The Later Proterozoic Torridonian Rocks of Sc
their Sedimentology, Geochemistry and Ori

Geological Society reviewing procedures

The Society makes every effort to ensure that the scientific and production quality of its books matches that of its journals. Since 1997, all book proposals have been refereed by specialist reviewers as well as by the Society's Books Editorial Committee. If the referees identify weaknesses in the proposal, these must be addressed before the proposal is accepted.

Once the book is accepted, the Society has a team of Book Editors (listed above) who ensure that the volume editors follow strict guidelines on refereeing and quality control. We insist that individual papers can only be accepted after satisfactory review by two independent referees. The questions on the review forms are similar to those for *Journal of the Geological Society*. The referees' forms and comments must be available to the Society's Book Editors on request.

Although many of the books result from meetings, the editors are expected to commission papers that were not presented at the meeting to ensure that the book provides a balanced coverage of the subject. Being accepted for presentation at the meeting does not guarantee inclusion in the book.

Geological Society Special Publications are included in the ISI Index of Scientific Book Contents, but they do not have an impact factor, the latter being applicable only to journals.

More information about submitting a proposal and producing a Memoir or Special Publication can be found on the Society's web site: www.geolsoc.org.uk.

It is recommended that reference to all or part of this book should be made in the following way.

STEWART, A. D. 2002. *The Later Proterozoic Torridonian rocks of Scotland: their Sedimentology, Geochemistry and Origin*. Geological Society, London, Memoir 24.

GEOLOGICAL SOCIETY MEMOIR NO. 24

The Later Proterozoic Torridonian Rocks of Scotland: their Sedimentology, Geochemistry and Origin

A. D. Stewart

Postgraduate Research Institute for Sedimentology, University of Reading, PO Box 227, Reading RG6 6AB, UK

2002
Published by
The Geological Society
London

THE GEOLOGICAL SOCIETY

The Geological Society of London (GSL) was founded in 1807. It is the oldest national geological society in the world and the largest in Europe. It was incorporated under Royal Charter in 1825 and is Registered Charity 210161.

The Society is the UK national learned and professional society for geology with a worldwide Fellowship (FGS) of 9000. The Society has the power to confer Chartered status on suitably qualified Fellows, and about 2000 of the Fellowship carry the title (CGeol). Chartered Geologists may also obtain the equivalent European title, European Geologist (EurGeol). One fifth of the Society's fellowship resides outside the UK. To find out more about the Society, log on to *www.geolsoc.org.uk*.

The Geological Society Publishing House (Bath, UK) produces the Society's international journals and books, and acts as European distributor for selected publications of the American Association of Petroleum Geologists (AAPG), the American Geological Institute (AGI), the Indonesian Petroleum Association (IPA), the Geological Society of America (GSA), the Society for Sedimentary Geology (SEPM) and the Geologists' Association (GA). Joint marketing agreements ensure that GSL Fellows may purchase these societies' publications at a discount. The Society's online bookshop (accessible from *www.geolsoc.org.uk*) offers secure book purchasing with your credit or debit card.

To find out about joining the Society and benefiting from substantial discounts on publications of GSL and other societies worldwide, consult *www.geolsoc.org.uk*, or contact the Fellowship Department at: The Geological Society, Burlington House, Piccadilly, London W1J 0BG: Tel. +44 (0)20 7434 9944; Fax +44 (0)20 7439 8975; E-mail: *enquiries@geolsoc.org.uk*.

For information about the Society's meetings, consult *Events* on *www.geolsoc.org.uk*. To find out more about the Society's Corporate Affiliates Scheme, write to *enquiries@geolsoc.org.uk*.

Published by The Geological Society from:
The Geological Society Publishing House
Unit 7, Brassmill Enterprise Centre
Brassmill Lane
Bath BA1 3JN, UK

(Orders: Tel. +44 (0)1225 445046
Fax +44 (0)1225 442836)
Online bookshop: *http://bookshop.geolsoc.org.uk*

British Library Cataloguing in Publication Data
A catalogue record for this book is available from the British Library.

ISBN 1-86239-103-3

Typeset by Aarontype Ltd, Bristol, UK
Printed by Bell & Bain, Glasgow, UK.

Distributors
USA
AAPG Bookstore
PO Box 979
Tulsa
OK 74101-0979
USA
(*Orders*: Tel. +1 918 584-2555
Fax +1 918 560-2652
Email bookstore@aapg.org)

India
Affiliated East-West Press PVT Ltd
G-1/16 Ansari Road, Daryaganj
New Delhi 110 002
India
(*Orders*: Tel. +91 11 327-9113
Fax +91 11 326-0538)
Email affiliat@nda.vsnl.net.in)

Japan
Kanda Book Trading Co.
Cityhouse Tama 204
Tsurumaki 1-3-10
Tama-shi
Tokyo 206-0034
Japan
(*Orders*: Tel. +81 (0)423 57-7650
Fax +81 (0)423 57-7651)

Contents

Acknowledgements

The writer is grateful to the following researchers who have generously given time to read and comment on parts of the text: H. Emeleus, F. Fraser-Menzies, R. E. Holdsworth, E. Irving, B. E. Leake, P. Nicholson and G. M. Young. Particular thanks are due to the three referees, A. R. Prave, P. M. Smith and N. J. Soper, who had the onerous task of reading the entire text, and detected more than a few dubious assumptions and murky arguments. Acknowledgement is also due to the University of Chicago and R. C. Selley for permission to use the copyright material contained in Figures 38 and 107.

Chapter 1
Introduction

Torridonian is an informal stratigraphic name for the Proterozoic reddish-brown sandstones overlying the Lewisian gneiss complex of the NW Scottish mainland. These sandstones form one of the principal elements of British stratigraphy, comparable in volume (over 150 000 km^3) to the Lower Old Red Sandstone of eastern Scotland, or the Triassic of England. They form the majestic mountains of NW Scotland, but also extend westwards under the Minch basin (Fig. 1). The subcrop has been identified 20 km north of Cape Wrath on the MOIST seismic reflection profile (Blundell *et al.* 1985), and beneath Devonian strata in the west Orkney basin (Cheadle *et al.* 1987). It extends south for 330 km to the latitude of Iona (Binns *et al.* 1974; Evans *et al.* 1982). The Torridonian was deposited on the edge of the Laurentian shield, near the roughly contemporaneous Grenville orogenic belt. It lies just outside the Caledonian orogen and has consequently escaped appreciable deformation, except in the Moine Thrust zone. Dips are generally low and the thermal history reflects little more than burial, giving ample scope for studies of the sedimentology, geochemistry, palaeoclimate and palaeomagnetism. Combined investigations of the sedimentology and chemistry of the rocks by several workers over the last ten years, using a total of nearly 600 whole rock analyses, have been particularly fruitful despite the relative neglect of the petrography. The most surprising lacuna in Torridonian studies is the paucity of published work on the micropalaeontology.

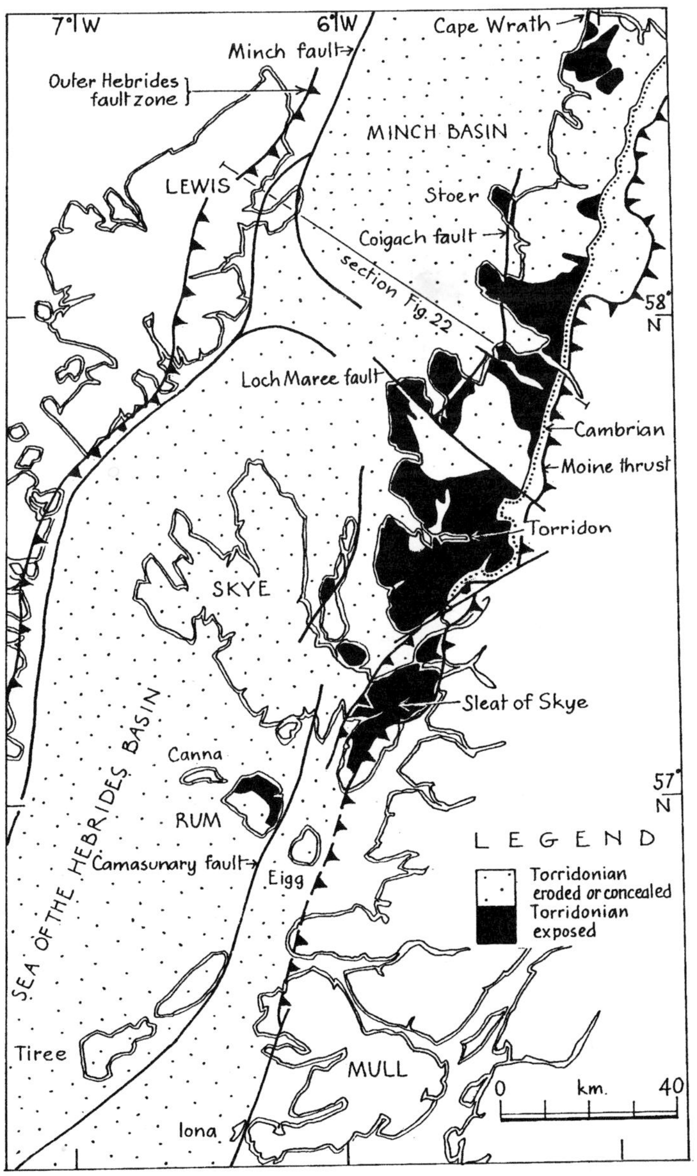

Fig. 1. Map of NW Scotland showing the present and former extent of the Torridonian, together with some major faults.

The main objects of this memoir are to provide a comprehensive field description of the Torridonian, and review, briefly, its origin and diagenesis. Special attention has been given to the stratigraphic framework, for it is clear from past researches that detailed studies of the rocks can be vitiated if their stratigraphic context is neglected.

The Torridonian can be divided into the Stoer, Sleat and Torridon Groups (Fig. 2 and folding Plate 1). The oldest is the Stoer Group which comprises fluviatile red sandstones and lake deposits unconformably covering the Lewisian gneiss complex on the foreland of the Caledonian orogenic belt. Although the Stoer Group is locally 2 km thick its areal extent is now limited to a narrow strip next to the Coigach fault shown in Figure 1.

The younger Sleat Group, not seen in contact with the Stoer Group, is confined to the Caledonian Kishorn nappe and best preserved in the Sleat of Skye. It consists of fluviatile sandstones with subordinate lacustrine or shallow marine shales, deposited unconformably on Lewisian gneiss. Caledonian deformation and lower greenschist facies metamorphism has affected most of the Sleat Group.

The youngest part of the Torridonian, and by far the most important volumetrically, is the Torridon Group. This also consists of mainly fluviatile sandstones, 6 or 7 km thick. Within the Kishorn nappe the Torridon Group conformably overlies the Sleat Group but on the foreland it covers a landscape unconformity that cuts across both the Lewisian gneiss complex and the westward-dipping beds of the Stoer Group.

The Torridon Group is truncated on its western margin by the Minch fault and is believed to have accumulated in a half graben. Reasons will be given later for thinking that the Minch fault, and others close to it, formed the western margin of the graben.

The Stoer and Torridon Groups were gently warped and tilted 5–6° westwards before being buried by at least 1.5 km of Cambro-Ordovician sediment.

A maximum age for the Torridonian as a whole comes from Rb–Sr and K–Ar biotite ages of about 1200 Ma in the underlying Lewisian gneiss complex (see pp. 21 & 42), and a minimum age of about 530 Ma is fixed by the Lower Cambrian fossils in the unconformably overlying Eriboll Formation. A maximum age for the Torridon Group is given by a zircon grain dated at 1046 ± 26 Ma by U–Pb (Rainbird *et al.* 2001).

Diagenetic ages for the Stoer and Torridon Groups have been obtained by Turnbull *et al.* (1996). They obtained dates of 1199 ± 70 Ma (Pb–Pb on limestone) for the Stoer Group and 994 ± 48 Ma (Rb–Sr on early diagenetic phosphate) for the Torridon Group. There are no age data for the Sleat Group. The palaeomagnetic pole positions for the Stoer and Torridon Groups compare closely with those for Laurentia at roughly the same time, confirming that the isotopic ages are not wildly wrong. The extensive biota in the grey shales of the Stoer and Torridon is consistent with the middle to late Riphean (Mesoproterozoic to early Neoproterozoic) ages given above.

The time gap between the Stoer and Torridon Groups, is much greater than that between, for example, the Old Red and New Red Sandstones of Britain, so that an all-embracing lithostratigraphic name such as 'Torridonian Supergroup' is undesirable. The term Torridonian is used in this book in its original sense, meaning all the sediments in Scotland west of the Moine thrust that were deposited after the formation of the Lewisian basement complex and before the Cambrian.

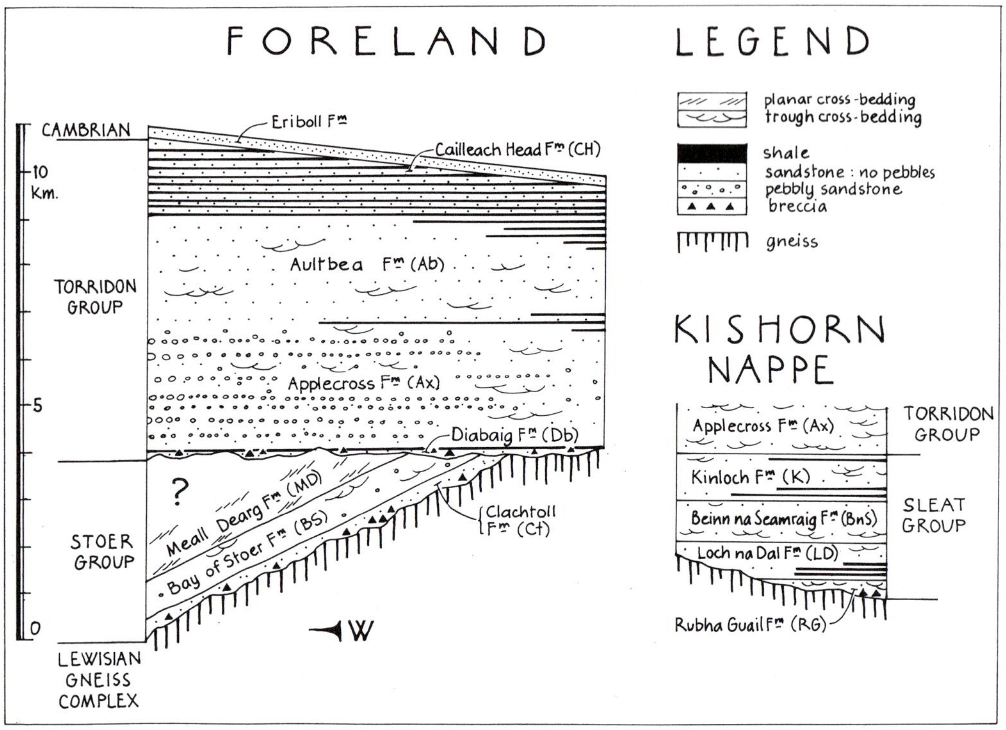

Fig. 2. Torridonian lithostratigraphy. All groups and formations are named. The formation codes used in Plate 1 and elsewhere in the memoir are given in brackets.

The foregoing sketch of the Torridonian is amplified in Chapters 2 to 5 which treat the regional stratigraphy, sedimentology, geochemistry and mineralogy of the sediments, and also their provenance, correlation, tectonic and climatic setting. Detailed field descriptions of the rocks in thirty-three sub-areas of the Torridonian will be found in the Directory, Chapter 6. The descriptions include definitions of both stratigraphic units and lithofacies, and the preferred environmental interpretation for each. Chapter 6 is intended to supply the field data needed to support the interpretations given in Chapters 2 to 4.

Some terms used in the memoir that may require clarification are listed below:

- *groups, formations* and *members* are defined following the Geological Society's *Guide to stratigraphical procedure* (Whittaker *et al.* 1991). Each formation has a two letter code (e.g. Ct = Clachtoll Formation) that is used in Plate 1 and elsewhere;
- *facies* is used to mean 'the sum of the primary characteristics of a rock' (Walther 1894, p. 989). The term is useful to designate lithologies that recur within a formation or formations. Each facies has a code consisting of two letters designating the formation from which it is first described, and a number, e.g. Ct1, the breccio-conglomerate facies of the Stoer Group, best developed at the base of the Clachtoll Formation;
- *lateral persistency of a bed* (p) is defined as the lateral extent of a bed divided by its maximum thickness. It is conveniently estimated from the expression $2(L/T)$ where T is the change in thickness of a bed observed in a distance L along it;
- θ is the vector mean direction of palaeocurrents obtained from n observations;
- *bearings* are from National Grid north;
- *grain size* is stated according to the Wentworth scale (Pettijohn *et al.* 1987, p. 72);
- *roundness* terms follow Pettijohn (1975, p. 57);
- *colour* is described by reference to the Geological Society of America rock-color chart (1963);
- *shale* is a clastic sediment with the modal grain size of silt, usually laminated;
- *sections* are drawn perpendicular to the strike of bedding using the construction of Busk (1929, p. 19).

History of research 1811–1969

The identification of the red sandstones of NW Scotland as a mappable stratigraphic unit is due to Dr John MacCulloch (1773–1835). He worked there intermittently between 1811 and 1818, travelling on horseback, by trading schooner and naval cutter, for at this time the Northern Highlands had no roads. The boundaries were plotted on Aaron Arrowsmith's quarter inch to the mile map of Scotland (1807), the best then available. MacCulloch showed that the pyramidal mountains had been carved out of a once continuous red sandstone unit resting unconformably on gneiss near present sea level. He also noted that the unconformity had considerable relief, associated with basal conglomerate and grey ripple-marked shale at several localities (MacCulloch 1819, vol. 1 p. 481 & vol. 2 p. 89–104).

Sedgwick & Murchison (1828) correlated the red sandstones of the NW coast with the Old Red Sandstone of the east of Scotland and they are thus shown on MacCulloch's geological map of Scotland, published in 1836 (Eyles 1937, 1939; Boud 1974). It is interesting that Hugh Miller, who quarried the Torridonian at Gairloch in 1823 while employed as a mason on the extension of Flowerdale House, also believed it to be Old Red Sandstone (Miller 1841) and maintained this view until his death in 1856.

Geological mapping of Sutherland by Cunningham (1841) showed that the quartzites, later shown to be Cambrian, step over the red sandstones onto the basement gneiss in an easterly direction, but it was James Nicol (1857*a*) who realized that the quartzites and red sandstones were separated by a regional angular unconformity. Nicol also provided the red sandstones with their first valid lithostratigraphic name; 'To these rocks as specially developed in Applecross and Gairloch, round Loch Torridon, the name of the Torridon Sandstone may well be given. It involves no theory and contradicts no fact.' (Nicol 1866, p. 29).

In 1880, Geikie made the startling suggestion that the ice-moulded 'mamillated' topography so characteristic of the Lewisian gneiss outcrop had been exhumed from beneath the Torridonian. However, similar topography occurs over the Moine schists near Loch Morar and over the Devonian lavas of Lorne. It also appears over the Lewisian gneiss near Cape Wrath, where the unconformity beneath the Torridon Group is featureless (see below). Geikie's suggestion is, therefore, wrong. More recently Godard (1957; 1965,

p. 564) has repeated Geikie's mistake, arguing that much of the present Lewisian surface has survived exhumation from beneath the Torridonian, unmodified by later erosion.

The British Geological Survey started systematic geological mapping of NW Scotland on a scale of six inches to a mile (1:10 560) in 1883. Regional variations in the palaeorelief of the gneiss–sandstone contact were soon detected. Near Cape Wrath, in the north, the contact was observed to be flat whereas in Assynt and farther south the palaeorelief was mountainous, reaching 600 m between Loch Maree and Beinn Dearg Bheag (Peach *et al.* 1888, p. 400–401; 1907, p. 275–277 & 311). The red sandstones were at first supposed to be Cambrian in age, but after the discovery of Lower Cambrian fossils in the unconformably overlying Fucoid Beds (Salter, *in* Murchison 1858), the time-stratigraphic term Torridonian was introduced (W. H. Hudlestone *in* discussion of Peach & Horne 1892; Geikie 1892; Peach *et al.* 1907, p. 32). Torridonian appears as a time-stratigraphic term on all Geological Survey maps issued after 1892 alongside a rock-stratigraphic name such as Torridon Sandstone. By 1893 the surveyors had completed the mainland mapping and were able to formulate a four-fold sub-division of the Torridon Sandstone, based on type sections at Diabaig, Applecross, Aultbea and Cailleach Head (Geikie 1894).

The nature of the Torridonian depositional environment was first considered by Goodchild (1897, 1898) who pointed out that the burial of fluvially eroded palaeotopography by locally derived detritus indicated a change from a humid to a semi-arid climate. He cited Pleistocene wadi deposits from Sinai as analogues for the Torridonian valley fill. Penck (1897, p. 149–160) reached a similar conclusion after a field trip to NW Scotland with John Horne in 1895. Penck, however, went on to draw a parallel between the cross-bedded red sandstones of the Applecross Formation and the fluvial sediments of the Indus and Ganges basins. He concluded presciently that the sediments must have formed in low palaeolatitudes, in the dry interior of a large continent and not in their present position on a continental margin. Penck rejected a lacustrine hypothesis because of the general absence of facies changes from coarse-grained red sandstone into fine-grained grey sediment, but also because Phanerozoic red beds generally have land faunas rather than lacustrine ones.

Penck (1897, p. 152) also proposed the ingenious hypothesis that the flat unconformity surface at Cape Wrath represented the remains of a plateau that farther south had been deeply eroded and was consequently covered by stratigraphically lower deposits (i.e. the Diabaig Formation).

Geochemical analyses of the Applecross sandstone by MacKie (1901) showed low values for Ca and Na but high K relative to the Lewisian gneisses which, he assumed, were the source of the sediment. MacKie concluded that the Ca and Na had been removed in solution during weathering and that consequently the climate was not arid. He also speculated that the atmospheric CO_2 concentration was high relative to present values, and chemical erosion thereby accelerated (MacKie 1926).

Mapping of the entire Torridonian outcrop was completed by the Geological Survey in 1896 and the full results published eleven years later in the monumental NW Highlands memoir (Peach *et al.* 1907, p. 269–362). Despite the mass of new data the basic stratigraphic framework remained essentially as Nicol had left it fifty years earlier – a single, conformable sandstone succession bounded unconformably below by gneiss and above by Cambrian quartzite. A lacustrine environment was tentatively suggested for the sediments (Peach *et al.* 1907, p. 273) on the basis of a perfunctory discussion that completely ignored the seminal ideas of Goodchild, Penck and MacKie. According to Penck the lacustrine hypothesis had been adopted by the Geological Survey in deference to the views of the former Director Sir A. C. Ramsay who believed that red beds formed in lakes (Ramsay 1871*a*, 1871*b*). Ramsay died in 1895 but the Survey continued to advance the hypothesis, which is identifiable in each of the first three editions of the Northern Highlands regional guide (Phemister 1936, 1948, 1960).

After the Geological Survey mapping was completed in 1896 active research on the rocks virtually ceased for sixty years. However, in 1948, H. H. Read emphasized the importance of Torridonian petrology during the discussion of a paper read before the Geological Society of London by P. Allen (1949). He returned to this theme in 1950 when his students Sutton and Watson read their paper on the evolution of the Lewisian basement to the Geological Society. For the discussion he wrote;

> The new interpretation [of the Lewisian] meant, again, that the Torridonian must be looked at properly. In it would be found the record of the Laxfordian cover at least. Samples of it were seen in the so-called foreign boulders in some of the Torridonian pebble beds. The sedimentary petrography of the Torridonian was a man-sized study of immense geological importance (Read *in* discussion of Sutton & Watson 1951).

The first fruits of Read's initiatives were seen in 1960 (Sutton & Watson 1960), swiftly followed by P. Allen and co-workers (Allen *et al.* 1960). Meanwhile, at Cambridge, Irving had demonstrated the existence of a dramatic shift in magnetization direction within what had been mapped as the Diabaig Formation, corresponding to a change in palaeolatitude from 18°N to 26°S (Irving 1954; Irving & Runcorn 1957). Irving's palaeomagnetic study was the first ever made of a Precambrian red bed sequence and also the first to show sequential polarity reversals in sediments. At roughly the same time Pavlovsky (1958), in a masterly but largely overlooked literature review of the Scottish Precambrian and Lower Palaeozoic, suggested that the Torridonian was deposited in a sedimentary basin bounded to the west by the Outer Isles fault, and to the east by a fault along the line of the later Moine thrust.

An initial attempt to put a maximum age on Torridonian sedimentation dates from 1955, when Holmes and co-workers produced the first K–Ar dates for potash feldspar in Lewisian pegmatites (Holmes *et al.* 1955). Argon leakage made the dates far too young, but the maximum age of 800 Ma proposed for the Torridonian (Holmes 1960) was fortuitously almost correct.

Microfossils were found by Teall in thin sections of phosphate nodules from the highest part of the Torridon Group (Peach *et al.* 1907, p. 288 & Plate LII). Later workers (Naumova & Pavlovsky 1961; Downie 1962; Diver 1980; Zhang Zhongying 1982; Zhang Zhongying *et al.* 1981) recovered organic walled microfossils, thought to be Riphean in age, from grey shales at almost all stratigraphic levels. The microfossils are unornamented spheroids, both isolated or arranged in clusters, and non-septate filaments.

Geological mapping during the early 1960s by students of P. Allen at Reading University disclosed a regionally extensive erosion surface within the Torridonian (Gracie 1964; Lawson 1965; Williams 1966*b*). This was soon found to be an angular unconformity that at Achiltibuie corresponded exactly to the palaeomagnetic break found by Irving (Stewart 1966*b*). It also became clear from the work of Selley (1965*a*) on Raasay, and Williams (1966, 1969*a*) at Cape Wrath, that the bulk of the Torridonian was fluviatile, as originally suggested by Penck. The source rocks of the Torridon Group were shown to be mainly 'Grenvillian' and late Laxfordian in age by Moorbath *et al.* (1967), and thought to be a basement complex located west of the Torridonian outcrop (Selley 1966; Williams 1969*a* & *b*).

The rocks at Torridon were first examined by the writer in 1960, and given their current group nomenclature nine years later (Stewart 1969). The strata beneath the newly discovered angular unconformity were called the Stoer Group. Those above the unconformity, that correspond exactly with Nicol's original definition of Torridon Sandstone, were called the Torridon Group. Strata 3.5 km thick in the Kishorn nappe of Skye originally assigned by the Geological Survey to the Diabaig Formation (the lowest formation of the Torridon Group) were renamed the Sleat Group.

Research from 1969 onwards is considered elsewhere in this memoir.

Chapter 2
The Stoer Group

The Group consists of alluvial red sandstones, interspersed with lake sediments, having a maximum exposed thickness of 2 km. The present extent of the Stoer Group is shown in Plate 1. It has survived only as a narrow strip next to the Coigach fault, apparently in a hanging wall roll-over (Stewart 1993*a*). Figure 3 shows the Stoer Group truncated by the Coigach fault, together with its unconformable relations with the Lewisian gneiss complex beneath and the Torridon Group above. The original extent of the group can only be inferred from the sediments. It has not been identified in the subsurface offshore to the west and it is unlikely that it ever existed at the present level of erosion east of the existing outcrop.

The general outlines of the sedimentary history are clear, but problems lurk in the details. For example, the oldest sediments of the group occupy palaeovalleys eroded in the gneiss complex, some of which were filled exclusively by alluvial deposits whereas others hosted swamps and temporary lakes. Another controversial topic is the origin of the volcaniclastic Stac Fada Member, and the amount of volcanic input to Stoer Group sediments generally.

Stratigraphy

The regional stratigraphy of the Stoer Group is shown in Figure 4. The stratotypes of the three constituent formations, originally defined at Stoer (Stewart 1991*a*), are described on pp. 57–70. The oldest is the Clachtoll Formation, overlying the Lewisian gneiss complex and identifiable by its clasts, virtually all of which can be traced to local basement lithologies. Next come the trough cross-bedded sandstones of the Bay of Stoer Formation, containing well-rounded pebbles of gneiss and quartzite. The alluvial Meall Dearg Formation completes the sequence. Unlike the Bay of Stoer Formation it lacks pebbles and is entirely built of tabular, planar cross-beds.

The Bay of Stoer Formation contains two members, the volcaniclastic Stac Fada Member and the lacustrine Poll a' Mhuilt Member. The stratigraphic section (Fig. 4) is hung from the Stac Fada Member, assumed to have been a horizontal time plane. The assumption is based on the absence of repetition of the Stac Fada lithology in any of the sections studied, and its stratigraphic position. At both Stoer and Enard Bay, for example, the Stac Fada Member is followed by the Poll a' Mhuilt Member and the Meall Dearg Formation. The Stac Fada Member does not appear randomly in the stratigraphic sequence. In the absence of the Stac Fada Member, the top of the Clachtoll Formation could be used as a datum, but the resulting stratigraphic section would not show the downwarping of the basement gneisses and the Clachtoll Formation at Stoer and Poolewe that is so evident in Figure 4.

Figures 5 & 6 are detailed stratigraphic profiles of the Stoer Group at Stoer and Poolewe, where it is best exposed. They give a good idea of sedimentary facies and environments, and form the basis of the following discussion.

Basement topography, drainage and weathering

The surface of the Archaean basement had relief of several hundred metres when Stoer Group deposition started (Figs 5 & 6). A possible explanation for the contrasting types of valley fill mentioned above, viz. river sands in some, but swamps in others, is drainage reversal, illustrated diagrammatically in Figure 7. Figure 7(a) shows a river flowing eastwards with uniform gradient, fed by two steeper tributaries. The contours shown are in arbitrary units. In Figure 7(b) the eastern edge of the map area has been raised by 250 units eliminating the gradient on the main river, which becomes a lake. In Figure 7(c) the eastern edge of the map area has been raised by a further 250 units so that the main river flows westwards. The southern tributary has no gradient and the valley is occupied by a swamp. In the last stage, Figure 7(d), the eastern edge of the map area has gone up by 1000 units altogether. The main river flows westwards and both tributaries have become swamps. Clearly, any section through the map area of Figure 7, especially in a north–south direction, will intersect valleys with contrasting fill, either alluvial or swamp. Recent drainage reversal of the kind described is well known from the area adjoining Lake Victoria in East Africa (Beadle 1981, p. 250).

The rivers that cut the valleys in the gneiss beneath the Stoer Group originally flowed eastwards, but were forced by regional tilting of the basement to reverse their flow direction. However, eastward-flowing palaeocurrents are absent from the lowest sediments of the Stoer Group (Clachtoll Formation); only westward-flowing ones are present.

No well-developed palaeosols now exist beneath the Stoer Group, though weathered gneiss can be detected locally. Ultrabasic gneiss at Clachtoll has been reduced to grus and rounded pebbles (p. 59) whereas at Enard Bay the gneiss is reddened and decomposed along cracks penetrating over a metre down from the unconformity (p. 71). More than this is hardly to be expected, for weathering-limited erosion was normal in Proterozoic hills, as it is today in the absence of natural vegetation. Well-developed soils over basement rocks only formed in exceptional locations such as plateaux and pediments, where slopes were gentle and free of sediment. The weathering products of basement rocks in upland areas were generally swept away as soon as they formed. Proterozoic

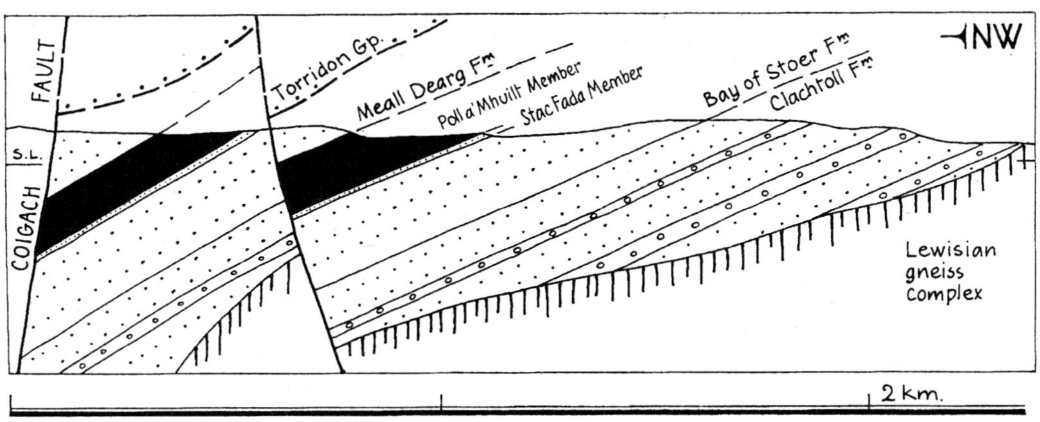

Fig. 3. True-scale section of the Stoer Group at Stoer showing its unconformable relationship to the Lewisian gneiss beneath and the Torridon Group above. The section extends from the Coigach fault at Cnoc Breac [NC 032317], southeastwards to Clashnessie [NC 054312]. The unconformity with the Torridon Group is exposed a short distance NE of the section.

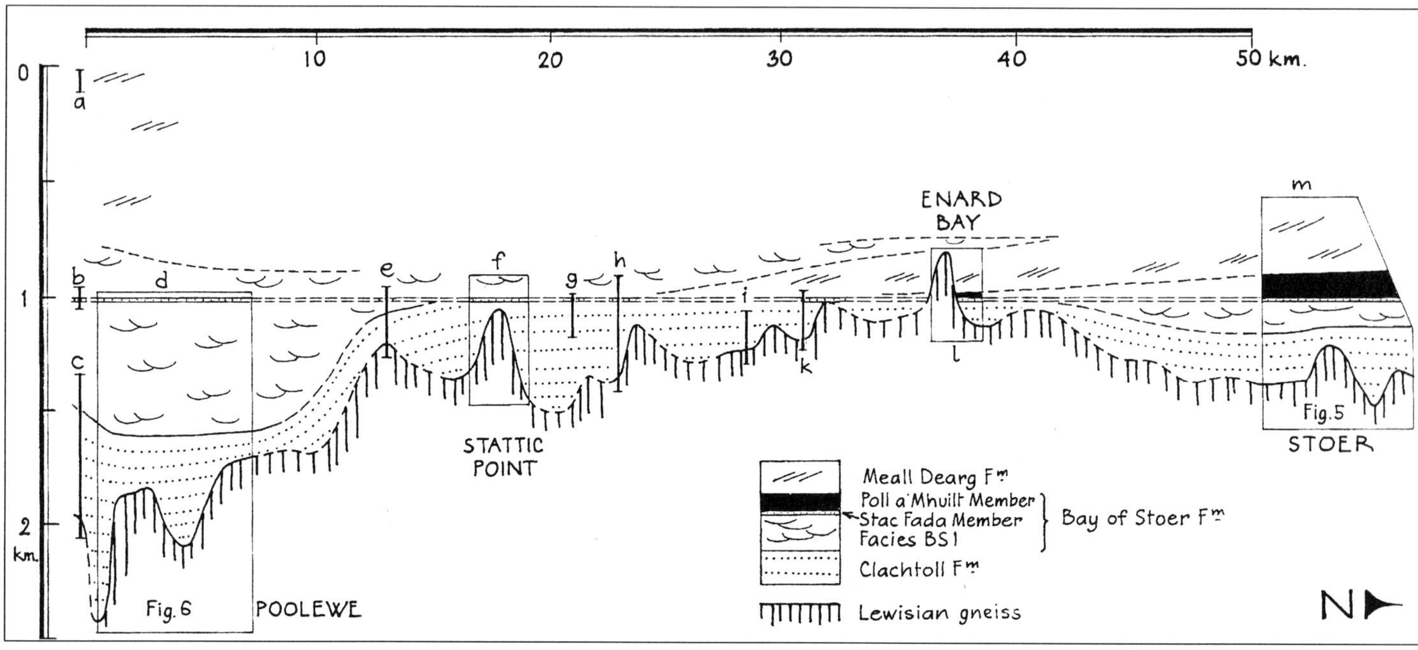

Fig. 4. Stratigraphic section of the Stoer Group, with the volcaniclastic Stac Fada Member forming the datum. The vertical exaggeration is ten times. The stratigraphy comes from the following sub-areas, detailed in the Directory, Chapter 6: **(a)** Rubha Reidh; **(b)** Bac an Leth-choin (Feadan Mór); **(c)** Bac an Leth-choin (Fig. 87); **(d)** Poolewe; **(e)** Gruinard Bay; **(f)** Stattic Point; **(g)** Cailleach Head (south side); **(h)** Cailleach Head (north side); **(i)** Achiltibuie (Horse Island); **(k)** Achiltibuie (Rubha Dunan); **(l)** Enard Bay; **(m)** Stoer.

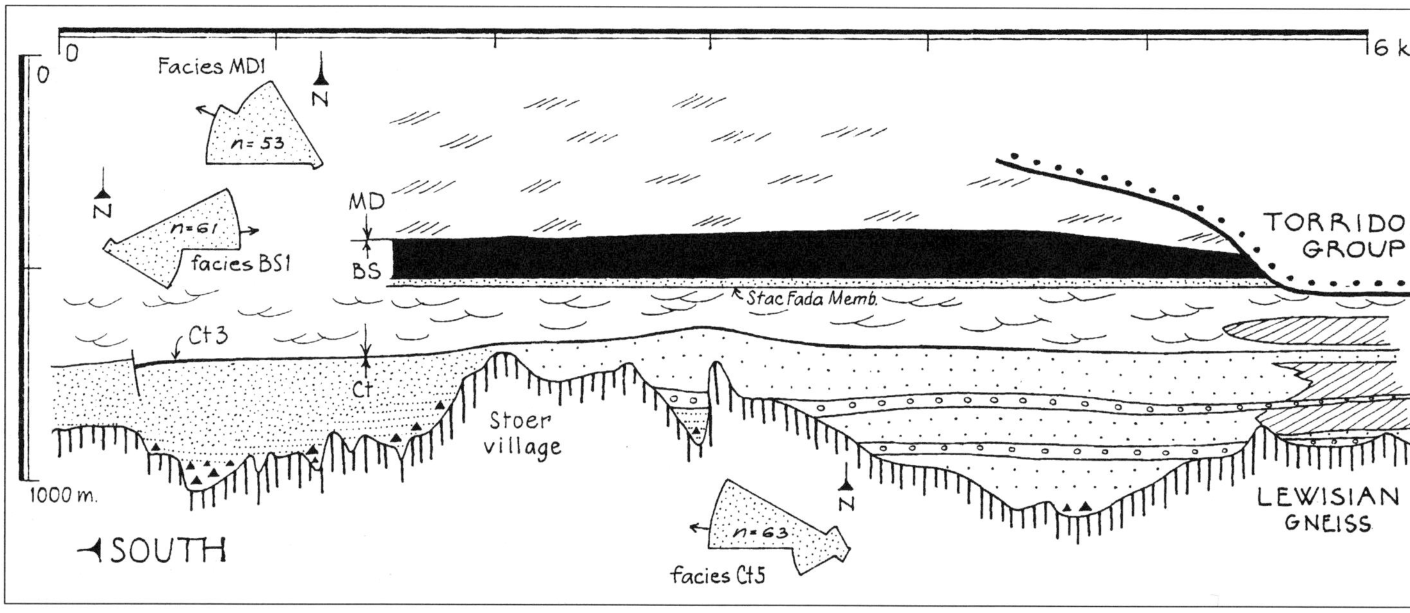

Fig. 5. Stratigraphic profile of the Stoer Group at Stoer. This is a down dip view of the stratigraphy with later faults removed, vertically exaggerated ×2. The rose diagrams show palaeocurrents for the Clachtoll Formation (Ct), Bay of Stoer Formation (BS) and Meall Dearg Formation (MD), deduced from trough cross-bedding. A key to the facies is given in Fig. 6.

palaeosols over sediments should be commoner, perhaps represented in the Stoer Group by the vertisol-like sediments in the Clachtoll Formation.

Facies and environments

Valley-confined alluvial fans

The lowest sediments in some palaeovalleys, for example the southern one in Figure 5 described in detail on pp. 57–59, are massive breccio-conglomerates (facies Ct1), that always overlie apparently fresh gneiss and contain a representative selection of local basement rocks, including ultrabasic types. Garnet, olivine, biotite and partly decomposed amphibole grains found in the matrix are all species common in the basement immediately east of Stoer (Cartwright *et al.* 1985). The blocks in the breccia are usually no more than half a metre in size, mainly subrounded in shape on the Pettijohn scale (Davison & Hambrey 1996, fig. 6), and clast-supported with a matrix of coarse sand and pebbles (Fig. 44). Stratigraphically upwards and away from the basement the size of the clasts diminishes and the breccio-conglomerate is interbedded with coarse pale-red sandstone. These tabular bedded pebbly sandstones, defined by containing more than 50% sandstone, constitute facies Ct2 (Fig. 45). Trough cross-bedding is present locally in this facies and may even become the dominant structure, in which case the rock is placed in facies Ct5. However, it is often difficult to map a boundary between Ct2 and Ct5.

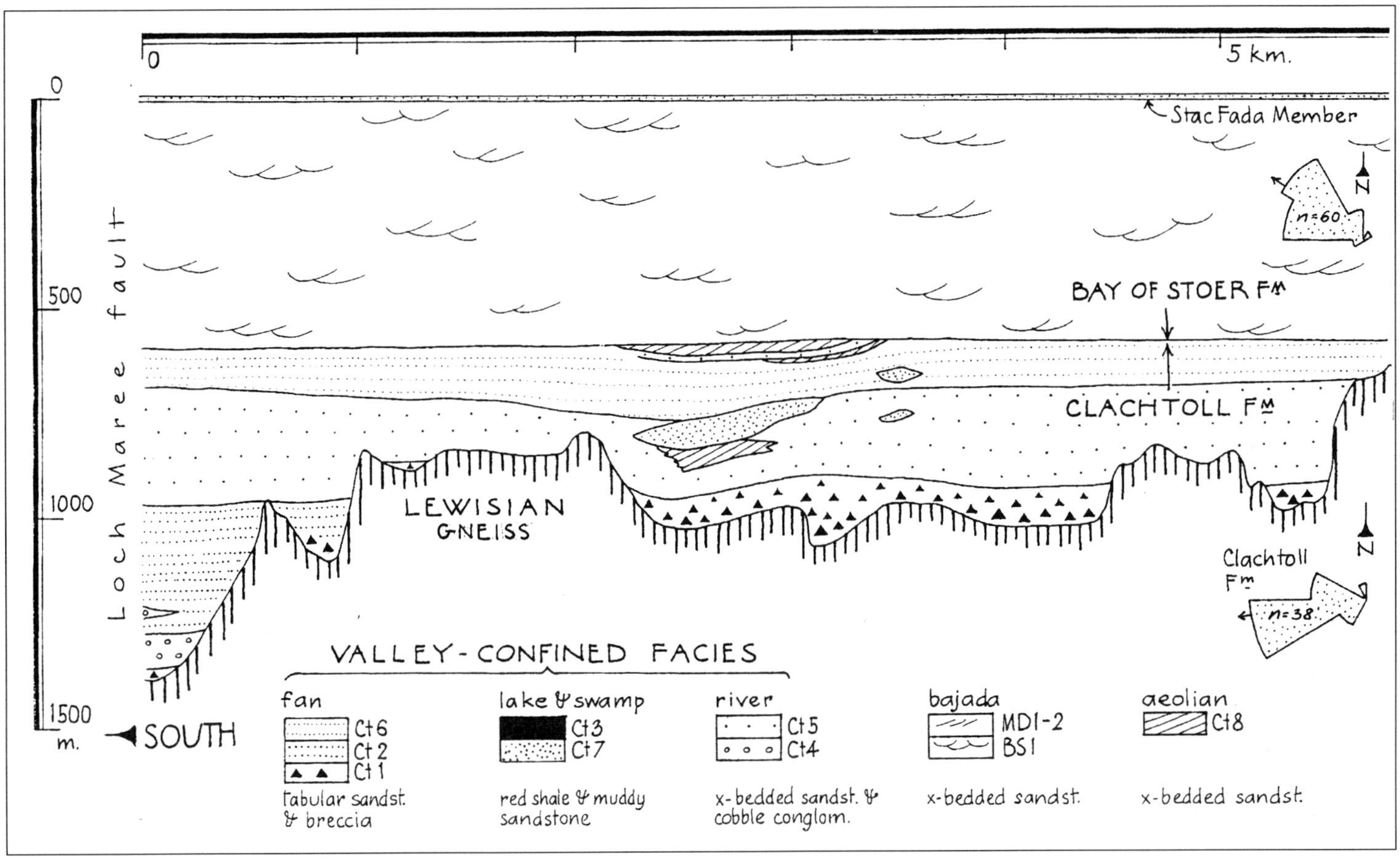

Fig. 6. Stratigraphic profile of the Stoer Group at Poolewe, vertically exaggerated ×2. The rose diagrams show palaeocurrents for the Clachtoll Formation (facies Ct5 & Ct2) and the Bay of Stoer Formation (facies BS1), deduced from trough cross-bedding.

The rounding of clasts in facies Ct1 suggests that it represents a fan-head conglomerate rather than talus or reworked talus material. The large size of the clasts probably means the sediment was supplied to the fan-head by a bedrock channel. Flood waters in such channels have boundary shear stresses that are orders of magnitude greater than in ordinary alluvial channels, and can easily transport cobbles in suspension and huge blocks as bed load (Baker & Kochel 1988). The huge elliptical acid gneiss blocks scattered through the breccia at Gruinard Bay (p. 87), and the 30 tonne block at Poolewe (p. 89), are probably reworked corestones. The tabular bedded breccio-conglomerates and sandstones are typical sheet flood deposits, like those commonly found in small alluvial fans, but not in river channels (Blair & McPherson 1994). The fining-upward from facies Ct1 to facies Ct2 is taken to indicate fan-head retreat up side valleys.

Facies Ct6, which is extensive at Poolewe, consists of fine to medium-grained sandstone with millimetre to centimetre lamination parallel to bedding. The maximum grain size is about 2 mm. The sandstone appears in the field to be poorly sorted, like the muddy sandstone facies (Ct7) into which it passes laterally, but thin sections show it has only about 15% of matrix. Low angle cross-bedding occurs very rarely. Desiccated red shale bands up to 3 m thick occur sporadically. The facies generally overlies either the trough cross-bedded facies (Ct5) or the tabular pebbly sandstone facies (Ct2). It is believed to represent upper flow regime sheet-flood deposition, though no current lineation has been seen. The sediments of the

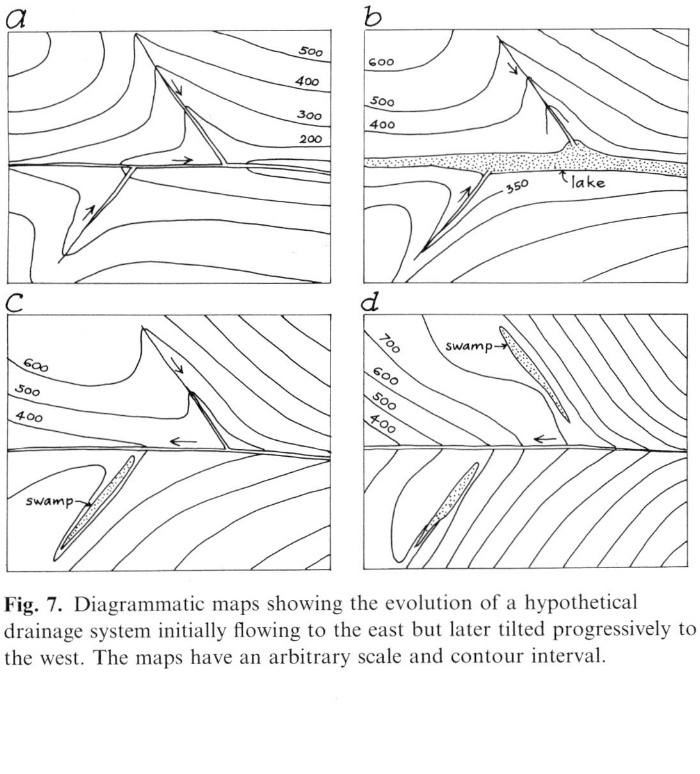

Fig. 7. Diagrammatic maps showing the evolution of a hypothetical drainage system initially flowing to the east but later tilted progressively to the west. The maps have an arbitrary scale and contour interval.

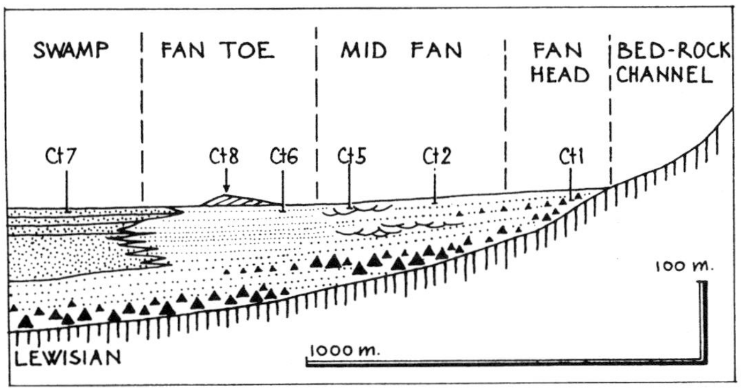

Fig. 8. Diagrammatic sketch of valley-confined alluvial fan facies and environments in the Clachtoll Formation

modern 'sandflat' environment, interposed between alluvial fans and saline lakes, may be comparable (Hardie *et al.* 1978).

A diagrammatic section of the valley-confined alluvial fan facies is shown in Figure 8.

Valley-confined swamps

The muddy sandstone facies (Ct7) occupies the centres of palaeovalleys, as can be seen from Figures 5, 6 & 8. At Stoer it is overlain by red shale (facies Ct3). The two facies also occur together at stratigraphically higher levels, Ct7 most prominently as the Stac Fada Member and Ct3 in the following Poll a' Mhuilt Member.

The muddy sandstone (Ct7) has several unusual characteristics, starting with the fact that the lowest 130 m of the facies at Clachtoll are completely devoid of bedding. Another peculiar feature is the diffuse pattern of relatively pale, discontinuous veinlets that ramify through the rock (Fig. 48). Graded beds 0.3 to 2 m thick are locally present at the base of the member (Fig. 46). The upper part of the facies is divided into beds decimetres to metres thick by thin desiccated limestone bands (Fig. 47). The limestones are up to 2 cm thick at the centre of the palaeovalley at Clachtoll, but only millimetres thick on the north side of Bay of Stoer, near the valley margin.

Petrographically the rock is a greywacke with 50% ferruginous matrix. The largest grains, up to 2 mm in size, are mainly quartz with angular or jagged shapes like those found in the Stac Fada Member and attributed by Sanders & Johnston (1989) to the explosive boiling of pore water in contact with magma. Normative calculations show that the muddy sandstone contains enough Mg and Fe (Table 4) for the original sediment to have been 40% smectitic clay. Much of the smectite could easily have been derived from weathering of the abundant basic and ultrabasic rocks in the Archaean basement nearby, and some also from the early diagenesis of fine-grained basic tephra.

The lateral continuity of the muddy sandstone with facies like the tabular-bedded pebbly sandstone (Ct2), well exposed on Horse Island (p. 76) and at Gruinard Bay (p. 87), shows that its massive nature arises from some post-depositional processes. A suitable process is suggested by the discontinuous veinlets, described above, which are sand-filled shrinkage cracks, partly assimilated into the surrounding sediment. The muddy sandstone was probably deposited in beds like those locally present at the base (Fig. 47), and then homogenized by repeated desiccation. The original content of smectitic clay immediately provides a possible mechanism; such clays are apt to shrink during a long dry season and swell during the following wet one. The cracks would tend to fill with sand, silt and possibly small flakes of mud at the start of the ensuing wet season, so that they could not close when the clay started to expand. Instead, the sediment between the cracks was deformed. Repetition of this process led to destruction of both the original bedding, and to some extent the cracks themselves. This process of pedoturbation is characteristic of modern vertisols. The muddy sandstone cannot be called a vertisol because it now lacks the complete set of definitive characters, but the process of homogenization is nevertheless applicable. Modern vertisols have, by definition, at least 30% clay, together with seasonally developed open, tortuous cracks at least a centimetre wide at a depth of half a metre from the surface. Vertisols are best developed on flat alluvial plains in warm climates with pronounced wet and dry seasons (Dudal & Eswaran 1988). They have also been described from seasonal lakes (Gustavson 1991). The depositional environment envisaged for the muddy sandstone would have been suitable for the growth of smectite, which is favoured by high pH and alkalinity, high partial pressure of CO_2, and surface runoff rather than groundwater input (Jones & Galen 1988, table 7).

As the palaeovalleys filled and the hill slopes progressively disappeared beneath sediment the rate of sedimentation diminished, permitting the formation of carbonate sheets by the evaporative pumping of dilute subsurface brines during the long periods without detrital influx. The carbonate sheets have not been examined for microbial structures. Channels are completely absent from the muddy sandstone facies but it would be rash to deduce that all storm water and suspended sediment was trapped in the valley. Most of it must have overflowed into interconnected palaeovalleys with trunk drainage. The surface water that remained would have been lost by evaporation or groundwater flow. Facies Ct7 therefore formed in a swamp rather than a lake.

The muddy sandstone is an unusual facies but has, nevertheless, a precise analogue in the Lower Jurassic East Berlin Formation of Connecticut (Demicco & Kordsch 1986; Hubert *et al.* 1992; Smoot 1991), where it forms massive beds 0.5 to 2 m thick, separated by thin bands of desiccated red shale. Dolomitic nodules and gypsum crystals are locally present in the sandstone. The muddy sandstone, described as 'flood-plain red mudstone' by Hubert *et al.* (1976), is marginal to grey mudstones deposited in perennial lakes. For an equivalent modern setting one might consider the 'ponded water mudflats' formed by the rapid deceleration and deposition of sediment-charged sheetwash in a temporarily expanded saline lake (Hardie *et al.* 1978).

Valley-confined rivers

The most evident sign of river deposition is provided by the cobble conglomerates (facies Ct4) that form sheets up to 40 m thick extending across the entire width of the palaeovalley north of Stoer (Fig. 5). They are multistorey and clast-supported, with a matrix grain size of 2 mm and cobbles generally less than 20 cm across (Fig. 54). Both the matrix grain size and the maximum cobble size diminish upwards through a given sheet. The cobbles are mostly made of coarse acid gneiss. The only basic clasts derive from the chilled margins of Scourie dykes. The cobbles are well-rounded, but this tells us little about the transport distance, for rounding is well known to develop rapidly to a maximum value in the first few kilometres of movement (Pettijohn 1975, p. 58–59). The arrangement of clasts into crude beds of 0.5–1 m thickness is significant, for this is comparable to the depth of channels at the base of the conglomerate. It is probable that the conglomerate beds formed in wide, shallow, braided channels of about this depth. Pebble sizes suggest current velocities somewhat over $1\,m\,s^{-1}$.

Trough cross-bedded red sandstones (Ct5) and tabular-bedded pebbly sandstones (Ct2) are interposed between the conglomerates. These, too, contain upward-fining cycles, generally about 40 m thick. The cyclicity is absent from the valley-confined alluvial fan and swamp sequences, suggesting that it was due to episodic uplift of the source area that supplied the cobble conglomerates, possibly tens of kilometres distant. Such uplift would have no effect on erosion in the purely local basin around a swampy valley.

Cobble conglomerates near the valley margins frequently contain a mixture of well-rounded and sub-angular clasts, presumably because the trunk streams reworked the locally-derived breccias washed down the valley sides.

Unconfined bajadas

By the time the local relief had been buried, much of it by its own waste, the fringes of large alluvial fans had advanced from the west and from the east to reach the position of the present day outcrop. They formed two alluvial wedges, or bajadas, the lower corresponding to the Bay of Stoer Formation and the upper to the Meall Dearg Formation. The total thickness of bajada deposits presently exposed is about 600 m at Stoer and 700 m at Poolewe. The bulk of this is made up of two facies. The first is trough cross-bedded sandstone (facies BS1) that only differs from facies Ct5 in containing a population of durable pebbles, including quartzites, and in being slightly contorted (Fig. 56). This facies is described in detail on p. 64 and the pebbles on p. 17. The second facies (pp. 68–70) is planar cross-bedded sandstone (facies MD1). These two facies were deposited, respectively, by linguoid dunes and relatively straight-crested transverse bars. The subordinate facies MD2, dominated by wave ripples, may represent the tops of the transverse bars reworked during falling stage (Fig. 64). The linguoid dunes are like those deposited by powerful floods in central Australia (Williams 1971), whereas the transverse bars of the Meall Dearg Formation resemble the deposits of the sluggish Platte River in Nebraska (Smith 1970, 1971; Miall 1996, p. 234).

The water streaming off the eastern edge of the lower bajada (facies BS1) at Stoer formed a perennial lake in which the shale facies (Ct3) was deposited. The shale directly overlies various valley-confined facies as the bajada toe, and the lake, advanced eastward. The lowest half metre is grey, but the rest is red, suggesting that the lake was only temporarily stratified. A genetic connection with the overlying alluvial sand facies BS1 is shown by thin ripple-bedded sandstone beds within the shale, supplied like the sands from a westerly source (Fig. 50). At Poolewe the Bay of Stoer Formation was deposited by currents coming from the east, like those in the underlying valley-confined facies of the Clachtoll Formation, and no such lake formed.

The alluvial, trough cross-bedded sandstones forming the bajada at Stoer contain cycles of muddy sandstone (Ct7) and red siltstone with ripple-laminated fine sandstone bands (Ct3), described in detail on pp. 64–65. A graphic log of a typical cycle is shown in Figure 9. Their persistency factor (p) is about 10 000. Seven such cycles can be mapped along strike for at least 6 km, so originally they must have covered tens, or even hundreds of square kilometres. Their origin is intriguing. Deposition of the muddy sandstone was preceded by a violent sheet flood that planed off the underlying bed forms and deposited quartz pebbles up to a centimetre. Almost immediately afterwards a muddy sandstone bed was deposited. A few of these muddy sandstone beds contain matrix-supported pebbles up to 2 cm in size. The muddy sandstones must have been deposited from a hyperconcentrated sheet flood or a mudflow. Then followed a period of quiet lake sedimentation during which ripple-laminated and desiccated fine sands and silts were deposited. Flat bedded sands near the tops of the shaly intervals probably indicate the proximity of the fluvial sands of facies BS1. Ripple-drift lamination in the shales indicates eastward-flowing palaeocurrents, as in the rest of the formation.

It is not evident from what direction the sheet floods came, but if it accorded with the slump direction in the Stac Fada Member, that moved down slope from the ENE (Fig. 61), then the source of the sediment lay in that direction, where Scourian gneisses and valley-confined facies were still exposed on the rift floor. The overflow of sheet floods onto the alluvial plain may have been due either to westward tilting of the rift floor, or an intra-rift normal fault west of the present outcrop, downthrowing to the east. Either could have temporarily arrested the eastward advance of facies BS1 and permitted shallow lakes to form. The sequence Ct7 > Ct3 > BS1 in the cycles is just like that seen at the base of the bajada sequence, i.e. the base of the Bay of Stoer Formation on the type section (Fig. 50).

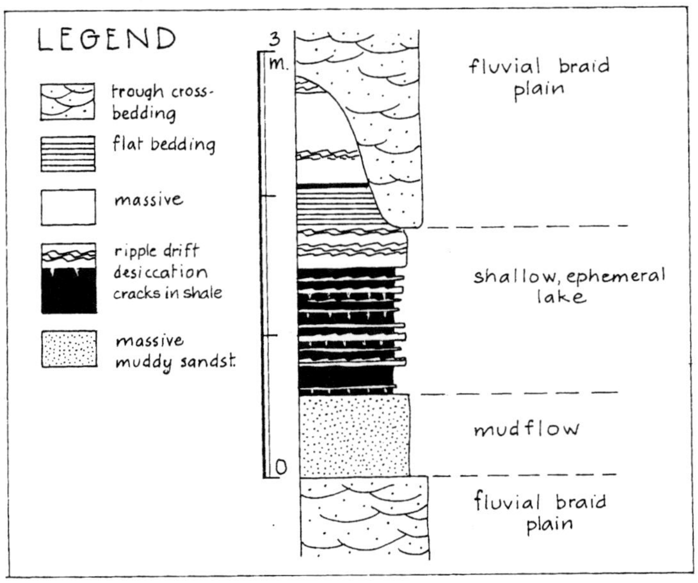

Fig. 9. Graphic log of a lacustrine cycle in the Bay of Stoer Formation at Clachtoll. For the location of this and similar cycles see Fig. 58.

Aeolian sands

The sandstones of the laminated sandstone facies (Ct8) are not characteristically aeolian for the grains are generally sub-angular and mica is a noticeable component. They are mineralogically quite like those of adjacent facies, from which they have presumably been winnowed, but are much better sorted. The most striking feature of the facies is the sharply defined and laterally persistent lamination (Fig. 86), with $p \approx 3000$. Grading within the laminae has not been noted. Surfaces exposed to erosion were evidently cohesive for blocks of the laminated sandstone up to 20 cm in size are incorporated in pebbly sandstone beds deposited by floods that frequently invaded the dune field. Good examples of such reworked sand blocks can be seen at Stoer (p. 63) and Achiltibuie (p. 76). The cementation may, perhaps, have been due to calcite precipitation from hard pore water. Another possibility is consolidation by cyanobacterial mats, like those commonly found in present-day deserts (e.g. Garcia-Pichel *et al.* 2001) and probably present also in the Proterozoic (Campbell 1979). Thin sheets of desiccated red siltstone are a common feature of the laminated sandstone, perhaps representing mud-drapes deposited on the dunes during floods (Fig. 55). Similar modern deposits are described by Glennie (1970, p. 48–49). As would be expected, the aeolian sands are in contact with virtually every other facies and are found at all stratigraphic levels.

Grainfall lamination (Hunter 1977) formed on the slip faces of small dunes appears to be the characteristic feature of facies Ct8. Climbing translatent stratification, usually the dominant type of lamination in modern dunes less than a metre high, should also be present, but has not been definitely identified. The wind-blown dunes presumably blanketed inactive parts of the coarse, shifting alluvial plain. The plain was invaded laterally by small alluvial cones that descended from bare gneiss hills. Landscapes of this kind can be found in many present-day arid areas (e.g. Moseley 1971) but their similarity to that described above arises from the absence of vegetation rather than aridity. The absence of talus (scree) at the base of the Stoer Group suggests reworking by substantial run-off.

The Stac Fada sequence

The sequence consists of the Stac Fada Member and the overlying Poll a' Mhuilt Member, together about 100 m thick (Fig. 10). The two members can be regarded as belonging to the muddy sandstone (Ct7) and in part to the shale facies (Ct3), respectively, but show peculiarities sufficient to warrant separate description.

The Stac Fada Member is about 10 m thick everywhere except at Enard Bay, where it reaches 30 m. It consists of a muddy sandstone (facies Ct7) containing abundant vesicular, glassy lapilli. Some of the larger quartz grains in the matrix are very angular and contain mosaic cracks. Accretionary lapilli are found abundantly in the topmost 10 m of the member at Enard Bay, separated from the lower part of the member by a shale horizon. Accretionary lapilli are sparsely present in the topmost few metres of the Stac Fada Member at Stoer. The member contains matrix-supported blocks of gneiss and sandstone up to about half a metre in size at several localities. At Stoer, the lower half of the Stac Fada Member contains rafts of sandstone up to 15 m long.

Lawson (1972) originally proposed that the Stac Fada Member was an ash flow resulting directly from a hydroclastic eruption. According to Sanders & Johnston (1989, 1990), basic magma penetrated the Stoer Group sediments when they were still wet,

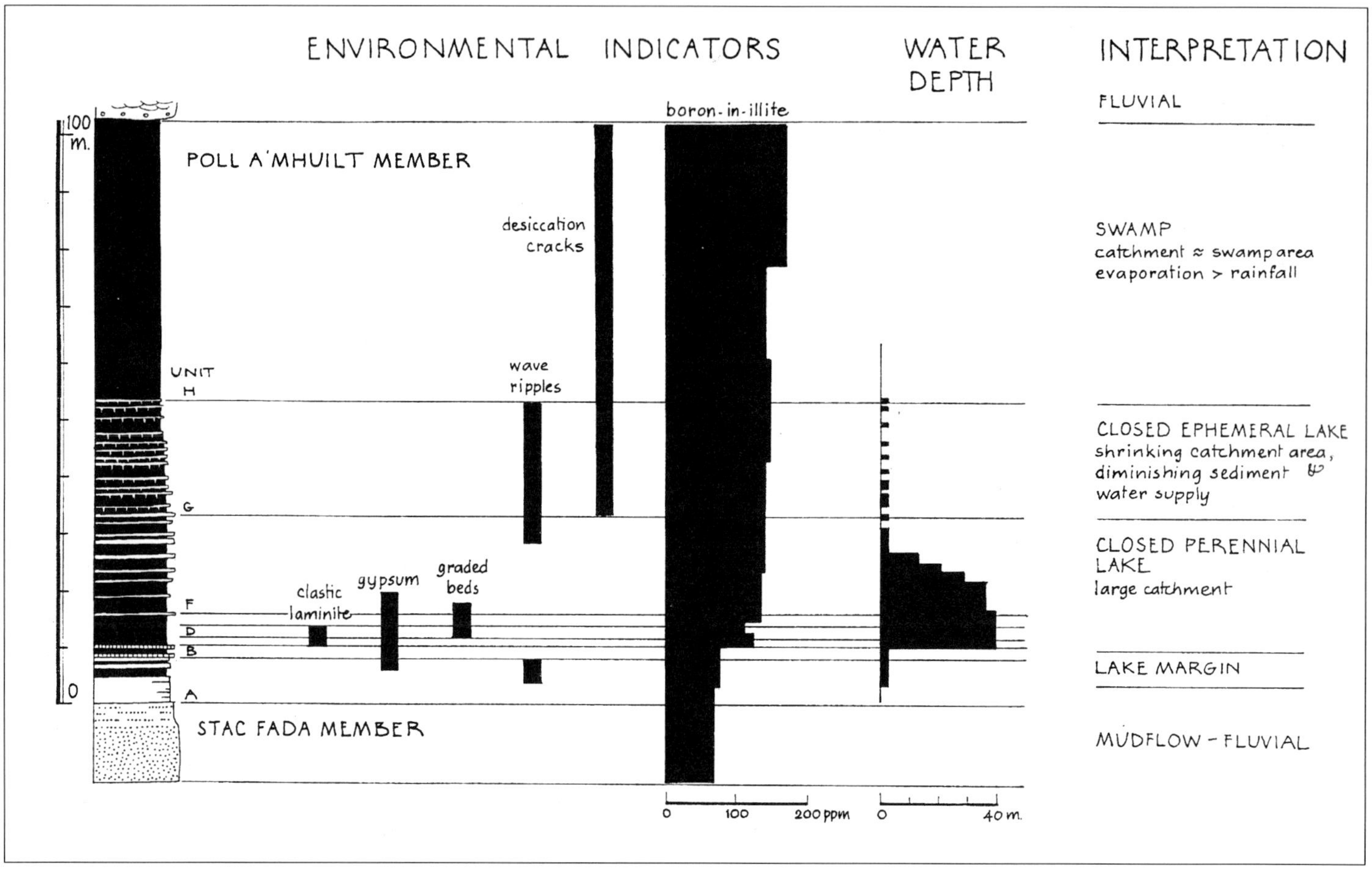

Fig. 10. Graphic log of the Stac Fada and Poll a'Mhuilt Members at Stoer, showing the boron content of illite, estimated water depth and sedimentary environments.

bringing the pore fluid to boiling point, chilling the magma to a glass and fragmenting it. The resulting slurry, they argued, was then extruded to form a hot mudflow, although they later admitted that the feeder pipes are nowhere seen. Geochemical studies suggested to Young (2002) that the original magma was basaltic. In brief, the Stac Fada Member represents a mudflow, or mudflows, incorporating both local siliclastic sediment and the products of a hydroclastic eruption. A more precise picture of its sedimentology is developed in the following paragraphs.

The base of the Stac Fada Member is generally flat, with only sand beneath it, but at Stoer the mudflow moved across sands *and* muds which it was able to intrude to a depth of about 2.5 m (Fig. 60). The resulting structures are described in detail on pp. 66–67. The folds (Fig. 61) and upturned beds show that the mudflow was moving westwards. If the sand beds were mainly dry they could easily have had a bulk density as low as 2000 kg m^{-3}, as compared with 1800–2300 kg m^{-3} for a mudflow (Costa 1988), making intrusion easy. The large gneiss clasts in the member come from the Clachtoll Formation, either picked up along with sand during the eruption or eroded while the mudflow was moving.

The accretionary lapilli are particularly interesting for they provide evidence of an eruptive centre near Enard Bay (Lawson 1972; Young 2002). The lapilli are about 4 mm in diameter, so the eruption could have been within 15 km of the bay, depending on the size and orientation of the eruption cloud from which the lapilli fell (Moore & Peck 1962; Fisher & Schmincke 1984, fig. 6-36). However, allowance has to be made for the fact that the lapilli lie in a non-volcanic matrix, i.e. they have been reworked and may have been carried beyond the fall-out zone. The shale horizon within the member at Enard Bay shows that the eruption that produced the accretionary lapilli was a distinct and relatively late event in the depositional history. Sanders & Johnston (1989) also concluded that the accretionary lapilli resulted from a fundamentally different eruption process than the rest of the Stac Fada Member. According to Young (2002, Fig. 8) there were two eruptive centres, each responsible for a series of separate flows that together constitute the Stac Fada Member. One of these centres, near Enard Bay, generated the mudflow containing accretionary lapilli, but convincing evidence for more than one other flow, or another identifiable eruptive centre is lacking. Young divided the Stac Fada Member at Stoer into three sub-units using geochemical differences that are of doubtful statistical significance. The only two sub-units in contact with each other are separated by a contact that Young interprets as erosional, but which could equally well be intrusional (Young 2002, fig. 5d). The evidence that any of these sub-units comes from a source north of Stoer, different to that which produced the accretionary lapilli, depends on directional structures (Young 2002, fig. 5a–c) that, in the writer's opinion, are ambiguous.

The genesis of the Stac Fada Member as a mudflow resulting from a phreatomagmatic eruption may seem plausible but three thorny problems still remain:

(1) Why is the member so extensive? It stretches for 50 km along strike. If the mudflow or flows originated from a single volcanic centre they should have moved downhill like lava streams rather than spreading over hundreds of square kilometres.
(2) Where did the water come from? The present volume of the Stac Fada Member can be conservatively estimated at about 5 km^3. If 50% by volume of the mudflow before compaction was water (Costa 1988) some 2.5 km^3 of water needs to have been available – equivalent to a medium-sized crater lake.
(3) If the Stac Fada Member was caused by a volcanic eruption why is it associated with an abrupt change in palaeoslope? It will be recalled that the mudflow moved west at Stoer. Lawson (1972, p. 359) reported a southwesterly direction of movement for the Stac Fada Member at Stattic Point. These

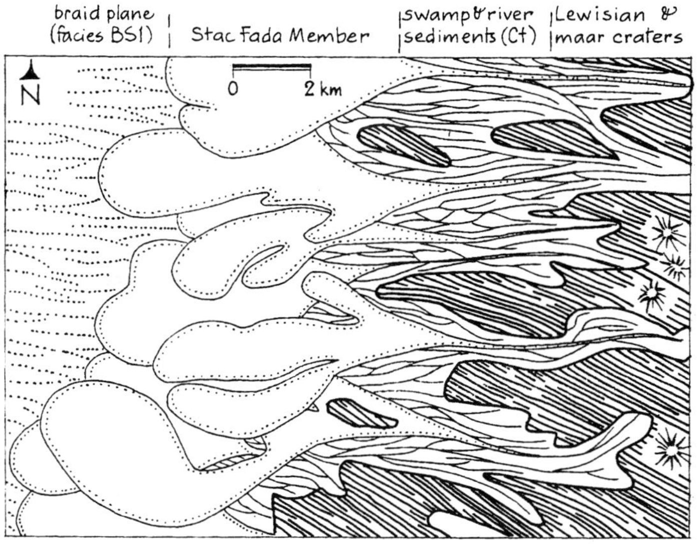

Fig. 11. Map and section showing how the mudflows forming the Stac Fada Member might have emerged from basement valleys containing the Clachtoll Formation and spread across the Bay of Stoer alluvial plain.

directions imply that the eastwardly inclined alluvial plain upon which the Bay of Stoer alluvial sandstones (facies BS1) formed had been tilted in almost the opposite direction immediately prior to deposition of the Stac Fada Member.

The three problems outlined above suggest an alternative hypothesis – that the Stac Fada Member originated in a dry, hilly region east of Stoer, where a group of maar volcanoes had deposited tephra. A dry, hilly source is indicated because mountains with an annual rainfall under 500 mm, such as those of central Asia (Rickmers 1913, p. 193–199), form the natural habitat of mudflows. Maar volcanoes result from hydroclastic eruptions, for example on rift floors where magma and ground-water are likely to come into contact, and could produce glassy tephra of basic composition like those in the Stac Fada Member (Fisher & Schmincke 1984, p. 257–262). Possibly in response to a major earthquake a large volume of rain-soaked sediment and tephra was dislodged and moved westwards through the valleys in which the Clachtoll formation was still accumulating and out across the coeval Bay of Stoer alluvial plain (Fig. 11). A later, similar event redeposited the accretionary lapilli. The fault movement that caused the earthquake also initiated the change in palaeoslope direction described above. The only shortcoming of this tentative hypothesis is that the basement now exposed east of Stoer is not known to be cut by late Precambrian volcanic pipes.

The Poll a' Mhuilt Member that follows the Stac Fada records the history of a perennial lake (Fig. 10). There is a detailed description on pp. 66–68. The basal unit **A** comprises fine to medium-grained sandstones with sabkha-like carbonate nodules, deposited in shallow water round the lake. The sandstones contain a substantial component of volcanic ash and have a correspondingly high Ni content of 130–150 ppm. The limestones of unit **B** may mark the lake shore. The water depth deepened abruptly to about 40 m following deposition of this unit so that black carbonaceous shale (unit **C**) with cryptarchs (p. 67) immediately follows the limestone. The water depth can be gauged roughly by decompacting the total thickness (about 20 m) of the permanent lake sediments forming units **C**–**F**. The gypsum and boron contents of units **C**–**E** show that the lake was hydrologically closed, i.e. evaporation was greater than inflow. Calcite should have preceded gypsum as a primary precipitate (Drever 1997, fig. 15-3) but is difficult to identify separately from that produced by the albitization of feldspar during diagenesis. The boron content of illite roughly doubled during deposition of these units as shown in Figure 10, suggesting a concentration factor of ten for boron in the lake water (Stewart & Parker 1979).

An increasing clastic input mainly from a westerly source is evident in units **E** and **F**, and by the start of unit **G** the lake ceased its permanent existence: the shaly units **G** and **H** are all desiccated. The disappearance of the permanent lake was probably due to aridity rather than lack of space, for the ephemeral lake sediments of units **G** and **H** are together over 50 m thick. If this supposition is correct the absence of any trace of sodium salts (e.g. analcime) in the sediment suggests that the lake initially contained fresh water, rather than the sea water that Downie (1962) and Cloud & Germs (1971) thought a necessary environment for the cryptarchs in unit **C**. The lake sediments were finally overwhelmed by river sediment (Meall Dearg Formation, facies MD1 & MD2) derived from the east. The top of unit **H** contains no sand beds to presage the approach of a fluvial system, so the abrupt appearance of Meall Dearg pebbly sediment may record a tectonic upheaval like that preceding the Stac Fada Member.

The depression of the area was probably tectonic, due perhaps to intermittent downward movement adjacent to a fault, though there is no direct evidence for this. The extent of the depression was quite limited for the thickness of the Poll a' Mhuilt Member is only 25 m at Enard Bay, 15 km south of Stoer, and 10 m at Stattic Point, 33 km south of Stoer (Fig. 4). Moreover the deep-water phase of lake history (units **C**–**E**) is missing at Enard Bay.

The setting for the Poll a' Mhuilt lake is shown in Figure 11. Just prior to the arrival of the Stac Fada mudflow the valley-confined sediments of the Clachtoll Formation were still being deposited by streams flowing from the east, while the toe of the Bay of Stoer bajada was advancing from the west. The area was then raised to the east and the Bay of Stoer alluvial plain warped or faulted down. The down-warp not only trapped the Stac Fada mudflow when it arrived but also defined the basin in which the Poll a' Mhuilt Member accumulated.

Geochemistry

Before embarking on a detailed examination of the rocks it is useful to look at their overall composition on a graph of soda against potash (Fig. 12). The main features that emerge are:

- The sodic nature of the Scourian gneisses beneath the Stoer Group, with average Na_2O well above that for average Archaean crust.
- An antipathetic relationship between Na_2O and K_2O in the sediments.
- High Na_2O in the shales, relative to either average Archaean or post-Archaean shale.

The same rocks plotted on a graph of K against Rb show that although some of the basal sediments have unusually high K/Rb ratios like those in the underlying basement most of the sandstones and all the shales have ratios that are much lower. The sediments with low K/Rb ratios also have concentrations of K and Rb so much higher than in the basement that it appears that these elements must have been contributed by some additional source. According to Stewart (1991*a*) the extra source was potassic volcanic material like that in the Stac Fada Member. Young (1999*a*), however, has proposed that the K and Rb were added metasomatically during burial diagenesis. The antipathetic relationship between K_2O and Na_2O has been shown by Van de Kamp & Leake (1997) to arise from the incomplete albitization of plagioclase and K-feldspar.

The metamorphic basement at Stoer

The basement immediately to the east of the Stoer Group belongs to the Scourian gneiss complex, of Archaean age. The rocks were extracted from the mantle at about 2900 Ma and metamorphism

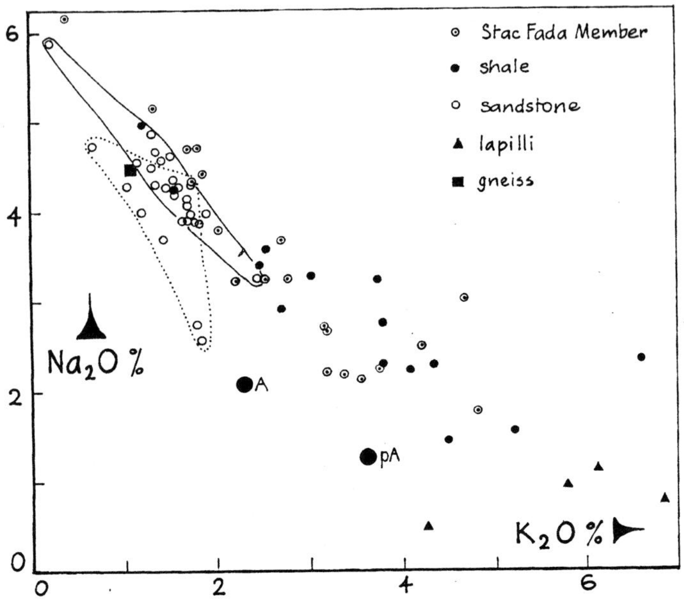

Fig. 12. The chemistry of Stoer Group sediments in terms of K_2O and Na_2O. The average composition of local Scourian gneisses and dykes (from Table 1B) is shown by the black rectangle. The average composition of Archaean (A) and post-Archaean (pA) shales are shown by large dots (from Taylor & McLennan 1985, tables 7.8 & 2.9, respectively). The field occupied by sandstones of the Bay of Stoer and Meall Dearg Formations is outlined by a solid line; the sandstones of the Clachtoll Formation (facies Ct2) are outlined by a dotted line (Donnellan 1981). Data for the volcaniclastic Stac Fada Member are from Young (2002, table 1) and the shales of facies Ct3 from Young (pers. comm.). The four lapilli analyses are from Young (2002, table 1), Stewart (1991*a*, table 2A) and Lawson (1972, table 1).

reached its peak at about 2700 Ma. The complex is well exposed and has been the object of prolonged research and frequent review (Sheraton *et al.* 1973; Cartwright *et al.* 1985; Park *et al.* 2002). The dominant rock type is orthogneiss of granodioritic or tonalitic composition, but basic and ultrabasic rocks are also common. Metasediments form about 10% of the Scourian near Stoer. The complex is notable for the almost complete removal of the heat-producing elements U, Th, Rb and K during crust formation (Tarney & Weaver 1987; Rollinson 1996). As a consequence, the K/Rb ratio in these rocks ranges between 500 and 3000, compared with an average upper crustal value of about 250 (Taylor & McLennan 1985, table 2.15). The depletion of thorium is reflected by the ratio La/Th = 27 compared with a value of around 3 or 4 in average crust of any age (Taylor & McLennan 1985, tables 2.15 & 7.10).

The Archaean basement was intruded by a major dyke suite, the Scourie dykes, over the period 2000–2400 Ma (Park *et al.* 1994). The dykes are predominantly dolerite and range in thickness up to 100 m. The dykes and the Scourian were severely deformed and migmatized by the Laxfordian orogeny at about 1700 Ma. The Laxfordian complex differs fundamentally from the Scourian in having typical post-Archaean upper crustal chemistry (Table 1C), except for the absence of a well-defined negative europium anomaly (Rollinson 1996, fig. 3). The ratio Na_2O/K_2O is only 2.4 and K/Rb averages 195 (Bowes 1972). Thorium is at upper crustal levels, with La/Th = 3. The nearest Laxfordian rocks to Stoer are 20 km distant. They are also developed in the Outer Hebrides, about 100 km to the west.

The average chemistry of the Archaean basement and the early Proterozoic (Scourie) dykes that cut it, based on systematic sampling of an area of 150 km^2 east of Stoer, is shown in Table 1B. The modal mineralogy is dominated by plagioclase (Table 1A) which in the Scourian acid gneisses of the mainland is oligoclase with a normative composition near An_{35} (Peach *et al.* 1907, p. 66; Bowes 1972). In the equivalent rocks of the Outer Hebrides the plagioclase is stated to average An_{27} (Fettes *et al.* 1992, p. 17).

Table 1. *Average chemistry of basement rocks near Stoer*

A Scourian mode			**B** Scourian chemistry $n = 208$	**C** Laxfordian chemistry $n = 219$
Quartz	25	SiO_2	64.20	69.4
Plagioclase	53	TiO_2	0.61	0.4
K-feldspar	4	Al_2O_3	15.66	14.7
Biotite	9	t.Fe_2O_3	5.96	3.2
Pyroxene	4	MnO	0.07	0.05
Hornblende	3	MgO	2.65	1.6
Other	2	CaO	5.10	3.1
		Na_2O	4.46	4.4
Total	100	K_2O	1.12	2.0
		P_2O_5	0.16	0.1
		Total	99.99	98.95
		Ba	830	795
		Ce	41	65
		La	≈19	55
		Ni	43	25
		Rb	13	85
		Sr	528	530
		Th	0.7	12
		Y	9	n.d.
		Zr	190	135
		K/Rb	715	195
		Rb/Sr	0.02	0.16
		Eu/Eu*	1.02	–
		$(La/Yb)_N$	6.5	–
		A	47.3	49.8
		CN	49.0	42.9
		K	3.6	7.3

Modes and major element data are per cent, the latter recalculated to total 100% volatile free. Traces are in ppm. Not determined = n.d. Number of analyses averaged is *n*. Column **A** is an estimate of the average mineralogy of the basement east of Stoer, based on a mode for the Scourian quartzo-feldspathic gneisses of the Outer Hebrides (Fettes *et al.* 1992, table 2) and Scourian dolerite dykes (Tarney 1973), combined in the ratio 9 : 1. Column **B** is an estimate of the chemistry of the basement, based on an average of 154 Scourian gneisses collected on a kilometre grid from the Assynt area, immediately east of Stoer (Sheraton *et al.* 1973, table 2B), and 54 Scourie dolerite dykes from the same area (Tarney 1973, table 2A & B), combined in the ratio 9 : 1. The Scourian gneiss average lacks La so the table above gives instead the average of 254 La determinations from similar gneisses at Drumbeg, about 7 km from Stoer (Sheraton *et al.* 1973, table 2A). Column **C** is an estimate of the composition of the Laxfordian complex by Bowes (1972). The original analysis included $H_2O = 1.1\%$ and $CO_2 = 0.2\%$. Thorium is a weighted average from Sheraton *et al.* (1973, table 4C–E). A, CN and K are mol percentages of Al_2O_3 (equivalent to the CIA index), CaO and Na_2O, and K_2O, respectively. The ratio Fe_2O_3/FeO for the Laxfordian is 0.4 (Bowes 1972).

The breccio-conglomerate and tabular sandstone facies

The lowest part of the breccio-conglomerate forming the well-known outlier at Clachtoll (Hambrey *et al.* 1991, p. 113, loc. 1D) consists of rounded blocks of pale grey picrite up to about a metre in size, in a matrix of mafic sand, in basement hollows 1–2 m deep along the northern edge of the outcrop. The breccia above, which is about 5 m thick, is mainly composed of subangular basic to acid gneiss clasts about 40 cm in size, in a sandy matrix. In the highest part of the deposit, clast size diminishes to about 4 cm and thin red and green sandstone bands appear. The source of the igneous material is almost certainly the Scourie picrite dyke, over 100 m wide, that crops out about 50 m NE of the breccia (Barber *et al.* 1978, fig. 12). Although it appears obvious in the field that the breccia was derived from local basement, the possibility mentioned above that potassium and other elements have been added to the basal sediments, either by metasomatism or in fine-grained tephra, makes a closer look at the chemistry advisable.

The chemistry of the breccia matrix and a sandstone near the stratigraphic top of the breccia are shown in Table 2. The matrix of the lowest, mafic, breccia (Table 2B) has a ratio Fe/Al suggesting that it is composed of a mixture of 75% picrite and 25% average basement (Table 2A). Upwards the proportion of mafic material falls to around 25% (Table 2C). The source rocks (Table 2A & E), it will be noted, are poor in Rb, K, Th and Y like the average basement (Table 1). They have much more Mg, but less alumina. Rubidium and K_2O values in the breccia are even lower than in the putative source rocks, almost certainly due to albitization. The Ca displaced by this process is partly in calcite (Table 2B) or epidote that, from the chemistry, forms about 30% of the green sandstone (Table 2D).

There is not the slightest evidence in these data that anything but Archaean basement has contributed material to this breccio-conglomerate.

Table 2. *Chemistry of sandstones forming the matrix of the breccio-conglomerate facies Ct1 in the Stoer Group outlier at Clachtoll, and possible sources*

	A Model source 1	B Breccia matrix $n=1$	C Breccia matrix $n=1$	D Sandstone band $n=1$	E Model source 2
SiO_2	51.8	32.5	53.2	58.5	58.9
TiO_2	0.4	0.34	0.73	0.75	0.5
Al_2O_3	7.7	5.93	12.28	12.72	13.2
t.Fe_2O_3	10.1	7.66	9.73	7.84	7.4
MnO	0.2	0.18	0.13	0.12	0.1
MgO	22.3	9.13	9.81	5.73	9.7
CaO	5.1	20.60	3.55	7.92	5.5
Na_2O	1.8	1.24	3.06	2.14	3.6
K_2O	0.5	0.05	0.26	0.05	0.9
P_2O_5	0.1	0.09	0.26	0.25	0.2
LOI	–	19.68	4.58	3.40	–
Total	100.0	97.40	97.59	99.42	100.0
CO_2	–	17.70	0.41	0.40	–
Rb	12	0	2	0	11
Sr	223	76	288	2286	454
Ba	304	50	125	49	606
Ni	1516	1082	545	304	544
Cr	2077	1705	1570	741	751
Zr	85	81	264	322	162
Y	7	8	14	14	8
K/Rb	346	<800	1079	<800	664
Rb/Sr	0.05	<0.01	0.01	<0.0004	0.02
A	37.7	61.8	54.5	43.5	44.4
CN	59.5	37.6	44.2	56.3	52.4
K	2.8	0.6	1.2	0.2	3.2

Major element data (in per cent) are ICP–MS analyses by Salim Malik, and the traces (in ppm) are XRF analyses by F. Street, both in the Postgraduate Research Institute for Sedimentology, University of Reading. The number of samples analysed is *n*. Column **A**, the model source for the lowest sediments, is 75% average picrite (Tarney 1973, table 1A–C) and 25% average Scourian gneiss (Sheraton *et al.* 1973, table 4A). Column **B** is an analysis of sandstone forming the breccia matrix 20 cm from the unconformity on the northern flank of the outlier [NC 04112666]. Column **C** is sandstone forming the breccia matrix about 10 m laterally from the gneiss [NC 04112665]. Column **D** is medium-grained green sandstone about 20 cm from the unconformity on the southern flank of the outlier [NC 04112664]. Column **E** is a model source for the uppermost sediments, composed of 25% picrite and 75% Scourian gneiss. A, CN and K are mol percentages of Al_2O_3 (equivalent to the CIA index), CaO & Na_2O, and K_2O, respectively.

The tabular-bedded pebbly sandstone (facies Ct2) at Stoer, always found laterally equivalent to the breccio-conglomerate facies, shows similar features. The chemistry of two samples about 60 m from the unconformity near Stoer village are shown in Table 3B & C. The Ba/K ratio for this sandstone has been taken by Van de Kamp & Leake (1997, fig. 11) to indicate a granulitic provenance, as does the very high ratio for K/Rb. Detrital plagioclase in the sandstone is commonly cloudy, with chess-board twins, and can be identified as albite by microprobe analysis (P. Van de Kamp, pers. comm.). Normative calculations show that about 85% of the original detrital andesine and K-feldspar has been albitized. Van de Kamp & Leake (1997) suggest that a proportion of the plagioclase in the basement was destroyed by weathering but that the Na released was regained by albitization under closed basin conditions. The sandstones in Table 3, indeed, have Na well in excess of what was supplied by the basement. The present ratio Na_2O/K_2O is about 6, similar to what it was before albitization and much greater than the value of 4 in the basement.

Samples of coarse sandstone, facies Ct2, from the type section on the shore at Stoer, about 350 m from the nearest gneiss outcrop, still show very low values for heat-producing elements (Table 3C). The ratio La/Th = 26, like that in the basement, is particularly significant for it is most unlikely to have been altered by sedimentary differentiation. On the other hand, a set of six samples of the same facies Ct2 at Achiltibuie, analysed by Donnellan (1981) and published by Van de Kamp & Leake (1997, fig. 5b), are quite unlike those at Stoer in having K/Rb ranging from 200 to 300 and average La/Th = 4. The Achiltibuie sandstones are texturally like those at Stoer and are adjacent to ultrabasic Scourian basement (Peach *et al.* 1907, p. 182; Bowes *et al.* 1964) but nevertheless contain a substantial component from elsewhere.

The shales and muddy sandstones

The two facies of the Clachtoll and Bay of Stoer Formations that contain an important clay component, the shale (Ct3) and the muddy sandstone (Ct7) are both notably enriched in elements generally lacking from the basement (Table 4). This is also shown by all the published analyses of shales and muddy sandstones from the Stoer Group (Stewart 1991*a*, table 1C & E; Van de Kamp & Leake 1997, table 2C; Young 1999*a*, table 1).

The most important difference is in uranium that forms only 0.05 ppm of the Scourian (Moorbath *et al.* 1969) but 3 ppm in the shales (Young 1999*a*, table 1). This is unexpected, for under oxidizing conditions uranium takes the soluble form U^{6+} and should therefore be depleted in the sediment rather than enriched (Drever 1997, p. 184). Thorium and Rb are both ten times more abundant in the fine-grained sediment than they are in the basement. In the basement La/Th = 27 whereas in the shales and muddy sandstones it is only about 4, a value near the average for shales (Taylor & McLennan 1985, fig. 2.16).

Potassium is up to five times more abundant in the shales and muddy sandstones (K/Rb ≈ 250) than in the basement or the tabular sandstone (K/Rb ≈ 700). The relatively high K values for the shales can be attributed to three processes. The first is the universal tendency for K-rich micas and clays winnowed from sands to be concentrated in laterally contiguous shales. This also concentrates Rb in the shales and somewhat reduces the K/Rb ratio relative to the sands. The second process is the probable transfer of K and Rb to the shales during the albitization of K-feldspar in the basement-derived sands. Despite albitization the K/Rb ratio of the sandstones has remained high, so K and Rb would have to have been removed from the sands and transferred to the shales in the same ratio, i.e. K/Rb ≈ 700. Such a process, however, would give very little Rb to the shales and do nothing to produce the observed ratio K/Rb = 250. The third process, believed to be dominant, is the addition to the shale of a detrital component relatively rich in K and Rb, with a ratio K/Rb in the region of 200. The possibility of K-metasomatism is discussed below.

Table 3. *Mineralogy and chemistry of the tabular-bedded pebbly sandstone facies Ct2 in the Clachtoll Formation at Stoer*

	A Modal mineralogy $n=2$		B Sandstone chemistry $n=2$	C Sandstone chemistry $n=2$
Quartz	25.4	SiO_2	60.99	64.23
Plagioclase	54.9	TiO_2	0.69	0.65
K-feldspar	0.4	Al_2O_3	14.44	12.76
Muscovite	1.0	Fe_2O_3	4.95	5.28
Biotite	0.7	FeO	0.99	n.d.
Chlorite	1.2	MnO	0.11	0.12
Clay	2.7	MgO	3.57	3.88
Calcite	2.4	CaO	4.90	5.01
Hematite	5.7	Na_2O	4.66	4.38
Opaque	2.0	K_2O	0.75	0.83
Zircon	0.1	P_2O_5	0.11	0.14
Epidote	3.3	H_2O+	1.96	n.d.
Dolerite	0.2	CO_2	1.00	n.d.
Total	100.0		99.12	97.28
		Ba	320	623
		Ce	–	43
		Co	18	n.d.
		La	–	26
		Ni	61	53
		Rb	8	14
		Sr	425	324
		Th	n.d.	1
		Y	20	11
		Zn	51	60
		Zr	290	209
		K/Rb	778	492
		Rb/Sr	0.02	0.04
		A	49.3	46.7
		CN	47.9	50.0
		K	2.8	3.3

Major elements are per cent and traces in ppm. Where FeO has not been detemined (n.d.) total iron is expressed as Fe_2O_3. Number of samples analysed is n. Columns **A** & **B** show the average mineralogy and chemistry for the same two samples of facies Ct2 from an outcrop just south of the old graveyard at Stoer [NC 040283]. The data are from Van de Kamp & Leake (1997, tables 1 & 2). The lithology is medium to very coarse sandstone with poor to moderate sorting and angular to sub-angular grain shapes. Mineralogy was determined by counting 400–600 points using the Gazzi–Dickinson method, after staining K-feldspar and plagioclase. Authigenic albite amounts to 3.0% of total feldspar. The chemistry is by XRF, except that an aliquot of the powder was used to determine FeO by titration. Column **C** is an averge of two XRF analyses of coarse sandstones from the Clachtoll Formation by Donnellan (1981), published by Stewart (1991*a*, table 1C). They are exposed on the shore at Bay of Stoer [NC 038283]. A, CN and K are mol percentages of Al_2O_3 (equivalent to the CIA index), CaO & Na_2O, and K_2O, respectively.

Table 4. *Chemistry of shales (facies Ct3) and muddy sandstones (facies Ct7) in the Stoer Group at Stoer*

	A Shale (Ct3) $n=4$	B Shale (Ct3) $n=4$	C Muddy sandstone (Ct7) $n=2$	D Muddy sandstone (Ct7) $n=2$
SiO_2	60.21	61.43	56.67	59.59
TiO_2	0.68	0.70	0.76	0.80
Al_2O_3	15.78	14.90	13.21	12.76
Fe_2O_3	7.22	5.28	8.43	6.60
FeO	n.d.	1.02	n.d.	n.d.
MnO	0.08	0.07	0.11	0.11
MgO	4.23	3.58	7.91	7.73
CaO	0.76	2.33	2.88	4.55
Na_2O	2.52	3.97	2.46	3.19
K_2O	4.13	2.40	3.24	1.57
P_2O_5	0.15	0.11	0.15	0.17
LOI	3.94	2.85	3.65	n.d.
Total	99.70	98.64	99.47	97.06
Ba	803	413	n.d.	472
Ce	64	n.d.	57	48
La	32	n.d.	44	23
Ni	93	68	47	64
Rb	128	73	109	42
Sr	138	185	196	248
Th	8	n.d.	n.d.	6
Y	21	28	24	19
Zr	192	238	185	215
K/Rb	268	273	247	317
Rb/Sr	0.93	0.39	0.56	0.17
A	61.6	53.2	56.1	53.7
CN	20.9	37.5	29.0	39.1
K	17.4	9.3	14.9	7.2

Major elements are in per cent and traces in ppm. Where FeO has not been determined (n.d.) total iron is expressed as Fe_2O_3. Number of samples analysed is n. Column **A** is an average of four samples of shale from couplets in the Bay of Stoer Formation, just beneath the Stac Fada Member (Young 1999*a*, table 1). The method of analysis of these samples is not stated. Column **B** is an average of four samples of shale, three from the Clachtoll Formation [NC 039292; 042271; 037271] and one from the Poll a'Mhuilt Member [NC 032286], published by Van de Kamp & Leake (1997, table 2C). Analyses are by XRF except for FeO, determined by titration. Loss on ignition (LOI) comprises H_2O+ (2.26%) and CO_2 (0.59%). Column **C** is an average of two XRF analyses of muddy sandstone by F. Street (Postgraduate Research Institute for Sedimentology, Reading University). One sample is from a road-side quarry at Clachtoll [NC 04232727] and the other from the base of a couplet on the type section of the Bay of Stoer Formation [NC 034285]. Column **D** is an average of two XRF analyses of muddy sandstones from the Clachtoll Formation by Donnellan (1981), published by Stewart (1991*a*, table 1C). They are exposed on the shore at Bay of Stoer [NC 038283]. A, CN and K are mol percentages of Al_2O_3 (equivalent to the CIA index), CaO & Na_2O, and K_2O, respectively.

The difference in geochemistry between the shales and the Scourian gneisses is emphasized by the chondrite-normalized rare-earth patterns shown in Figure 13. The pattern for the gneiss east of Stoer shows an insignificant negative europium anomaly (Eu/Eu* = 1.02) together with light rare earth enrichment, features typical of Archaean basement (Taylor & McLennan 1985, fig. 9.1). By contrast there is a marked negative europium anomaly in the shales (Eu/Eu* = 0.73) and the Stac Fada Member (Eu/Eu* = 0.79), like that generally found in post-Archaean upper crust and clastic sediment. Thus the Clachtoll shales and muddy sandstones contain a post-Archaean detrital component unrelated to the local basement.

A clue to the nature of this component is provided by the work of O'Nions *et al.* (1983) on samarium and neodymium in the Poll a' Mhuilt Member. Two shale samples from the member have crustal residence ages (τ) of 2.00 and 2.33 Ga, respectively. These ages are much less than for the Scourian, that has $\tau = 2.9$ Ga (Hamilton *et al.* 1979). However, if roughly a third of the analysed shales were composed of a volcanic component, derived in two stages from the mantle during deposition of the Stoer Group, they would have the right crustal residence age (Stewart 1991*a*). Mantle-derived Laxfordian material (if such ever existed) is not present because the Laxfordian lacks the required negative europium anomaly, and also because zircon grains in the Stac Fada Member and the subjacent Bay of Stoer sandstones are almost entirely of Scourian age (Rainbird *et al.* 2001). Grains giving Laxfordian ages form only a few percent of the suite, making it difficult to appeal to *any* basement source for the europium anomaly in the shale.

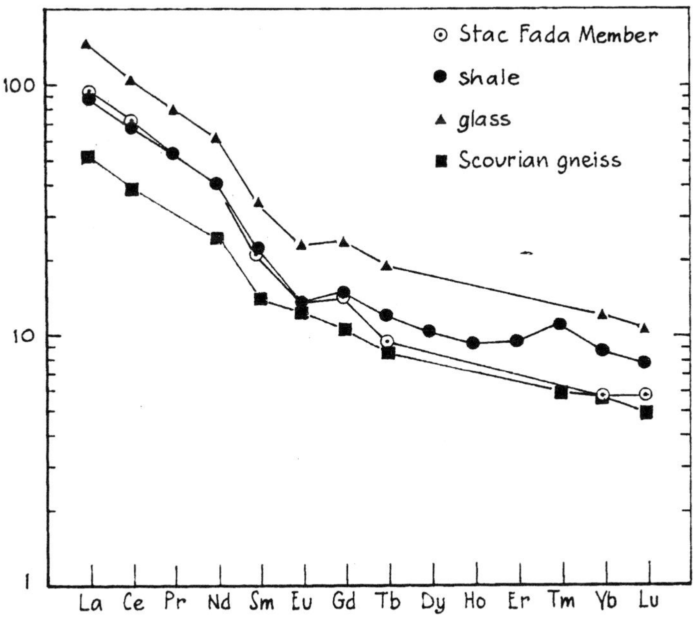

Fig. 13. Chondrite normalized rare-earth element patterns for the Stoer Group (shales, lapilli glass and Stac Fada Member), and the predominantly Scourian source rocks. Note the negative europium anomaly in the sediments and the glass, but not in the gneisses. The patterns for the Stac Fada Member and for the shales are each based on ten samples from Stoer (Young 2002, table 1 and 1999*a*, table 1). The lapilli glass is a single sample from Stoer (Young 2002, table 1). The pattern for the Scourian is a composite of 65% tonalitic gneiss and 10% basic gneiss, both from Weaver & Tarney (1980), 10% dolerite dyke (Weaver & Tarney 1981), and 15% metasediment (Fettes *et al.* 1992, fig. 8, analysis S 63088). All patterns are normalized to carbonaceous chondrite (Taylor & McLennan 1985, p. 298).

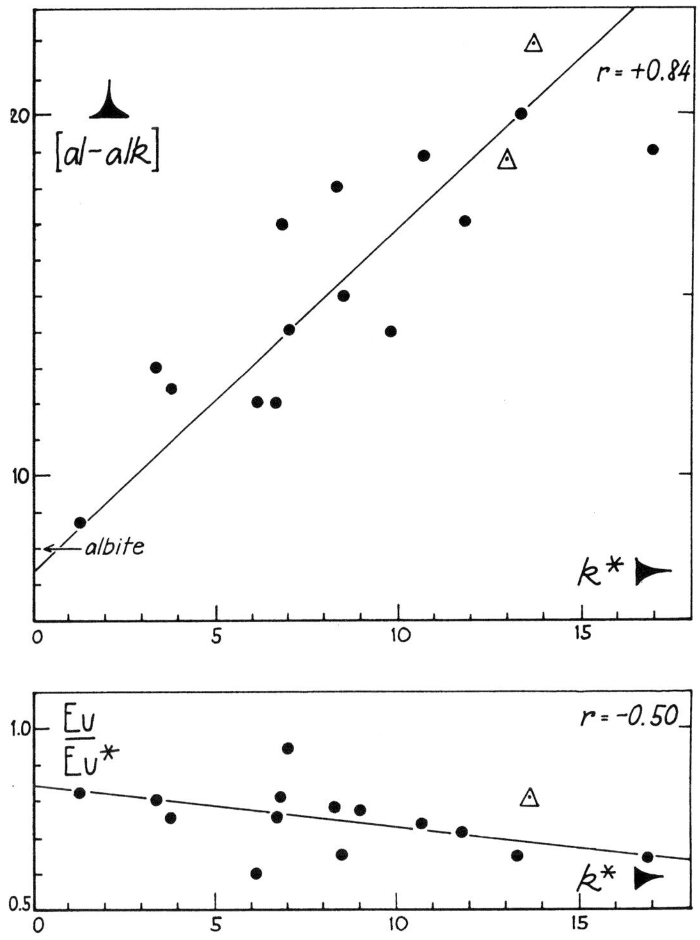

Fig. 14. The potassium content (k^*) of shales in the Stoer Group in relation to Niggli [*al–alk*] and Eu/Eu*. The parameter k^* is the product of the Niggli values *al* and *k*, equivalent to mol per cent potassium in the rock analysis with silica excluded. The Niggli term [*al–alk*] measures alumina available to form minerals such as muscovite, illite and chlorite (Niggli 1954, p. 15). Eu/Eu* is the europium anomaly (Taylor & McLennan 1985, p. 299). The fourteen shale analyses (black dots) were kindly provided by G. M. Young. Volcanic glass analyses (triangles) are from Young (2002, table 1).

Young (1999*a*) has argued that the relatively high K and Rb concentrations in Stoer Group sediments as a whole, compared with local basement rocks, are due to metasomatising fluids that permeated the sediments during burial diagenesis and enriched the clays in these elements. Regional potassic metasomatism of this kind is well known (e.g. Fedo *et al.* 1997; McLennan *et al.* 2000). There are, however, two objections to the application of the hypothesis to the Stoer Group. First, the sandstones of the group contain scarcely any illite – most of the K and Rb is in the cores of the detrital K-feldspar grains that have survived albitization. It is therefore the composition of the feldspar, not the illite, that is relevant. As pointed out above, some of these sandstones have low K and Rb contents, with high ratios for K/Rb, whereas others, although sedimentologically similar, are rich in K and Rb, with low ratios for K/Rb. If the metasomatism has generated K-feldspar it is difficult to see why it should have affected some sandstones but not others. Secondly, although the K in the shales is associated mainly with clay (Fig. 14) it also correlates (weakly) with increasingly negative Eu/Eu* which suggests that it originated in tephra rather than a metasomatizing fluid.

The significance of the glassy tephra in controlling sediment composition can be viewed graphically by using the chemical index of alteration (Nesbitt & Young 1982, 1984). This index uses the sediment chemistry to measure the conversion of original feldspar to clays, by combining the molecular proportions of Al, Ca, Na and K as follows:

$$\text{Chemical index of alteration (CIA)} = 100 \times Al_2O_3/(Al_2O_3 + CaO^* + Na_2O + K_2O)$$

Note that CaO* is calculated from the CaO in silicates only. The index is about 100 for highly aluminous sediment such as clay, but as low as 45 for some the feldspar-rich basement rocks. The molecular proportions can also be plotted on a triangular diagram

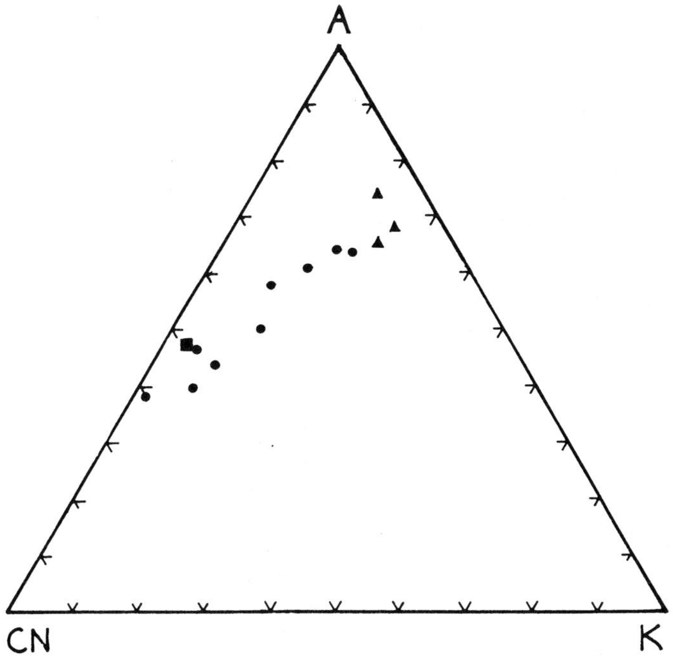

Fig. 15. Stoer Group shales (black dots) plotted in A–CN–K space (Young 1999*a*, Fig. 5a). The predominantly Scourian source rocks (from Table 1) are shown by a rectangle and three samples of volcanic glass from the Stac Fada Member (Table 5 and Stewart 1991*a*, table 2A) by triangles.

with apices Al_2O_3, ($CaO^* + Na_2O$) and K_2O, called A, CN and K for short. The A–CN–K diagram in Figure 15 shows that the trend in shale composition attributed by Young (1999*a*) to metasomatism could just as well arise from the mixing of two end members – the basement and the volcanic glass. Figure 15 shows the present composition of the glass, not the original composition which would have been calcic. This calcium has now been lost from the silicate system, and probably the sediment as well (cf. Table 5A), so that it has no relevance to the A–CN–K diagram in Figure 15.

The Stac Fada Member

The member is composed of coarse muddy sandstone containing plagioclase, microcline and quartz grains averaging about 0.1 mm in size set in a fine-grained ferruginous matrix. Grains over 0.5 mm in size, mainly microcline and gneiss fragments, are rare. The most striking feature of the rock, however, is the abundance of vesicular pale-green glassy fragments ranging in size up to about 70 mm, properly called lapilli (Fisher & Schmincke 1984, p. 90). Analysis of the lapilli by XRD and XRF shows that chlorite and illite are the main components.

Table 5. *Chemistry of the Stac Fada Member at Stoer*

	A Stac Fada whole rock $n = 1$	B Volcanic glass $n = 1$	C Volcanic glass $n = 1$	D Stac Fada (model)
SiO_2	62.28	50.32	43.20	63.0
TiO_2	0.60	1.16	1.44	0.8
Al_2O_3	14.66	19.30	16.79	15.0
$t.Fe_2O_3$	6.50	7.42	16.07	6.1
MnO	0.10	0.11	0.34	0.1
MgO	4.64	6.82	13.92	4.8
CaO	1.01	0.81	0.59	4.1
Na_2O	3.69	0.73	0.49	3.6
K_2O	2.56	6.38	4.12	2.3
P_2O_5	0.19	0.30	0.17	0.2
LOI	3.59	6.39	n.d.	–
Total	99.82	99.74	97.13	100.0
Ba	750	754	n.d.	655
Ce	68	100	44	57
La	34	54	30	33
Ni	154	558	1266	179
Rb	76	288	199	82
Sr	178	46	42	254
Th	6	9	n.d.	3
Y	16	38	21	17
Zn	60	85	169	66
Zr	175	396	353	256
K/Rb	280	183	172	233
Rb/Sr	0.43	6.3	4.7	0.3
A	58.9	68.1	73.8	–
CN	30.0	7.5	6.6	–
K	11.1	24.4	19.6	–

Major element data are per cent and traces in ppm. Not determined = n.d. Number of samples analysed is *n*. Column **A** is the average of ten analyses of the Stac Fada Member at Stoer analysed by ICP (Young 2002, table 1). Column **B** is an analysis of the volcanic glass chiselled out of the Stac Fada Member at Stoer and analysed by ICP (Young 2002, table 1). Column **C** is an XRF analysis of volcanic glass hand picked from a crushed sample of the member at Stoer (Stewart 1991*a*, table 2). Column **D** is a model composition for the Stac Fada Member based on 75% facies Ct2 (this volume, Table 3C) and 25% volcanic glass (this Table, column **B**). A, CN and K are mol percentages of Al_2O_3 (equivalent to the CIA index), CaO & Na_2O, and K_2O, respectively.

The glass is poor in silica but rich in Mg, Fe and Ni (Table 5B & C), suggesting a highly mafic original composition. The high K and Rb values, both thought to be original constituents for reasons given earlier, together with the contents of Si, Al and Ti, suggest a Type III potassic lava (Foley *et al.* 1987) rather than the Type II (East African) lava previously suggested (Stewart 1991*a*). Unlike ultra-potassic rocks the glass is poor in CaO and Na_2O, both lost during diagenesis (Stewart 1991*a*). The Ca may be represented by the abundant calcite in the Stac Fada Member and adjacent shales; evidence for the movement of Na out of the member is given below in the section on Na metasomatism.

The whole-rock chemistry of the Stac Fada Member is shown in Table 5A, together with that of the glassy lapilli (Table 5B & C). The chemistry of the member can be modelled satisfactorily (Table 5D) by combining 25 parts of volcanic glass ($K/Rb = 172$) with 75 parts of sandstone derived from the local, mainly Archaean basement (Table 3C). The sandstone has relatively little K and Rb, and $K/Rb = 500$. Laxfordian detritus is largely absent from the Stac Fada Member for, as mentioned earlier, 95% of the zircons obtained from it give Archaean ages when dated by U–Pb (Rainbird *et al.* 2001).

The sandstones of the Bay of Stoer and Meall Dearg Formations

The sandstones are mainly composed of quartz and altered, ruddy feldspar, including microcline and perthite. The detrital grains are enveloped by a ferruginous pellicle, and the pore space is filled by quartz. The chemistry of these sandstones (facies BS1 & MD1) is summarized in Table 6A & B. The relatively low ratios for $La/Th \approx 3.5$ and $K/Rb \approx 300$ show that, like parts of the Clachtoll Formation, the sandstones were not eroded solely from the Scourian complex. Zircons from both formations have concordant $^{207}Pb/^{206}Pb$ ages that are mostly Archaean, with about 20% of the grains in the Meall Dearg giving Laxfordian ages (Rainbird *et al.* 2001). A source for the Meall Dearg that was 60% Scourian and 40% Laxfordian would give the correct K/Rb ratio and a suitably low value for La/Th. The greater abundance of zircon in the Scourian would also explain the high proportion of Archaean ages. A source area extending 40–50 km east of Stoer would have intersected the southeastward prolongation of the Laxfordian injection zone, now exposed near Loch Laxford, and provided a suitable mixture of Laxfordian and Scourian detritus to form the Meall Dearg Formation. The Bay of Stoer Formation contains much less Laxfordian material, for only a single zircon grain out of a sample of thirty has an age of 1900 Ma (Rainbird *et al.* 2001). The source of the Rb and Th in this formation is unknown.

Additional evidence about the provenance of the alluvial sandstones comes from the included pebbles, half of which are microcline-bearing gneiss and the rest quartzite. The latter are mostly fine-grained, undeformed or little deformed fuchsite-bearing sandstones, together with some metaquartzites and a few silica-cemented arkoses. A lack of sand-sized quartzite and gneiss grains in the alluvial sandstones indicates that the source area did not include siliceous sediments as once suggested by the writer (Stewart 1991*a*), but rather a pre-Stoer Group conglomerate capable of yielding the gneiss and quartzite pebbles.

The ratio Na_2O/K_2O in the sandstones lies between 2.3 (facies BS1) and 2.9 (facies MD1). The negative gradient on the graph of Na_2O against K_2O (Fig. 12) is consistent with albitization of both plagioclase and K-feldspar. Before albitization the ratio Na_2O/K_2O would have been lower. A norm for average Meall Dearg sandstone indicates about 41% plagioclase (almost entirely detrital) and 9% of K-feldspar. Knowing the ratio of albite (28%) to oligoclase (13%) in the norm, and assuming a similar extent of albitization in both plagioclase and K-feldspar, it follows that about half the feldspar has been albitized. The sediment had an original ratio Na_2O/K_2O of about 0.7 so that roughly three-quarters of the Na in the source rock was lost during weathering but more than replaced during diagenesis.

Table 6. *Chemistry of alluvial sandstones forming the Bay of Stoer and Meall Dearg Formations*

	A Bay of Stoer facies BS1 $n=7$	**B** Meall Dearg facies MD1 $n=14$
SiO_2	76.22	77.86
TiO_2	0.56	0.35
Al_2O_3	10.25	10.82
t.Fe_2O_3	3.50	2.63
MnO	0.05	0.04
MgO	1.57	1.69
CaO	2.13	0.69
Na_2O	3.95	4.40
K_2O	1.73	1.51
P_2O_5	0.09	0.08
Total	100.05	100.07
Ba	555	480
Ce	57	32
La	31	16
Ni	12	8
Rb	48	42
Sr	127	100
Th	8	5
Y	10	7
Zr	181	135
K/Rb	299	298
Rb/Sr	0.38	0.42
A	46.0	52.1
CN	45.6	40.0
K	8.4	7.9

Major elements are in per cent and traces in ppm. Number of samples analysed is n. Columns **A** & **B** are average compositions based on XRF analyses by Donnellan (1981), publishd by Stewart (1991*a*, table 3). A, CN and K are mol percentages of Al_2O_3 (equivalent to the CIA index), CaO & Na_2O, and K_2O, respectively.

The chemistry of the two sandstone facies (BS1 and MD1) is remarkably similar, the only difference being the greater concentration of heavy minerals in facies BS1, with consequently higher values for Fe, Zr, Y and rare earths. The low levels of Ca and Sr relative to basement values are probably the result of the dissolution of calcic plagioclase during albitization. The relatively high ratio for SiO_2/Al_2O_3 in the sandstones reflects the removal of plagioclase from the source rock during weathering.

Sodium metasomatism

Van de Kamp & Leake (1997) have stressed the importance of albitization, arguing that perhaps a quarter of the plagioclase in the basement was destroyed during weathering of the basement surface and subsequent sediment transport, and that the sodium thereby released was trapped in the sediment and later consumed in transforming detrital andesine and K-feldspar to albite. Apart from the petrography the main evidence for albitization is in Figure 12 which shows a clear antipathetic relationship between Na_2O and K_2O in the alluvial sandstones (facies BS1 and MD1), some of which have an even higher Na_2O/K_2O ratio than that in the local basement. The gradient of -1 defined by the data points in the figure requires that andesine was originally twice as abundant as K-feldspar and that both species in a given sample have been albitized to the same extent.

In order to explore Van de Kamp & Leake's model it is worth attempting to construct a budget for sodium in the mainly locally derived Clachtoll Formation, as shown in Table 7. The first problem that arises is estimating the relative proportions of sandstone, shale and dissolved load. This can be easily done if the chemistry of the sandstone, shale and source rocks are *accurately* known (Stewart 1993*b*), but the original composition of the shale in the Clachtoll is not, and there are few analyses of the sandstones. The best that can be done is to adopt proportions like those for present-day river systems in the warm, dry climates capable of yielding sediment with the low CIA indices apparent in Tables 4 & 5.

Table 7. *Geochemical budget for sodium in the Clachtoll Formation at Stoer*

Na_2O in a kilogram of basement rock		Na_2O in a kilogram of sediment and dissolved load, as supplied		
53% Plagioclase (An_{33})	41.3	*Sand (294 g)*		
4% K-feldspar	2.4	42% Plagioclase (An_{33})	9.1	
9% Biotite	0.4			9.1 g
4% Pyroxene	0.3	*Silt (588 g)*		
3% Hornblende	0.3	30% Plagioclase (An_{33})	13.8	
		14% K-feldspar	0.5	
Total	44.7 g	20% Illite	1.2	
				15.5 g
		Total dissolved load (118 g)		
		17.0% Na_2O	20.1	
				20.1 g
		Total		44.7 g
		Na_2O in a kilogram of sediment now		
		Sandstone (110 g)		
		38% Albite	4.2	
		4% Plagioclase (An_{33})	0.3	
				4.5 g
		Shale (379 g)		
		19% Albite	10.4	
		18% K-feldspar	0.9	
		20% Illite.	0.3	
				11.6 g
		Total		16.1 g

The assumed proportions of sand, silt and total dissolved load are typical of present-day warm, dry climates. Sand/silt = 0.5 (Stewart 1993*b*); silt/dissolved load = 5 (Meybeck 1976, 1988 annex). The present ratio of sandstone to shale is that observed in cross-sections. The modal mineralogy of the basement is repeated from Table 1. The dissolved load does not include atmospheric CO_2. The present plagioclase content of the sandstone (38% An_0, 4% An_{33}) is determined from the Al_2O_3 and Na_2O now in the rock (Table 3A), assuming that similar proportions of detrital K-feldspar and plagioclase have been albitized. The mineralogy of the shale is based on a norm for shale in the Poll a'Mhuilt Member and probably includes a tephra component.

A few general conclusions can be reached from Table 7. The first is that half the plagioclase and the Na supplied by the source rocks were missing from the sand and mud *before* they were deposited. The plagioclase had been weathered to clays and the Na removed as dissolved load. Van de Kamp & Leake (1997) obtained the much lower figure of 25% plagioclase loss because they considered only the composition of the sandstones and shales, but not their relative masses, or the dissolved load. Secondly, Table 7 predicts far more shale in the swampy palaeovalley immediately south of Stoer than is actually shown in the profile, Figure 5. Most of the shale-forming detritus, together with the storm water and the dissolved load it contained, overflowed into an adjacent valley and disappeared. There is nothing surprising about this for the palaeovalley was no more capable of containing all the water and sediment poured into it than a silted-up reservoir.

Compared with the 20.1 g of Na_2O lost in solution by weathering of a kilogram of basement rock, the 5.5 g used in albitization of a kilogram of the existing sediment (2.2 g for sandstone and 3.3 g for shale) is small, and is quite insensitive to the ratio of sandstone to shale. For these Clachtoll sediments, therefore, the suggestion of Van de Kamp & Leake (1997) that some Na lost during weathering has been regained from Na-rich groundwater during diagenesis appears perfectly reasonable, but is much less convincing for the Bay of Stoer and Meall Dearg alluvial sandstones that were not deposited in any kind of hydraulically closed system.

Albitization is particularly evident close to the Stac Fada Member, where virtually all the original andesine and K-feldspar have been made into albite, and the Ca displaced to form the conspicuously eroded calcite-cemented pods seen on the outcrop. Abundant Na may have come from the diagenetic alteration of the volcanic glass in the Stac Fada Member. An indication of the mobility of Na, K and Ca in the member is provided by the feldspar–quartz–calcite–illite veins, estimated to form about 1% by volume of rock. Lawson (1972) has shown that the feldspar is two-thirds albite and one-third orthoclase. Sanders & Johnston (1989) have proposed that the veins were created by the passage of steam, escaping from the freshly deposited, boiling hot Stac Fada mudflow, and later compacted by 40%. The feldspars, however, are undeformed.

Albitization of feldspar is a process that is well-known from sandstones, starting when the burial temperatures approach 100°C and ending at about 150°C. It is often accompanied by the breakdown of detrital smectite (liberating Na) and the formation of illite (requiring K). At least a third of the Na needed for albitization in the Clachtoll Formation could have come from this source. The rest of the Na, and most of the K for illitization and authigenic K-feldspar, must have come from elsewhere. The K may have been derived from the breakdown of volcanic glass, as suggested earlier.

Palaeomagnetism

Palaeomagnetism can give a palaeopole position that helps to date the rocks, and also a palaeolatitude that is relevant to the palaeoclimate. A brief look at the data available is therefore appropriate.

At first sight it looks certain that the magnetization of the Stoer Group is detrital in origin (DRM). There are three steps in the argument. First, thermal demagnetization curves yield unblocking temperatures mainly between 650°C and 700°C showing that the remanent magnetism resides in specular hematite (Stewart & Irving 1974; Torsvik & Sturt 1987). Second, the direction of magnetization of ten sandstone blocks (facies BS1) incorporated in the Stac Fada Member (Fig. 60) is random (Irving & Runcorn 1957, fig. 14). Lastly, these ten sandstone blocks are no different to other sandstones of the same facies, and like them almost certainly have their magnetization held by hematite. It follows that the remanent magnetization of facies BS1 is held by hematite and was randomized by the movement of the Stac Fada mudflow almost contemporaneously with sedimentation.

Unfortunately, some additional facts are difficult to reconcile with this simple story. The source rocks for the Stoer Group were the Scourian basement that has $Fe_2O_3/FeO \approx 0.5$ (Beach 1976) and would have supplied mainly magnetite (Beckmann 1976; Piper & Poppleton 1991; Piper 1992). Any hematite derived from the basement could have come only from shear zones that are quite limited in area (Beach 1976). Polished sections of Stoer Group sandstones show that at least half of the hematite has been mobilized to form shreds lying parallel to the bedding, or girdling detrital silicate grains. Martite textures are common, indicating that the hematite was originally magnetite. The presence of magnetite is suggested by unblocking temperatures between 500°C and 600°C in the Stac Fada Member (Stewart & Irving 1974). Traces of magnetite have also been detected in unspecified facies of the Stoer Group by Torsvik & Sturt (1987) and it is dominant in the matrix of basal conglomerates at Enard Bay (Piper & Poppleton 1991). All these data suggest that most of the magnetite derived from the Scourian basement was converted to hematite, giving present whole rock Fe_2O_3/FeO in the range 5–9 (Van de Kamp & Leake 1997, table 2).

Most magnetite probably converted to martite during soil formation or shortly afterwards (Van Houten 1968, tables 2 & 3; Steiner 1983) but complete conversion requires millions of years (Van Houten 1968, table 4). The slow oxidation of magnetite to hematite may explain some of the variation in palaeomagnetic inclination (I) between sites of virtually the same age, especially bearing in mind that the magnetization vector is inclined in the direction of dip of the beds. For example the Bay of Stoer Formation immediately below the Stac Fada Member at Poolewe gives $D = 310°$, $I = +43°$ (Smith *et al.* 1983 supplementary publication table 2 site 136) while the same horizon at Stoer, 55 km away, gives $D = 310°$, $I = +30°$ (Stewart & Irving 1974, table 1 site SS10). The data from Poolewe, in fact, give an average inclination of +51°, as compared with +29° at Stoer (Smith *et al.* 1983, p. 35). Both sequences are thick and it seems most unlikely that compactional dips are involved. Most of the Poolewe rocks are greyish green rather than pink, the result of hot formation water coming from the nearby Loch Maree fault, but Smith *et al.* nevertheless found that hematite was responsible for the magnetization. Their explanation for the discrepancy is a difference in age between Poolewe and Stoer, during which interval there was rapid polar wandering. However, there is no evidence for a significant age difference. Piper & Poppleton (1991) claim that the steeper inclinations result from the growth of hematite during the Mesozoic and consequent magnetic overprinting. It is probable that more careful thermal demagnetization of the Stoer rocks would clear up some of these problems.

Overall the data from Stoer appear the more reliable, giving mean $D = 315°$, $I = +27°$ (16 sites) after blanket thermal cleaning at 500°C, and a palaeolatitude of about 14°N (Stewart & Irving 1974, table 1). The mean pole for the Stoer Group is 234°E, 35°N, with $A_{95} = 7°$. The results before thermal cleaning are very similar. Torsvik & Sturt (1987) calculated $D = 318°$, $I = +16°$ (7 sites) at Stoer after careful thermal cleaning, which gives a palaeolatitude of about 8°N (assuming normal polarity). These palaeolatitudes agree closely with those for adjacent Laurentia and Baltica at about 1150 Ma, as shown in Figure 39.

Palaeoclimate

Source rock weathering tends to spare quartz but destroy feldspar, especially plagioclase. In consequence the resulting sediment has a higher concentration of quartz, but less feldspar, than the parent rock. The chemistry of the Stoer Group sediments is of little use in quantifying the degree of weathering because of the changes wrought by diagenesis, but also because of the likelihood of added tephra. Van de Kamp & Leake (1997), however, have pointed out that the detrital components of the Stoer Group sandstones are still recognizable, despite albitization. For sandstones of the Clachtoll Formation (facies Ct2) directly derived from the basement gneisses, average quartz/(quartz + feldspar) = 0.3 and plagioclase/total feldspar = 1.0. These ratios can be compared with those for recent sandstones derived directly from tonalitic source rocks (like the Scourian gneiss), for there is no evidence of secular changes in the chemistry of post-Archaean clastic sediments to show that Proterozoic weathering products differed from those of today (Taylor & McLennan 1985, p. 98). On this basis Van de Kamp & Leake proposed that the climate was Mediterranean (Köppen type Cs). From Figure 16, however, it could just as well have been arid, or even glacial. Davison & Hambrey (1996, 1997) have indeed suggested a glacial origin for the unconformity beneath the Stoer Group and the immediately overlying breccias – a claim that has been rejected both by Stewart (1997) and Young (1999*a*). Davison & Hambrey's conclusion was based on the following five observations:

(1) The existence of *roches moutonnées* at Enard Bay, illustrated in their figures 3a, 3b and 4. The whale-back form of the basement hill in their Figure 3a is clear, but the 'smooth' unconformity surface they describe (Stewart 1997, fig. 2) is not remotely smooth. It has, on the contrary, joint-controlled relief of several metres amplitude with signs of rounding by chemical weathering (Stewart 1997, fig. 2). In their reply (Davison &

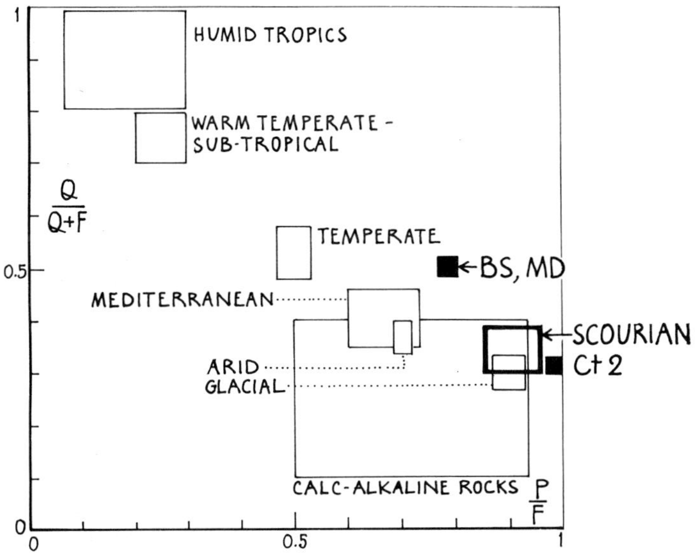

Fig. 16. The modal mineralogy of Stoer Group sandstones (facies BS, MD & Ct2), compared with recent first cycle sands derived from calk-alkaline basement rocks (Van de Kamp & Helmold 1991). Minerals used are quartz (*Q*), plagioclase (*P*) and total feldspar (*F*). The modal mineralogy of Scourian gneisses is also shown. Sandstones Ct2 derived from the gneisses have been relatively enriched in plagioclase due to Na metasomatism.

Hambrey 1997) they admit that the surface is not as smooth as it might be, but contend that it is still much smoother than the western side of the ridge. But the western side of the ridge has been severely sheared and the unconformity is unrecognizable. The smooth surface they figure from another supposed Precambrian *roche moutonnée* at Enard Bay (top right-hand part of their fig. 3b) is a glacially scoured surface of Pleistocene age cutting Stoer Group basal breccia.

(2) Dropstones at Enard Bay (Fig. 67). These are metre-sized blocks made of acid gneiss indistinguishable in the field from that which forms the local basement. They are not found more than 10 m laterally from the basement ridge (another supposed *roche moutonnée*) that now stands some 6 m above them, or at any other locality. The laminated sediments that envelop the dropstones have low lateral persistency – a centimetre thick band typically dies out within a metre. The coarser bands have an average grain size of about 0.5 mm as compared with 0.1 mm for the finer ones. Most of the extensive bedding surfaces are covered by wave ripples, and locally by films of shale cut by complete desiccation polygons. These sediments were deposited in shallow water, so the inference that the dropstones fell from melting icebergs (Davison & Hambrey 1996, fig. 8) is surprising. More probably the stones tumbled onto an ephemeral lake shore from the basement ridge above them.

(3) Massive sandy diamictite at Clachtoll and Enard Bay. A diamictite, as originally defined by Flint *et al.* (1960) should be poorly sorted with matrix-supported clasts, but Davison & Hambrey (1997) are content to accept the clast-supported breccia at Clachtoll, shown in their figure 5b, as glacial. The lowest breccia, which occurs in basement hollows 1–2 m deep which the northern edge of the outcrop, consists of rounded blocks of pale-grey altered picrite up to about a metre in size, in a matrix of picritic sand. The breccia above, which is about 5 m thick, is mainly composed of subangular basic to acid gneiss clasts about 40 cm in size in a sandy matrix. In the highest part of the deposit clast size diminishes to about 4 cm and thin red sandstone bands appear. The source of the picrite is almost certainly the ultramafic Scourie dyke, over 100 m wide, that outcrops about 50 m NE of the breccia (see pp. 59 & 61). The deposit, therefore, has all the characters of a locally derived, fining-upwards fan-head breccia rather than the basal till claimed by the authors.

(4) Clast roundness in the basal breccias. The variation of clast roundness was initially attributed by the authors to glacial action. They later admitted that the ultrabasic pebbles at Clachtoll, mentioned above, are well-rounded whereas the acid ones are subangular, but suggested that the rounding was due to modern weathering (Davison & Hambrey 1997). Modern weathering can be excluded, however, for at some points the outcrop has been smoothed by recent erosion so that the original pebble shapes are obvious. It seems that the clast roundness described from the basal Stoer Group by Davison & Hambrey is adequately explained by alluvial processes.

(5) Glacial hydrofracturing at Clachtoll. The sediment-filled dilatational fractures penetrating the basement gneisses at Clachtoll, thought by Davison & Hambrey (1996, 1997) to be of glacial origin, are due mainly to horizontal extension (Beacom *et al.* 1999, fig. 8) rather than the vertical extension to be expected from glacial hydrofracturing.

In short, there is no evidence for glaciation. This is not surprising, for nowhere on the planet are there signs of glaciation between 2250 Ma and about 950 Ma (Crowell 1999, p. 65).

The palaeolatitude (from palaeomagnetism) was 10°–20°N when the Stoer Group was being deposited. The climate in these latitudes today is like that deduced from the desiccation of muddy sediments in the Clachtoll Formation, viz. annual rainfall 300–1200 mm and a dry season 4–8 months long (Stewart 1991*a*), equivalent to the Köppen climatic zone BSh. The associated precipitation of carbonate is also consistent with such a climate (Mack & James 1994), as is the sandstone mineralogy (Van de Kamp & Leake 1997).

No palaeosols are preserved in the gneiss beneath the Stoer Group but the presence of detrital magnetite in the sediments (p. 19) suggests that source area weathering was not intense enough to have turned all the magnetite into hematite. This merely indicates vigorous erosion rather than any lack of atmospheric oxygen.

Basin analysis

Tectonism

Several lines of evidence indicate active deformation during Stoer Group sedimentation.

(1) Extensional faults with throws up to 30 cm developed in only partly consolidated sediment. Such faults are are commonly seen in the Clachtoll Formation and also at other levels (Fig. 64). A fault that was active during deposition of the Clachtoll Formation at Clachtoll had a notable effect on the palaeocurrents and stratigraphy (p. 64).

(2) Local downwarping of the basin floor during deposition of the Bay of Stoer Formation (Fig. 4).

(3) Sealed microcracks penecontemporaneous with the Stoer Group occur throughout much of the Lewisian gneiss complex, including Clachtoll and Enard Bay where the group is exposed (Hay *et al.* 1988). The cracks are steeply dipping, with a poorly defined north–south strike. They are filled by four distinct mineral assemblages, of which the earliest is calcite + K-feldspar. This earliest assemblage is cut by hematite-coated cracks descending from the basal Stoer Group at Clachtoll, which suggested to Hay *et al.* that the calcite + K-feldspar assemblage was contemporaneous with or even slightly earlier than the Stoer Group. Later crack-fill assemblages are prehnite + calcite + albite, pumpellyite + calcite + quartz, and stilpnomelane. At Enard Bay, pumpellyite occurs in clasts within the basal Stoer Group, in sediment-filled veins cutting gneiss below the Stoer Group and in veins cutting the Stoer Group (p. 71). The sealed microcracks thus indicate the existence of a roughly east–west tensile stress during Stoer Group deposition.

(4) Dilatational cracks at Clachtoll (pp. 60–63) opened under tension in an east–west direction just beneath the unconformity and allowed sediment to enter (Beacom *et al.* 1999). The

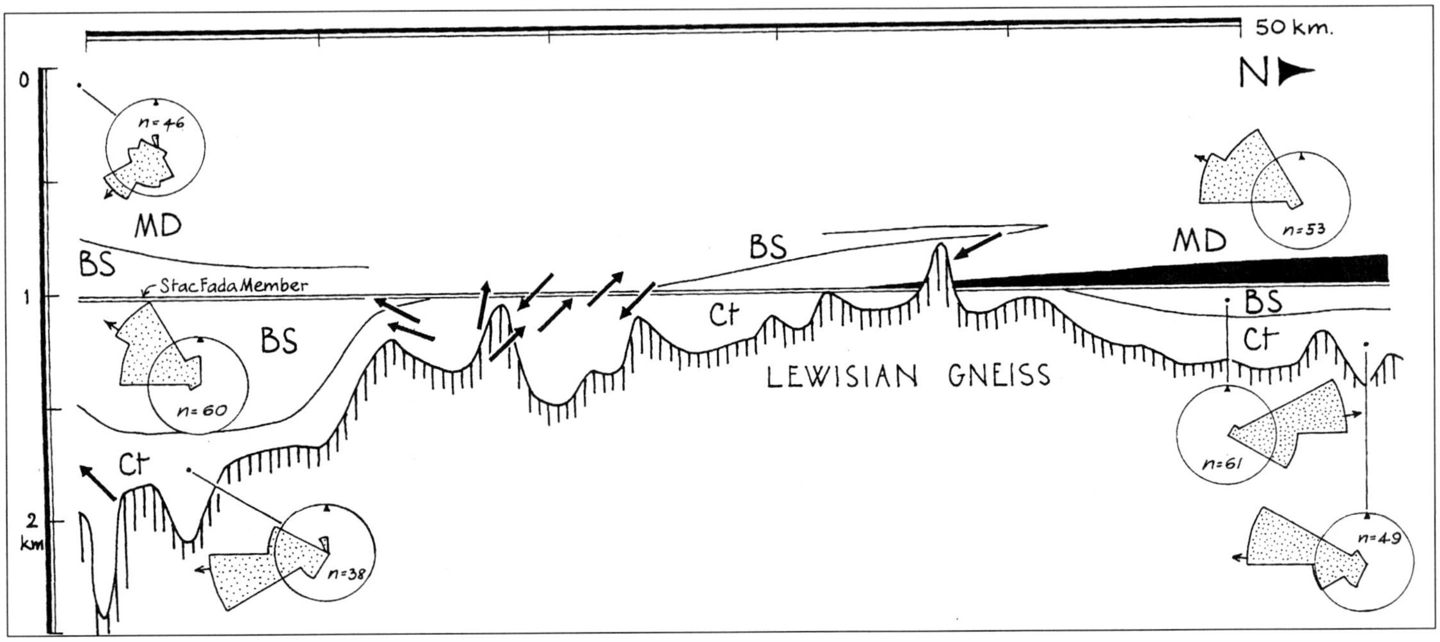

Fig. 17. Stoer Group palaeocurrent directions derived from trough cross-bedding, plotted on a stratigraphic section like Figure 4. Vectors are plotted with north assumed to be pointing upwards. Circles represent 20% of the observations. Black arrows are means of 2–20 observations. Formation codes; Ct = Clachtoll, BS = Bay of Stoer, MD = Meall Dearg.

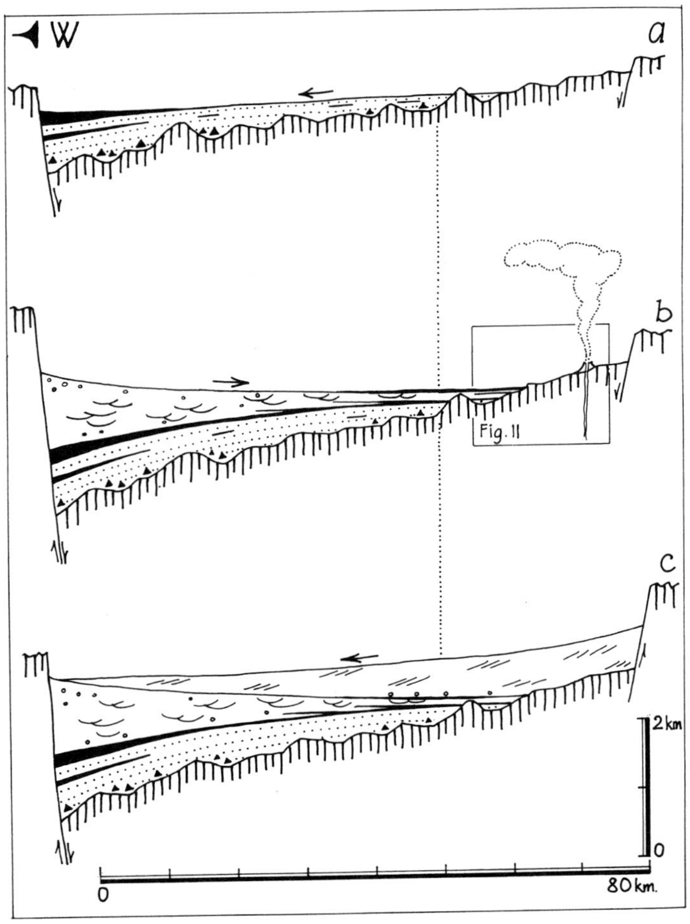

Fig. 18. Diagrammatic sections showing the evolution of the Stoer rift. The longitude of Stoer is shown by the dotted line. The topmost section (**a**) shows the Clachtoll Formation, confined to palaeovalleys at Stoer, but envisaged to thicken westwards. Section (**b**) shows the Bay of Stoer Formation, including the volcaniclastic Stac Fada Member and the following lake sediments. The lowermost section (**c**) shows the Meall Dearg Formation. The rift boundary faults are hypothetical.

sediment is laminated parallel to the bedding in the contiguous Stoer Group, showing that extension preceded development of the adjacent Coigach fault and the related tilting of the group (Stewart 1993*a*). The dilatational cracks are generally lined by hematite and are probably cognate with those described in the previous paragraph. The sediment in the interior of the cracks locally contains pods of K-feldspar and calcite, but it is uncertain if these minerals belong to the K-feldspar + calcite assemblage of Hay *et al.* (1988). The dilatational cracks at Clachtoll evidently developed prior to sediment cementation, probably during a phase of high pore-fluid pressure, when the unconformity was deeply buried by Stoer Group sediment. Cementation is likely to have started at around 80°C, inhibiting sediment migration into the cracks. The contemporaneous formation of pumpellyite suggests that the cracks were forming at temperatures in the range 125–230°C (Hay *et al.* 1988). In a rift environment (see below) these temperature would be expected at 3–6 km depth, assuming a geothermal gradient of 40°C km^{-1} (Allen & Allen 1990, fig. 9.18).

Palaeocurrents

Stoer Group palaeocurrents are summarized in Figure 17. Those above the Stac Fada Member (Meall Dearg Formation), came from the east and NE in sediments apparently deposited by a single distributary system. The beds immediately beneath the Stac Fada Member (Bay of Stoer Formation) came from the west at Stoer and the east at Poolewe, as if they were deposited on two giant alluvial wedges expanding from opposite directions. The beds that overlie the basal unconformity (Clachtoll Formation) show a more complicated story, but the great majority come from the east. The palaeocurrents suggest that the basin axis was oriented roughly north to south and that sediment came in from the sides.

Basin evolution

The contemporary stress regime, reversals of palaeoslope and the basic volcanism are taken to indicate sedimentation in a rift, but the position of the margins is quite uncertain. The complete lack of coarse, Scourian input from the west indicates that the Coigach fault was not a boundary fault. Indeed, there are no suitably oriented major faults to the west before reaching the Minch fault, or to the east before the Moine thrust zone. The detritus forming the Meall Dearg Formation probably came from as far away as the Moine thrust zone. Tentatively, therefore, the basin boundaries are assumed to coincide with the positions of the Moine thrust and the Minch fault at the present level of erosion – the level that also brings the Stoer Group to the surface.

The suggested evolution of the rift in the latitude of Stoer is shown in the series of cross-sections in Figure 18. In figure 18a the Clachtoll Formation, which at Stoer comprises sediments confined to valleys in the Lewisian gneiss complex, is envisaged thickening westwards into a major clastic wedge. Maar volcanoes are shown on the rift floor, where they could encounter sediment, rather than on the flank beyond the boundary fault. In Figure 18b bajada sediment (Bay of Stoer Formation) advance from the west. Upward movement of the rift floor on the east arrested the growth of the alluvial wedge and formed a depression that trapped the Stac Fada mudflow and the lake sediments constituting the Poll a' Mhuilt Member that follows. In Figure 18c alluvial fans advance from the east (Meall Dearg Formation). The eastern boundary of the rift is shown in Figure 18 as a fault, but its existence is quite hypothetical and the basin could just as well be a half graben.

Age and correlation

Limits on the age of the Stoer Group are provided by the age of the underlying basement, the age of clasts within the sediment, he timing of sediment diagenesis and the position of the palaeomagnetic pole.

Basement uplift started about 1700 Ma and was largely completed by 1400 Ma (Park *et al.* 1994). Nevertheless, K–Ar ages of about 1150 Ma have been obtained from biotite in gneisses near Torridon (Moorbath & Park 1972), and Rb–Sr ages from biotite in Laxfordian gneisses in the northern part of the Outer Hebrides average $1186 \pm 203\,(1\sigma)$ Ma (Cliff & Rex 1989). Biotite from a block of gneiss in the Stac Fada Member has given a K–Ar age of $1187 \pm 35\,(1\sigma)$ Ma (Moorbath *et al.* 1967). These ages of around 1190 Ma are generally attributed to Grenville-related shearing and local uplift. The youngest zircon ages in the Stoer Group at 1800 Ma are probably Laxfordian (Rainbird *et al.* 2001) and are of little help in dating the sediments.

The diagenesis of limestone in the Poll a' Mhuilt Member (unit **B**) has been dated at $1199 \pm 70\ (2\sigma)$ by the Pb–Pb method (Turnbull *et al.* 1996). By contrast an Rb–Sr isochron age from the $<0.2\,\mu$m clay fraction of Poll a' Mhuilt shales dates a much younger diagenetic event at about 689 Ma. The whole-rock Rb–Sr age of 1009 ± 130 Ma obtained by Turnbull *et al.* (1996) from Poll a' Mhuilt shale is of uncertain significance. About 20% of the shale is K-feldspar, possibly detrital, and about 20% albite, probably metasomatic, so the 'isochron' used to obtain the age may be a mixing line.

The palaeomagnetic directions for the type section of the Stoer Group, corrected for the later opening of the Atlantic (Fig. 38), give a pole position near those of Laurentian rocks in the age range 1050–1150 Ma. The older age agrees with the limestone age quoted above.

Although cryptarchs (Diver & Peat 1979) occur in units **B** and **C** of the Poll a' Mhuilt Member they are of little use for dating. They were thought by Cloud & Germs (1971) to be Palaeoproterozoic, though Downie (in Stewart 1975) once suggested they were Middle Riphean (i.e. 1050–1350 Ma).

The adopted age for the Stoer Group, considering the results from the limestone in the Poll a' Mhuilt Member, the clasts and the palaeomagnetic pole positions, is 1150 ± 50 Ma, contemporaneous with, or slightly earlier than, the Grenville orogeny, in which collisional events started at 1200 Ma and culminated at about 1000 Ma (Davidson 1995). The group is almost the same age as the earliest lavas and sediments filling the Mid-continent rift of north America, an idea considered further in Chapter 5.

Chapter 3
The Sleat Group

The group consists of up to 3500 m of coarse, grey fluviatile sandstones, with subordinate grey shales, resting unconformably on the Lewisian basement complex. Although the beds have been thrust tens of kilometres from the east as part of the Caledonian Kishorn nappe (Ramsay 1969) their stratigraphic position is secured by a conformable relationship with the overlying Torridon Group, presumed to be only slightly younger (Fig. 19). The Stoer Group does not form part of the Kishorn nappe and its relationship to the Sleat Group is unknown. Lithostratigraphic correlation of the two groups is thought unlikely, however, for reasons given later. Slight Caledonian deformation and very low grade metamorphism have had little effect on sedimentary structures in the Sleat Group between Loch na Dal and Kylerhea, in Skye. The metamorphism has, however, dissuaded attempts at isotopic dating.

The lithology, framework mineralogy of the sandstones, palaeocurrents and stratigraphic nomenclature of the Sleat Group are summarized in Figure 20. No formal facies nomenclature is proposed. The present outcrop is shown in Plate 1.

Stratigraphy

Sleat Group lithostratigraphy was originally established by the Geological Survey (Peach *et al.* 1907, p. 349) and revised by Sutton & Watson (1964). Stratotypes are described in detail on pp. 109–112. The dominant feature of the group is an upward decline in grain size, from very coarse sandstones in the basal Rubha Guail Formation to very fine sandstone in the Kinloch. The two lowest formations are easily distinguished by their coarse grain size, shale content, relatively thin bedding and lack of contortions, but the higher formations are more problematic. The reappearance of grey shales was used by the Geological Survey to trace boundaries between the Kinloch Formation and its neighbours, rather than any great difference in the nature of the sandstones. The Kinloch, however, has an average grain size of only 0.10 mm, significantly less than either the Beinn na Seamraig Formation beneath, or the Applecross Formation above, both of which have an average grain size of 0.15 mm. The Kinloch Formation also contains much more small-scale cross bedding, especially ripple-drift lamination, than its neighbours.

The poorly exposed boundary between the Sleat and Torridon Groups in Skye was said to be conformable by Clough (in Peach *et al.* 1907, p. 359), who mapped it. According to Sutton & Watson (1964) the shaly units in the uppermost Kinloch Formation (Sleat Group) interfinger with Applecross sandstones (Torridon Group), indicating a lateral facies change. None of these workers record a difference in dip between the two groups, and there are no sandstone pebbles in the Torridon Group to record erosion at their common boundary. Moreover, the pre-Caledonian magnetization vectors for the uppermost Kinloch Formation and the Applecross are similar to those for the Torridon Group on the mainland (Potts 1990, fig. 7). The boundary between the Sleat and Torridon Groups is, therefore, assumed to be conformable.

Facies and environments

Basement relief

The Geological Survey recorded relief of about 200 m on the mainland north of Skye (Peach *et al.* 1907, p. 343), where both the Rubha Guail and Loch na Dal Formations are cut out against the Lewisian gneiss. Thick breccias are lacking from the Rubha Guail Formation, however, suggesting that 200 m of relief is near the maximum. The Sleat Group is only 1400 m thick on the mainland, in Lochalsh, compared with twice that thickness in Skye. Whether the difference is due to burial of original relief, tectonic warping during sedimentation, or Caledonian deformation, is unknown.

Fan-delta and lake deposits

The Rubha Guail Formation which overlies the Lewisian gneiss complex consists almost entirely of coarse sandstone, with gneiss breccia close to the unconformity. Details are given on p. 109. Trough cross-bedding is typical of the coarser beds. Fine-grained, banded sediments, which become more abundant towards the top of the formation, contain wave ripples and polygonal desiccation cracks (Sutton & Watson 1960, fig. 7; Stewart 1962, fig. 11). They are followed above by 200 m of laminated dark grey siltpstones with millimetre-thick phosphatic bands, belonging to the Loch na Dal Formation.

The upward fining through breccia, coarse sandstone and into grey shale suggests that the lowest 250 m of the Sleat Group shown in Figure 108 originated as small alluvial fans, pro-grading from basement hills or a fault scarp into a lake or ria. The frequent occurrence of coarse or very coarse sandstone bands in the grey siltstones (Fig. 110) suggests steep basement topography near the lake shore. Lake transgression eventually covered the fans.

The upper part of the Loch na Dal Formation is a regressive sequence dominated by coarse and sometimes pebbly trough cross-bedded sandstones, often filling metre-deep channels (Fig. 109). The channel sands represent the top of a fan-delta which had prograded into the lake or ria and ultimately filled it. Palaeocurrent directions for this selquence, allowing for tectonic tilting are uniformly towards the ENE (Fig. 20).

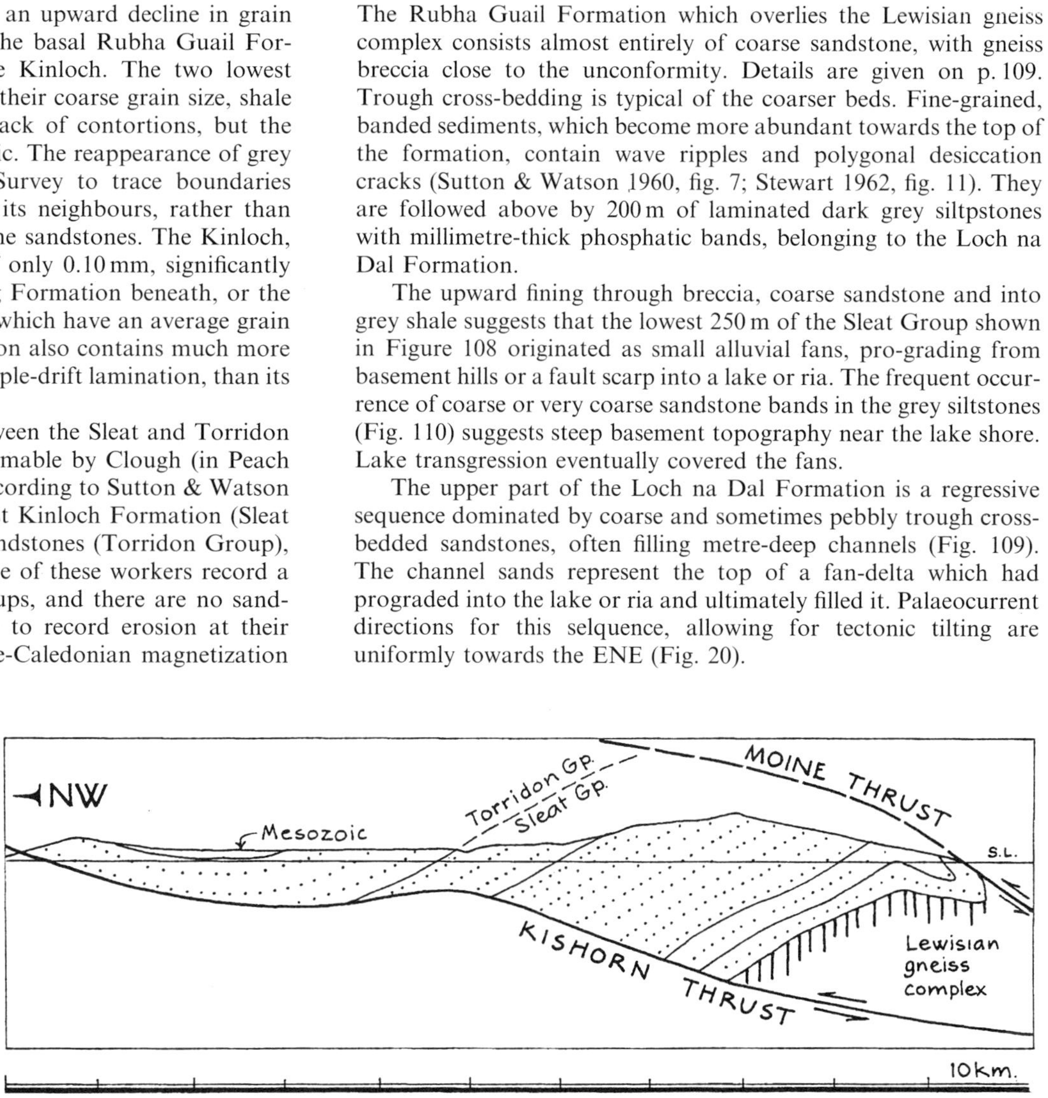

Fig. 19. True-scale section of the Sleat and Torridon Groups (stippled) in the Kishorn nappe of Skye, between Ben Suardal [NG 639295] in the NW and Rubha Guail [NG 750163] in the SE, based on Peach *et al.* (1907, Figs. 60 & 61) and Bailey (1955, Fig. 4). Beneath the Mesozoic it is probable that Cambrian strata unconformably overlie the Torridon Group as they do a few kilometres to the SW. Rocks beneath the Kishorn nappe, exposed to the SW in the Ord window, belong to the Sleat and Torridon Groups, and to the Cambrian.

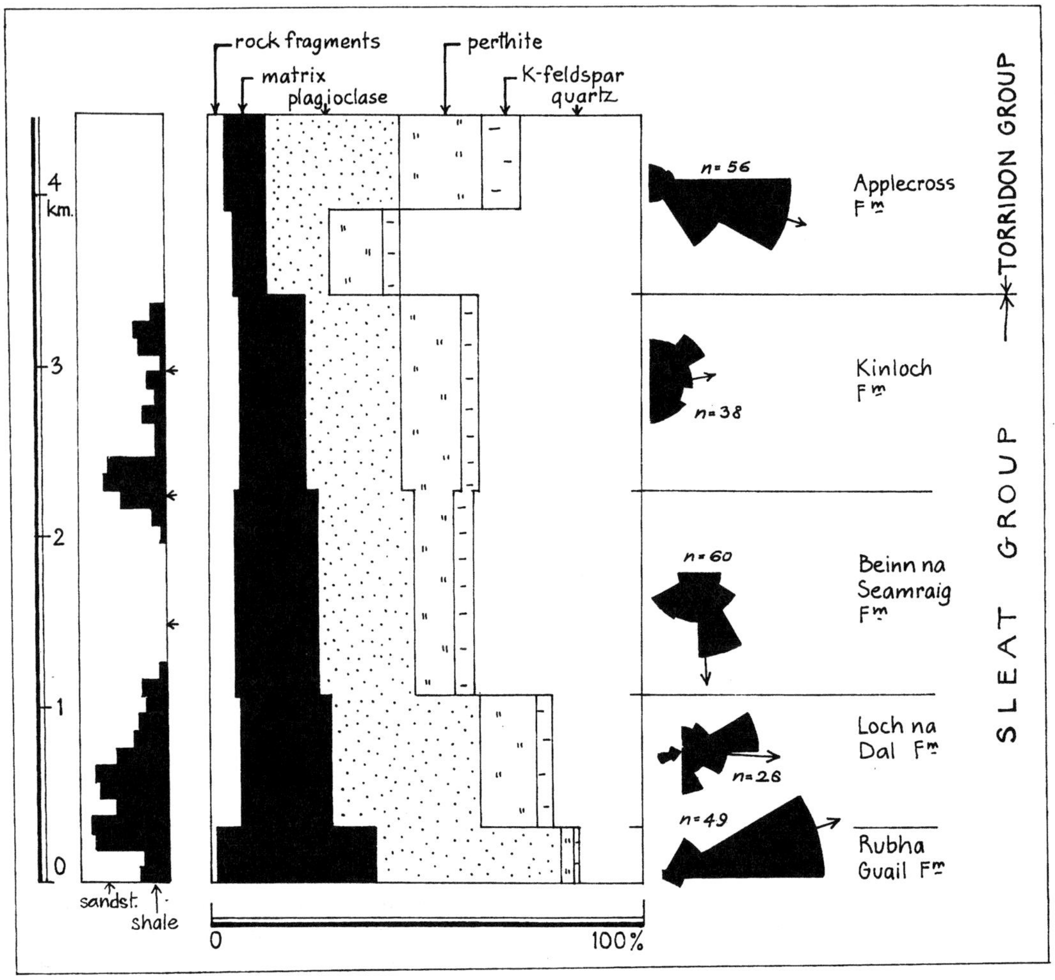

Fig. 20. Lithology, sandstone mineralogy, palaeocurrents and stratigraphy of the Sleat and Torridon Groups in Skye. The lithology column shows the average proportion of sandstone (white) to shale (black) in 100 m stratigraphic intervals. The column has been compiled from measured ections marked by arrows. The extremities of the four sections are (base) NG 738158–NG 704158 (coast line); NG 718165–NG 705182; NG 741215–NG 735230; NG 661163–NG 638178 (top). The average mineralogy of the formations is based on stained thin sections (800 points counted) of samples taken at 100 m intervals (Byers 1972). Palaeocurrent rose diagrams are based on the data of Sutton & Watson (1960, 1964). The formations were established by Peach *et al.* 1907, p. 348–62. Groups are from Stewart (1969).

Braided rivers

The rest of the Sleat Group and the overlying Applecross Formation of the Torridon Group consists of strongly contorted, cross-bedded sandstones (Figs 111–113). Ripple-lamination forms metre-thick sequences, especially in the Kinloch Formation (Fig. 112). Less commonly there are grey, shaly intercalations like those forming the lower part of the Loch na Dal Formation. In the Kinloch Formation the shales form the upper parts of fining-upward cycles 25–35 m thick. The whole section represents braided river deposits with occasional transgressive lacustrine or shallow marine episodes. Palaeoslopes were variable in direction. The palaeocurrents in the Beinn na Seamraig Formation flowed south but in all the others, including the Applecross, flowed eastwards.

Geochemistry

The chemistry of the sandstones and shales is summarized in Tables 8 and 10; the sandstone mineralogy is given in Table 9 and Figure 20. Even a cursory glance at the data shows that the sandstone composition changes upwards through the section. The quartz content and the ratio SiO_2/Al_2O_3 increase steadily, along with the ratio K-feldspar/total feldspar, K and Rb. Meanwhile K/Rb ratios fall. I have shown previously that these trends result from changing source rock composition. Basic rocks were being eroded during deposition of the lowest sediments, but stratigraphically upwards an acid source becomes dominant. In addition, quartzose metasediments are an important component in the lowermost Applecross Formation (Stewart 1991*b*).

These upward changes in mineralogy and chemistry are accompanied by a decline in average grain size through the group, from over a millimetre at the base to only 0.1 mm at the top, suggesting increasing transport distance. The lowest sediments almost certainly come from the Lewisian gneiss complex in which the commonest lithologies, where it unconformably underlies the Sleat Group in nearby Lochalsh, are hornblende gneiss, biotite gneiss and massive pink pegmatite (Peach *et al.* 1907, p. 262–263). The gneisses are cut by Scourie dykes transformed into hornblende schists by Laxfordian reworking. Unfortunately, analytical data for these basement rocks are lacking.

Source rock compositions

The chemistry of interbedded sand and shale can be used to put more precise limits on the composition of the original source rock (Stewart 1995*b*). Suppose the source contains two elements with concentrations C_o and C'_o. The concentrations of the same elements in the bed load are C_b and C'_b, in the suspended load C_s and C'_s, and in the dissolved load C_d and C'_d. In addition, the total bed load is defined as T_b, the total suspended load as T_s, and the total dissolved load (without atmospheric inputs) as T^*_d. Then:

$$[T_s/T^*_d]C_o(C'_b - C'_s) + [T_s/T^*_d]C'_o(C_s - C_b) + C_o(C'_b - C'_d) + C'_o(C_d - C_b) = [T_s/T^*_d](C_sC'_b - C'_sC_b) - C_bC'_d + C'_bC_d \quad (1)$$

$$[T_b/T_s]C_o(C'_d - C'_b) + C_o(C'_d - C'_s) + [T_b/T_s]C'_o(C_b - C_d) + C'_o(C_s - C_d) = [T_b/T_s](C_bC'_d - C'_bC_d) - C'_sC_d + C_sC'_d. \quad (2)$$

These equations have to be solved to get values for C_o and C'_o knowing only the equivalent concentrations in the bed load (sand) and suspended load (shale). The composition of the dissolved load is completely unknown, so the elements chosen need to be insoluble, or nearly so. Iron and aluminium are ideal for they are

Table 8. *Chemistry of sandstones in the Sleat and Torridon Groups of Skye*

	Rubha Guail Formation $n=2$	Loch na Dal Formation $n=5$	Beinn na Seamraig Formation $n=11$	Kinloch Formation $n=16$	Lower Applecross Formation $n=7$	Upper Applecross Formation $n=7$
SiO_2	59.38	66.35	75.43	76.65	83.34	73.87
TiO_2	0.96	0.81	0.61	0.55	0.42	0.40
Al_2O_3	14.57	14.84	11.57	11.68	9.29	13.85
t.Fe_2O_3	7.80	6.04	3.39	3.13	1.81	2.38
MnO	0.13	0.10	0.07	0.06	0.03	0.05
MgO	4.28	2.25	0.64	0.52	0.29	1.01
CaO	6.06	2.14	1.74	1.07	0.10	0.17
Na_2O	3.66	3.19	2.34	2.61	2.23	3.89
K_2O	1.96	3.62	3.89	3.60	2.92	4.14
P_2O_5	0.27	0.19	0.28	0.26	0.18	0.10
Total	99.07	99.53	100.14	100.13	100.61	99.86
Ni	49	<15	<15	<15	<15	<15
Rb	32	65	87	92	74	106
Sr	420	419	240	177	79	162
Y	13	12	19	23	20	15
Zn	70	46	18	26	11	25
Zr	317	253	342	278	241	194
SiO_2/Al_2O_3	4.1	4.5	6.4	6.6	9.0	5.3
Na_2O/K_2O	1.9	0.9	0.6	0.7	0.8	0.9
K/Rb	508	462	371	325	327	324
Rb/Sr	0.08	0.15	0.36	0.52	0.94	0.65
A	53.3	55.6	51.6	53.5	57.0	55.3
CN	39.0	29.7	29.9	28.6	23.7	26.8
K	7.7	14.6	18.5	17.8	19.4	17.9

Major elements are in per cent and traces in ppm. Number of samples analysed is *n*, taken from the stratigraphic intervals stated. The modal mineralogy corresponding to the analyses is shown in Figure 20 and Table 9. The analyses were carried out using XRF, at the Postgraduate Research Institute for Sedimentology, Reading University, by F. Street. Partial wet chemical analyses of six sandstones and two shales by S. Malik in the same Institute show Si, Al, Fe and Na concentrations within 10% of those determined by XRF, but K by XRF is overstated by 14%. A, CN and K are mol percentages of Al_2O_3 (equivalent to the CIA index), CaO & Na_2O, and K_2O, respectively. Sample numbers follow: Rubha Guail Formation = T1 & T2, Loch na Dal = T6 & T8–T11, Beinn na Seamraig = T12–T22, Kinloch = T23–T37 & K2, lower Applecross = T38–T43, upper Applecross = T44–T50.

scarcely soluble in stream water (Drever 1997, table 9-1) and have the additional advantage of being virtually immobile during diagenesis. Other relatively insoluble elements which may be used are K and Si. Extreme values for T_b/T_s and T_s/T_d^* in modern rivers, and presumably for Precambrian ones, are $0.1 \le T_s/T_d^* \le 10^3$ and $0 \le T_b/T_s \le 1$. Substituting these ratios in equations (1) and (2), along with the concentrations of the chosen insoluble elements in the sandstones and shales gives four linear equations in C_o and C_o' which can be displayed as a fence diagram (Fig. 21). The fences bound an area within which the source rock composition must have lain.

A source for the lowest sediments (Rubha Guail Formation) which accords with the fence diagrams is 80% Scourian hornblende gneiss and hornblende–biotite gneiss (Sheraton *et al.* 1973, table 3

Table 9. *Modal mineralogy of Sleat and Torridon Group sandstones in Skye*

	Rubha Guail Formation $n=2$	Loch na Dal Formation $n=5$	Beinn na Seamraig Formation $n=11$	Kinloch Formation $n=15$	Lower Applecross Formation $n=6$	Upper Applecross Formation $n=7$
Quartz	15.0	21.2	38.6	38.1	56.0	28.4
K-feldpar	0.5	3.7	5.0	4.3	4.0	8.6
Perthite	2.8	12.7	9.4	13.1	11.9	18.9
Plagioclase	43.2	34.5	22.3	22.2	15.2	30.7
Matrix	37.0	21.8	19.0	15.2	6.7	8.6
Lithic grains						
volcanic	–	1.1	1.6	2.6	2.9	1.6
other	–	1.3	1.7	2.7	3.1	1.6
Total	98.5	96.3	97.6	98.2	99.8	98.4
P/F	0.93	0.68	0.61	0.56	0.49	0.53
$Q/(Q+F)$	0.24	0.29	0.51	0.49	0.64	0.33

Analyses are per cent, based on 800 points per slide from each of *n* samples (Byers 1972). The sample are about 100 m apart stratigraphically, and are identical to those which yielded the chemical data in Table 8 except for the Kinloch Formation where there is one sample (K2) less. Plagioclase was stained and includes antiperthite. Muscovite and heavy minerals account for the difference between the stated totals and 100%. *P* is plagioclase, *F* total feldspar and *Q* quartz.

Table 10. *Chemistry of shales in the Sleat Group of Skye*

	Rubha Guail Formation $n=2$	Loch na Dal Formation $n=4$	Kinloch Formation $n=2$
SiO_2	54.30	54.39	56.95
TiO_2	1.22	1.24	1.17
Al_2O_3	17.08	17.82	17.71
t.Fe_2O_3	10.98	10.74	9.59
MnO	0.16	0.13	0.11
MgO	5.84	4.58	3.71
CaO	3.45	1.48	0.66
Na_2O	2.69	2.05	1.50
K_2O	2.83	4.12	4.50
P_2O_5	0.31	0.39	0.17
Total	98.86	96.94	96.07
Ni	56	39	42
Rb	92	140	184
Sr	353	190	121
Y	12	29	28
Zn	107	121	133
Zr	227	215	257
SiO_2/Al_2O_3	3.2	3.0	3.2
Na_2O/K_2O	0.9	0.5	0.3
K/Rb	255	444	203
Rb/Sr	0.26	0.74	1.52
A	61.1	63.1	67.5
CN	28.0	21.1	14.0
K	10.9	15.8	18.5

Major elements are in per cent and traces in ppm. Number of samples analysed is n. Analyses are by XRF, done in the Postgraduate Institute for Sedimentology, Reading University, by F. Street. Partial wet chemical analyses of six sandstones and two shales by S. Malik in the same Institute show Si, Al, Fe and Na concentrations within 10% of those determined by XRF, but K by XRF is overstated by 14%. A, CN and K are mol percentages of Al_2O_3 (equivalent to the CIA index), CaO & Na_2O, and K_2O, respectively. Sample numbers: Rubha Guail Formation = T3 & T4, Loch na Dal = T5, T7, FL-1 & LD-3, Kinloch = FL-2 & FL-3.

F & G) and 20% Scourian pegmatite (Evans & Lambert 1974, assuming 50% microcline). The hornblende–biotite–gneiss suggested as a source by Stewart (1991*b*, table 5) contains insufficient Fe and K to be suitable alone. In the topmost Sleat Group (Kinloch Formation) the proportion of pegmatite needs to increase to 40%. Scourian source rocks of this kind also account for the rather high whole-rock K/Rb values in the Sleat Group, which range from 300 to 500.

Extra information on the source rocks comes from the pebbles in the various formations. Clasts in the Rubha Guail Formation are mainly quartz–feldspar or K-feldspar. Pebbles of acid igneous rock (Peach *et al.* 1907, p. 353; Bailey 1955, p. 123), fine-grained dolerite (probably from Scourie dykes) and metaquartzites are very rare. In the three higher formations, on the other hand, the vast majority of pebbles are porphyry, of rhyolitic or rhyodacitic composition; the remainder are acid gneiss, typically composed of quartz and plagioclase ± microcline ± biotite. The igneous pebbles are like those in the Applecross Formation of the Torridon Group, dated by Moorbath *et al.* (1967) as Laxfordian. In the Applecross Formation porphyry pebbles are still present but, in addition, so are sedimentary quartzites. The abundance of quartzose detritus in the lower half of the formation is responsible for the high content of Si, Zr, Y and P.

It is clear that the lowest Sleat Group sediments (Rubha Guail Formation) were derived little altered from local Scourian basement but that at higher stratigraphic levels distant sources contributed both acid volcanic pebbles and relatively more K-feldspars. The proposal that the volcanic pebbles in the Sleat Group and the quartzite pebbles in the Torridon Group came from tectonic slices in the basement (Stewart 1991*b*, fig. 7) is one possible scenario. But they could also have been derived from a sequence overlying the basement, or from earlier conglomerates. All that can be said with any certainty about the K-feldspars is that their K/Rb ratios are too high for them to have come from Laxfordian granites or gneisses. The Scourian pegmatite or something similar would be a more suitable source.

Sodium metasomatism

Plagioclase in the Sleat Group has been partially albitized and in the overlying Applecross Formation totally albitized. Multiple least squares analysis shows that Na_2O in the Sleat Group is carried exclusively by plagioclase and perthite, the former being mainly andesine. However, two samples from the Kinloch Formation with over 15% plagioclase have whole-rock CaO/Na_2O as low as 0.1, indicating albite. A sandstone from the Beinn na Seamraig Formation with over 20% plagioclase has a similarly low value for CaO/Na_2O, calculated on a carbonate-free basis (Stewart 1991*b*, table 1A). The secondary calcite seen in the Loch na Dal Formation and locally elsewhere is probably the by-product of albitization.

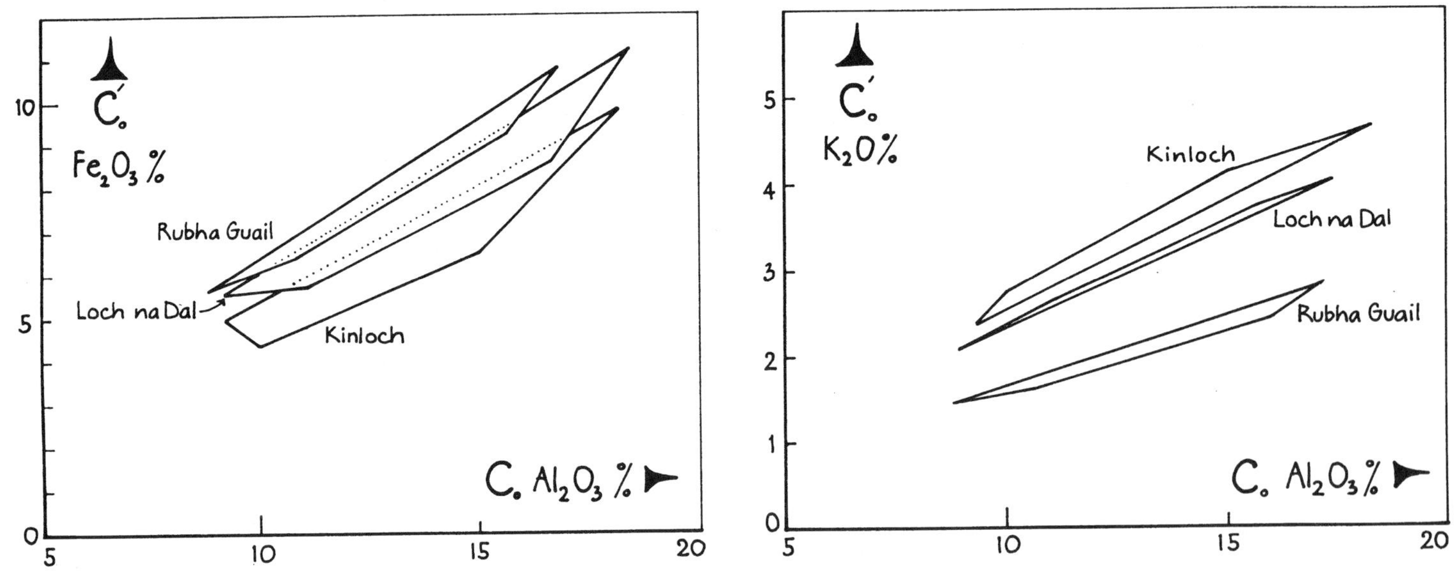

Fig. 21. Fence diagrams showing the composition of Sleat Group source rocks in terms of Al_2O_3, Fe_2O_3 and K_2O. The diagrams are based on analyses of interbedded sandstones and shales, available only from the Rubha Guail, Loch na Dal and Kinloch Formations (Tables 8 amd 10).

There are too few analyses from the Skye Applecross to get a reliable estimate of the feldspar composition, but after uniting the data with those from the same formation on the adjacent islands of Rum and Raasay least squares analysis indicates plagioclase with 8–11% Na_2O, i.e. albite or sodic oligoclase. This is confirmed by the very low calcium values in the Applecross Formation. In Skye the average whole-rock ratio CaO/Na_2O is only 0.04, indicating that the plagioclase is albite. The corrosive formation waters which attacked the plagioclase also dissolved much of the apatite, removing the Ca in solution but leaving the PO_4 ion attached to clays. There is no doubt, therefore, that plagioclase in the Applecross sandstones, which is mostly cloudy in thin section, has been albitized during burial diagenesis. The K-feldspar, though, survived intact, for perthite and microcline are still visible, unaltered, in thin section. Moreover the percentage of K-feldspars and K_2O in the Applecross Formation does not vary with stratigraphic level in Skye, Raasay or Rum. The relatively slight degree of albitization of the Sleat Group may depend on the greater clay content and frequent shale units which impeded circulation of the acid formation waters. The conditions for albitization were outlined on p. 18.

Weathering and palaeoclimate

The standard procedure or determining source rock weathering is to look at indices which measure the surviving proportion of relatively soluble plagioclase to relatively insoluble K-feldspars, or the proportion of relatively soluble feldspars (all types) to quartz. Suitable indices based on modal analyses are shown for the sandstones of each formation in Table 9. The upward decrease in Na-feldspars (P) relative to total feldspar (F), however, is due to the change in source area during deposition of the Sleat Group (pp. 24–26) and not to any increase in weathering intensity. The upward increase in the ratio of quartz (Q) to feldspar (F) is also unrelated to weathering. Total feldspar in sandstones remains much the same throughout Sleat Group and up into the overlying Torridon Group, while the detrital quartz content increases and matrix drops (Fig. 20). A better idea of weathering intensity can be had from the chemical index of alteration (CIA) shown in Tables 8 and 10. The index is quite constant for the sandstones at about 55 and for shales at about 65, as compared with 47 for average Scourian basement gneiss (Table 1).

Basin analysis

The Sleat Group predates the Torridon Group and occupies a large sedimentary basin which once existed east of Skye (Stewart 1982). Its length exceeds 80 km (Plate 1) but the width presently exposed in the Kishorn nappe is only about 10 km (Fig. 19), and no boundary faults are preserved. Nevertheless, some lines of evidence are available to infer the origin of the basin.

(1) The tripartite facies succession, consisting of fan breccias at the base (Rubha Guail Formation), followed by lake sediment (lower Loch na Dal Formation), and fluviatile sands (upper Loch na Dal, Beinn na Seamraig and Kinloch Formations) is typical of many continental extensional basins (Schlische & Olsen 1990; Lambiase 1991). It reflects the result of constant sediment input to a rift, or a pull-apart basin, subsiding at a decelerating rate (Allen & Allen 1990, figs 3.14 and 5.15). The basin was initially small and easily filled by sediment, but as it rapidly deepened and widened the sediment supplied was insufficient to fill the space available and a lake formed. Later still, the declining subsidence rate allowed the sediment supplied to more than fill the entire basin. The lake was invaded by deltas and replaced by an alluvial plain.

(2) The geochemistry of Sleat Group sandstones suggests that Scourian crust was probably a significant sediment source. The igneous pebbles appear to be like those in the Applecross Formation, which are Laxfordian (p. 41). The Sleat Group is thus very much younger than the detritus which forms it, which rules out a forearc basin like the Alpine molasse, or a Pannonian-type intermontane basin.

(3) There are a series of reasons for thinking that the Torridon Group formed in a rift (pp. 43–44), thus its conformable relationship to the underlying Sleat Group points to a similar origin.

The Sleat Group is therefore viewed as having filled the western part of a rift basin in Figure 36. Caledonian compression severed the sediments from the basement and thrust them over the rift margin in much the same way as the slightly younger Rivieradal Group was forced over the western edge of the Hekla Sund basin in east Greenland (Higgins *et al.* 2001, fig. 11c).

Age and correlation

Little can be said about the age of the Sleat Group except that it evidently post-dates the Scourian basement (2500 Ma), and predates the Cambrian strata which unconformably overlie the Applecross Formation in the Sleat of Skye. The Applecross Formation can be identified by the petrography of the pebble suite and palaeocurrent direction ($\theta = 106°$, $n = 56$), but also by its palaeomagnetism. The Applecross Formation in Skye, and also the immediately underlying sandstones in the Kinloch Formation of the Sleat Group, have directions of magnetization predating Caledonian deformation which are comparable with those in the Applecross elsewhere (Robinson & McClelland 1987; Potts 1990).

Although there is no doubt that the Applecross Formation on Skye correlates with the Applecross–Aultbea complex, the Sleat Group is more problematic. The hypothesis that it is laterally equivalent to the Applecross or Aultbea Formations is not supported by either the palaeocurrent directions or the absence of quartzite pebbles in the Sleat Group. Clough thought that the Sleat Group was merely 'an enormously expanded form of the Diabaig [Formation] of the region to the north'. (Peach *et al.* 1907, p. 349). Like the Sleat Group, the Diabaig is sandwiched between the Lewisian basement and the Applecross Formation so that Clough's suggestion is at first sight quite plausible. But the Diabaig contains no porphyry pebbles, so those in the Sleat Group would need to have been supplied from the west by trunk streams of which no trace now remains in the Diabaig Formation of foreland areas such as Rum and Raasay.

Since the Stoer Group occupies a position between the basement and the Torridon Group, the possibility of correlation between the Sleat and Stoer Groups needs consideration. Both groups are formed of highly feldspathic fluviatile sandstones derived mainly from the Scourian basement which they unconformably overlie. But here the resemblance ends. The clastic material forming the sandstones and shales of the Sleat Group is more weathered than that of the Stoer Group, with higher CIA indices. For example, sandstone in the Rubha Guail Formation, the lowest of the Sleat Group, has an average CIA index of 53 (Table 8), whereas that in the Clachtoll Formation, at the base of the Stoer Group, has an index of about 48 (Table 3). Another significant fact is the presence of porphyry pebbles in the Sleat Group and the Applecross of Skye, but not in the Stoer Group. Rock-stratigraphc correlation of the Sleat and Stoer Groups can therefore be excluded.

As pointed out in the section on Sleat stratigraphy, there is no obvious sign in Skye of an angular unconformity between the Sleat and Torridon Groups like that between the Stoer and Torridon Groups on the mainland. Indeed, the palaeomagnetic evidence noted above indicates the contrary. Even if there were an erosional break in Skye between the Sleat and Torridon Groups it would not, of course, be sufficient to demonstrate contemporaneity of the Stoer and Sleat Groups. For these reasons time-stratigraphic correlation of the Sleat Group with the Stoer Group is thought unlikely.

The remarkable resemblance between the sedimentary structures in the Beinn na Seamraig and Kinloch Formations (Figs 111

& 112), and those in the Moine psammites of Morar (Glendinning 1988, figs 2.12 & 2.15) is intriguing. The two successions are roughly contemporaneous (see p. 45) but Glendinning attributes the arkosic Upper Morar Psammite (5 km thick) to a tidal shelf environment because of the bimodal palaeocurrent directions. Furthermore, Prave *et al.* (2001) state that Grenville age zircons are present in the Morar succession but not in the Sleat Group.

At present the Sleat Group cannot plausibly be correlated with anything.

Chapter 4
The Torridon Group

The group unconformably overlies both the Lewisian gneiss complex and the Stoer Group, and is in turn unconformably overlain by the Lower Cambrian (Figs 2 & 22). The unconformity surface at the base is generally rugged, with relief reaching 600 m. The maximum thickness of the Torridon Group is about 7 km onshore and 6 km offshore in the Sea of the Hebrides basin (Stein 1988, fig. 11; 1992, fig. 2B), but albitization of the highest beds indicates that the original thickness was 3–4 km greater. Lake deposits at the bottom of the group occupying palaeovalleys in the gneiss are followed by kilometres of red sandstones, all deposited in a subsiding rift. Alluvial sands interfinger with lake sediments to form cyclothems at the top of the group. As mentioned in the Introduction the Torridon Group is by far the most extensive and voluminous part of the Torridonian (see Plate 1), but nevertheless poses fewer problems of interpretation than the Stoer and Sleat Groups.

Stratigraphy

The formal stratigraphy established by the Geological Survey (Geikie 1894) has been retained even though it is in some respects unsatisfactory. The sediments filling the palaeovalleys at the base of the group form a well-defined lithostratigraphic unit, the Diabaig Formation, characterized by breccias and sandstones derived from the immediately adjacent basement (Fig. 23). The Cailleach Head Formation at the top of the Group is also a valid lithostratigraphic unit, formed of coarsening-upward cyclothems of grey shale and red sandstone. The bulk of the Torridon Group, however, is red sandstone best regarded as a single unit, but divided by the Geological Survey into two formations: the coarse-grained, pebbly Applecross Formation below, and the fine-grained pebble-free Aultbea Formation above. These two formations are in reality mere facies. Three examples will suffice to demonstrate this. Firstly, the fine-grained red sandstones in Skye, some 2 km thick, believed by the Geological Survey to overlie the Diabaig Formation and therefore assigned to the Applecross Formation are, in fact, virtually indistinguishable from the Aultbea. Secondly, the lowest part of the Applecross Formation, up to 300 m thick, is generally formed of fine-grained, pebble-free red sandstone, much like the Aultbea Formation. Thirdly, the fine-grained red sandstones at Toscaig mapped as Aultbea Formation by the Geological Survey pass conformably up into pebbly sandstones over 500 m thick that are indistinguishable in all respects from the Applecross. Precise definitions of the Aultbea and Applecross Formations in their type areas will be found on pp. 88 and 101, respectively, but outside these areas they are useless.

The construction of a stratigraphic section for the Torridon Group is attended by two main problems: the lack of even one traverse spanning the entire group and the absence of a regionally extensive horizontal datum at the top. The first problem is due to the Toscaig and Coigach faults, which slice the group in half. Movement on these faults is poorly constrained so that the 7 km thickness of the group shown in Figure 23 could be a kilometre in error. The second problem is essentially due to lack of research. The upper part of the section (Fig. 23) is hung from the zone of palaeomagnetic reversals and thin grey shales exposed at Aultbea, the Summer Isles, Torridon and Toscaig. However, neither the reversals nor the shale have yet been located in Rum. The reversals probably formed during burial diagenesis (p. 43) and are therefore essentially time planes parallel to the depositional surface.

There is no comparable horizontal time plane that can be used to hang the lower part of the section (Fig. 23q). Both the Diabaig–Applecross boundary and the first appearance of abundantly pebbly sandstones in the lowermost Applecross Formation are potentially diachronous, at least in the direction of palaeocurrent flow. The basal unconformity is a time plane, but might not have been regionally horizontal when sedimentation started. However, a closer look at the Diabaig–Applecross boundary in a section perpendicular to the palaeocurrents (Fig. 24) shows no evidence of interfingering between the two formations. Only where the Applecross Formation approaches the basal unconformity can any evidence of lateral equivalence be detected (Fig. 27). Consequently, for want of a better, the Diabaig–Applecross boundary has been chosen as the datum. Selection of the horizon at which pebbly sandstone appears would produce an identical result.

The gradual thickening of shales in the Diabaig Formation from a feather edge in the latitude of Stoer (Fig. 23a) to 400 m in Rum (Fig. 23a) could, in principle, be due to several factors, such as toplap, onlap and offlap (Levorsen 1960, p. 24–29; Allen & Allen 1990, p. 145). However, the upper surface of the shales does not show erosion, i.e. toplap. Nor is there evidence for a change in basement relief along the section that could be used to argue that the shales were filling an original topographic low, i.e. onlap. Consequently the thinning of the shale is taken to indicate progressive southward tilting of a regionally flat (but locally hilly) unconformity surface during Diabaig sedimentation, i.e. offlap.

High-resolution heavy mineral analysis is a valid method of locating time planes in red-beds over distances of at least 150 km (e.g. Mange *et al.* 1999) and will no doubt one day be used to clarify Torridonian stratigraphy as it has other comparable sequences.

Basement topography and drainage

The unconformity at Cape Wrath is quite flat or has, at most, local relief of 150 m (Williams 1969*a*, 2001). The flat part, where exposed, is locally weathered to a depth of 3 m (Williams 1968; Retallack & Mindszenty 1994), suggesting a genetic difference from the hilly unconformity to the south which always cuts fresh gneiss. The hilly unconformity has relief of 300–400 m in any 100 km^2 area. There is insufficient evidence to show that the relief varies systematically from west to east as suggested by Stewart (1972).

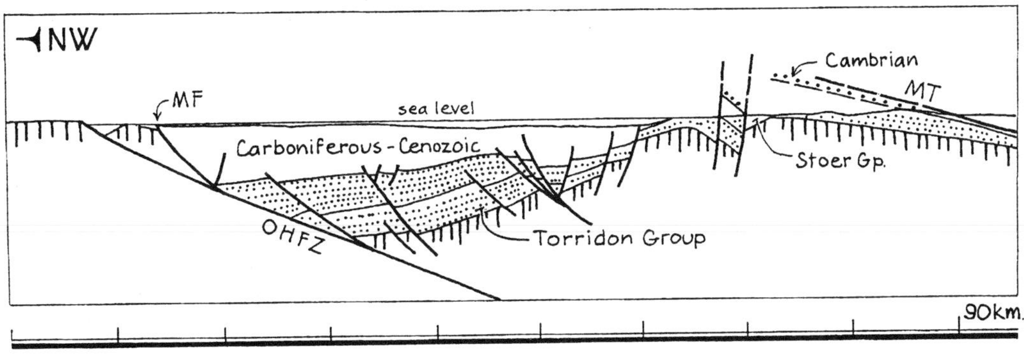

Fig. 22. True-scale section the Torridon Group (stippled) between the Outer Hebrides [NB 370420] in the NW and Dundonnell [NH 140850] in the SE. The offshore part of the section is based on Stein (1992, fig. 2). The section line is shown in Fig. 1. The Torridon Group is limited to the west by the Minch fault (MF) and the Outer Hebrides fault zone (OHFZ), and the east by the Moine thrust (MT). The significance of the reflector within the Torridon Group is uncertain.

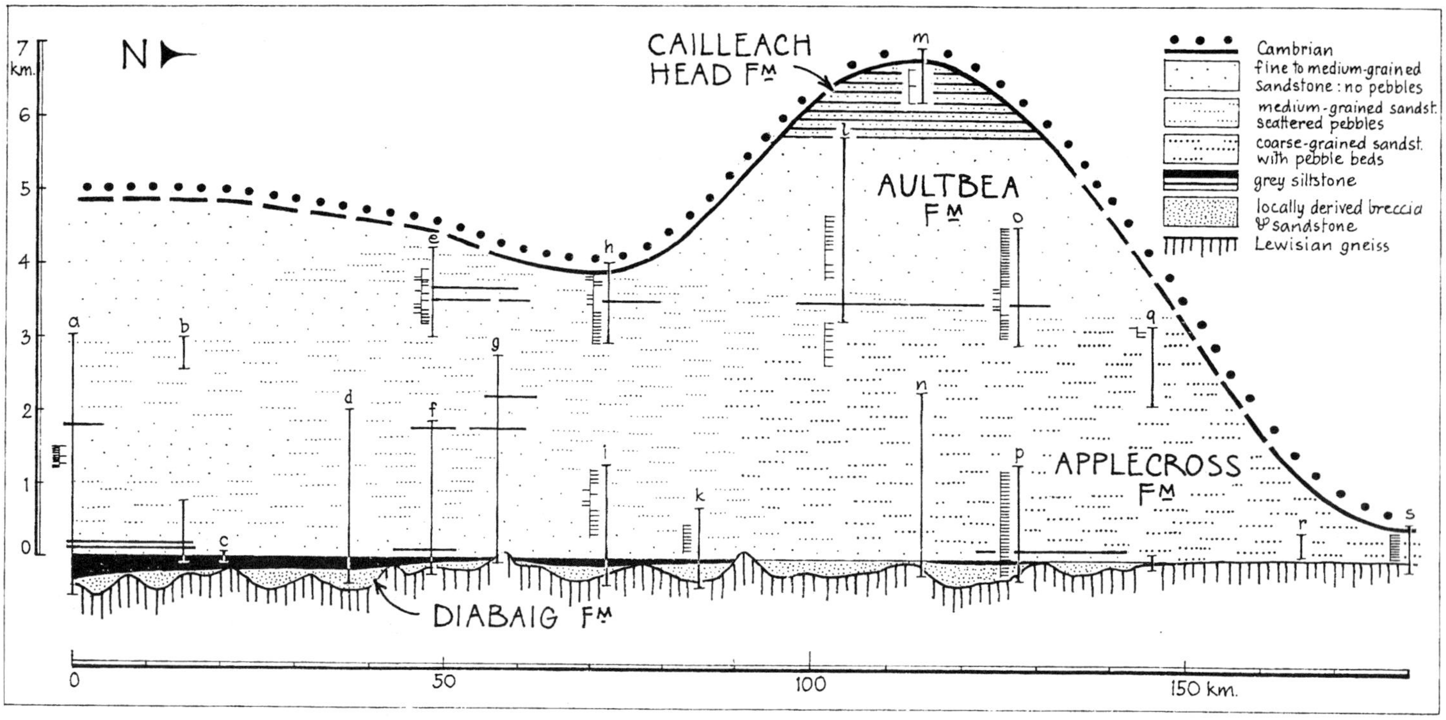

Fig. 23. Stratigraphic section of the Torridon Group along a line joining Cape Wrath and Rum (azimuth 028°), almost perpendicular to the palaeocurrents in the Applecross Formation ($\theta = 123°$). The vertical exaggeration is ten times. Sub-areas shown are: **a**, Rum; **b**, Soay; **c**, Camusunary; **d**, Scalpay; **e**, Toscaig; **f**, Raasay; **g**, Shieldaig to Applecross; **h**, Torridon, east of the Fasag fault; **i**, Torridon west of the Fasag fault; **k**, Gairloch; **l**, Aultbea; **m**, Cailleach Head; **n**, Scoraig to Dundonnell; **o**, Isle Ristol to Badentarbat; **p**, Achiltibuie to Strath Kanaird; **q**, Rubha Stoer, west and east of the Coigach fault; **r**, Handa; **s**, Cape Wrath. The palaeomagnetic stratigraphy is shown with SE positive polarity to the right of the vertical line and NW negative to the left (Irving & Runcorn 1957; Stewart & Irving 1974; Smith *et al.* 1983; Robinson & McClelland 1987; Williams & Schmidt 1997).

Outliers of the Torridon Group are abundant enough in three areas, each about 100 km^2, to allow reliable three-dimensional reconstruction of the unconformity. The first is between Quinag and Cùl Mòr (Stewart 1972, fig. 7). Dip-corrected relief is about 250 m at Quinag and 330 m between Cùl Mòr and Loch Veyatie. The second area is between Loch Maree and An Teallach where the relief is about 600 m (Stewart 1972, fig. 5 1988*b*, fig. 9.2). It has been aptly described as 'probably the most dramatic Precambrian landscape in Europe' (Butler 2000). The last area, with relief of about 300 m is between Gairloch and Loch Torridon (Fig. 25). In Raasay, some 450 m of two-dimensional relief is seen in a stratigraphic profile 6 km long (Fig. 106). The palaeovalleys are partly filled by the breccias, sandstones and shales of the Diabaig Formation, whereas the interfluves are buried by the overlying fluviatile Applecross Formation.

In general the palaeovalleys trend northwestwards, parallel to the foliation in the gneisses, but the original flow directions of the

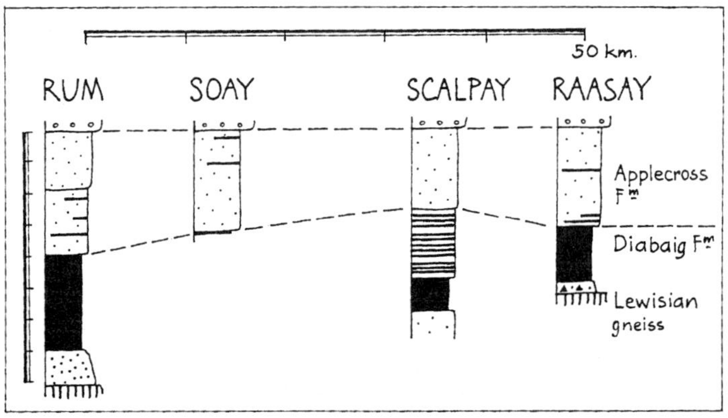

Fig. 24. Graphic logs through the lowermost Torridon Group along a line linking Rum and Raasay, perpendicular to the palaeocurrents. The scale bar is divided at 100 m intervals. The datum adopted for this diagram is the first appearance of *abundantly* pebbly sandstones. Grey shales are ornamented black. Further details will be found in Figs 105 & 114.

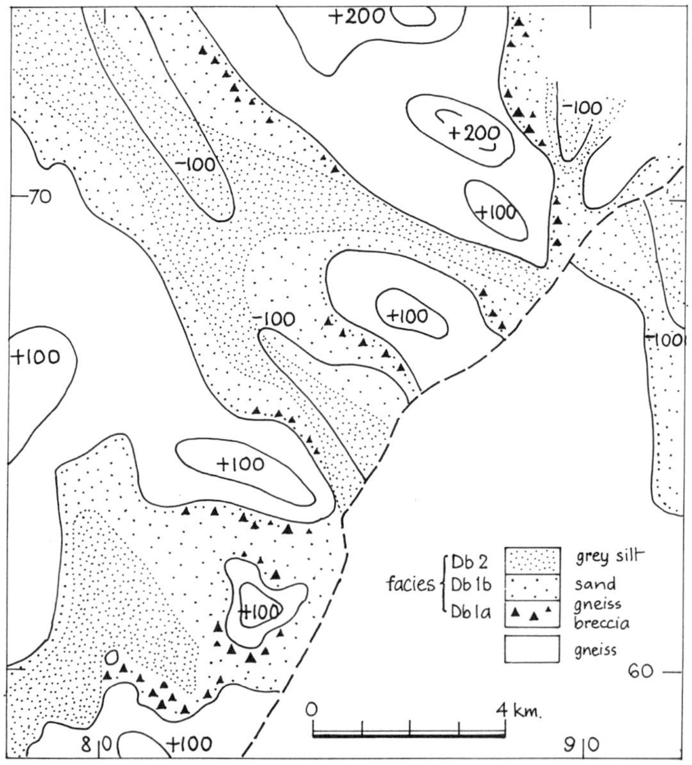

Fig. 25. Reconstructed unconformity relief and Diabaig facies in the area bounded by Lochs Diabaig, Gairloch and Maree. See Plate 2 for location. Heights are given in metres relative to the contact between the Diabaig Formation and the overlying Allt na Béiste Member of the Applecross Formation. The extension of the Applecross fault which crosses the map downthrows to the east so that the basement there is largely concealed. The extent of the shale facies encompasses known outcrops but is mainly conjectural. The map is based on Stewart (1972, fig. 3) and Rodd & Stewart (1991, fig. 2).

rivers that cut them cannot be deduced from the topographic reconstructions. However, the rivers were not simply stopped by the instantaneous elimination of the regional palaeoslope. Had this occurred, typical river deposits containing far-travelled pebbles ought to be covered by the local breccias and lake deposits of the Diabaig Formation – the lowest unit of the Torridon Group. But such fluvial sediments are noticeably absent, except in the area between Inverpolly Forest and Canisp. It is not difficult to devise an explanation for this. If the palaeoslope was towards the east, say, it may be suggested that the gneiss surface was first warped, increasing the slope in the area now exposed but diminishing it farther to the west. By this means the rivers were deprived of their upstream discharge and sediment supply, and induced to erode their floodplains so vigorously as to destroy them. Warping then resumed, reducing the local palaeoslope and permitting the formation of lakes and alluvial cones in the basement valleys.

The hilly unconformity beneath the Torridon Group was depressed towards the south during deposition of the Diabaig Formation (Figs 23 & 24). From Quinag southwards it was conserved by being buried, but to the north, around Cape Wrath, it seems to have been degraded by an erosional regime that produced transport-limited slopes. In other words, weathering intensified while erosion became less effective, so that slopes were progressively reduced and the original relief was softened. Williams (1969*a*) has attributed this to pedimentation that started near Quinag and advanced westwards hand in hand with alluvial fan formation. It could, however, simply be due to a change of climate. Modelling shows that slope reduction of the kind envisaged can be accomplished within a few tens of thousands of years (e.g. Armstrong 1976). Retallack & Mindszenty (1994) suggest several hundred thousand years were needed for the formation of the palaeosols at Cape Wrath.

Unconformity weathering

Weathering of the Lewisian gneiss for up to 3 m immediately below the Torridon Group was described by Williams (1968) from five localities near Cape Wrath, four of which are on sea cliffs. Inland only one exposure of weathered gneiss has been found, so the lateral extent of the weathering is uncertain. In all cases the weathered unconformity surface appears essentially flat and parallel to the overlying beds. Williams identified two distinct weathering sequences, or palaeosols, developed over biotite gneiss and amphibolite, respectively, prior to deposition of the overlying Applecross Formation. He concluded that the climate was warm and moderately humid, perhaps with a dry season. Later work by Retallack & Mindszenty (1994) has confirmed Williams' results. A palaeomagnetic study by Williams & Schmidt (1997) showed that the dominant iron mineral in the palaeosol is hematite, with a magnetization direction like that in the overlying Applecross Formation, reinforcing the original conclusion that the weathering was Precambrian and disproving the suggestion by Stewart (1995*a*) that it was Cenozoic. Williams & Schmidt (1997) also concluded that the weathering was capable of producing detritus like that forming the Applecross Formation. The Applecross, however, originally contained abundant detrital magnetite (see pp. 42–43), which is absent from the Cape Wrath palaeosols.

The palaeosol grades up through sandy claystone with corestones of gneiss (locally cut by the unconformity), into dusky red claystone. Retallack & Mindszenty (1994, 1995) claim that the palaeosols have only been compacted to 80% of their original thickness, basing their conclusion on the low density of the claystone relative to the parent gneiss, and the lack of ptygmatic folding in pegmatite sheets preserved within one of the palaeosols. However, their table 7 gives the average claystone density as 2700 kg m^{-3}, as compared with 2700 kg m^{-3} to 2900 kg m^{-3} for the gneiss, indicating that the palaeosol is almost as compact as it could be. Ptygmatic folding of the pegmatites is not to be expected for they are little altered and 10–30 cm wide, possessing substantial rigidity. They also approach the unconformity at low angles (cf. Williams 1968, figs 4 & 5) so that compaction would have tended to stretch most of them, not the contrary. The palaeosols contain pedogenic smectite, cut by clear diagenetic illite associated with K-metasomatism of the palaeosols (Retallack & Mindszenty 1994, figs 8 & 11; Young 1999*b*, fig. 6). Bulk samples of the claystone have Weber indices as low as 348 and 775. Retallack & Mindszenty consequently believe the palaeosol was buried to a depth of less than 1500 m and heated to no more than 120°C. But the evidence for compaction, cited above, and the geological setting of the palaeosols suggest otherwise.

Since the Torridon Group is at least 6 km thick, the proposed burial of the basal beds by 1500 m appears to be a gross underestimate. The Torridon Group at Cape Wrath was almost entirely eroded during the late Proterozoic, covered unconformably by some 1500 m of sediments during the Cambro-Ordovician period and overridden by the Moine nappe during the Silurian (Dallmeyer *et al.* 2001). Conodonts in the Cambro-Ordovician rocks at Durness, about 20 km east of the palaeosol localities, have a coloration index of 5 (Aldridge 1986), indicating heating to a temperature of 300–325°C (Rejebian *et al.* 1987). Illite crystallinity (2–6 μm fraction) in the Durness and Torridon Groups of Assynt, about 30 km south of the palaeosol localities, indicates a maximum temperature of between 250°C and 300°C (Johnson *et al.* 1985). These high temperatures resulted from the emplacement of the Moine nappe. The nappe is a gently dipping structure with a displacement of at least 80 km (Elliott & Johnson 1980). Thermal modelling requires it to have had a thickness of some 12 km (Johnson *et al.* 1985). Williams (pers. comm.) has suggested that erosion may have prevented the thrust sheet from reaching Cape Wrath, but this is improbable, for 5 km of vertical erosion (i.e. relief) over 20 km horizontally is the most that metamorphic rock can sustain without collapsing under gravity (Jeffreys 1970, p. 256). Fission track dating shows that there were still 3–4 km of overburden on the Torridon Group at Cape Wrath in Devonian times (Lewis *et al.* 1992).

These considerations suggest that the palaeosol has, in fact, been strongly compacted and heated. The high Weber indices obtained by Retallack & Mindszenty (1994) were on bulk samples containing relatively coarse illite crystals. It is possible that Weber indices obtained from the finer 2–6 μm fraction of the same samples might have much lower indices, like those found by Johnson *et al.* (1985, fig. 3). Nevertheless, the coarse, poorly crystallized illite and the survival of pedogenic smectite despite K-metasomatism are surprising. Isotopic dating of the illite is desirable to verify the Precambrian age of the clay suite.

The Cape Wrath weathered sequences that developed over biotite gneiss and amphibolite are like the well-known early Proterozoic Pronto and Denison palaeosols in Ontario (Gay & Grandstaff 1979), that Holland (1984, p. 277–291) used to put limits on the ratio of oxygen to carbon dioxide in the contemporary atmosphere. In the poorly drained Denison palaeosol the carbon dioxide and oxygen used in the weathering reactions were entirely supplied from the atmosphere, via the groundwater trapped in the soil, and the oxygen was insufficient to oxidize Fe^{2+} to Fe^{3+}. Consequently iron was lost in solution. The Pronto palaeosol, on the other hand, was well drained and constantly resupplied with atmospheric carbon dioxide and oxygen. There was sufficient oxygen to oxidize Fe^{2+} to Fe^{3+} and iron was retained in the weathered profile. These two palaeosols, which are of about the same age, enable limits to be placed on the ratio of oxygen to carbon dioxide in the atmosphere at that time. Unfortunately, both of the Cape Wrath palaeosols are like the Pronto palaeosol in that the ratio Fe_2O_3/FeO increases up the profile and total iron remains constant (Retallack & Mindszenty 1994, table 7). Holland's parameter R which is the ratio of oxygen demand to carbon dioxide demand for the parent rock is 0.045 for the amphibolite, which means that $P(O_2)/P(CO_2) > 0.7$, where P denotes the atmospheric partial pressure for the stated gases during weathering. This is not very illuminating, considering that the present value of $P(O_2)/P(CO_2) \approx 600$. All that can be said with certainty is that there was enough oxygen in the atmosphere to oxidize most of the magnetite in the weathered profile, a conclusion already reached by Retallack (1990, p. 337–9). The conclusion accords with the atmospheric model originated by Cloud (1968) and most recently elaborated by Rye & Holland (1998) and

Holland (1999), which requires a steep rise in oxygen content about 2200 Ma ago, from only 1% of the present atmospheric level (PAL) to 15% PAL and possibly much more. The model depends on the fact that before this time the iron in palaeosols is reduced, and the easily oxidized minerals uraninite, pyrite and even siderite (Rasmussen & Buick 1999) appear as detrital grains in sandstones. After 2200 Ma the iron in palaeosols is oxidized, red beds make their appearance, and detrital uraninite and pyrite are no longer found.

Facies and environments

Valley-confined alluvial fans

The palaeovalleys are usually fringed by a coarse, massive, clast-supported breccia (facies Db1a) that grades stratigraphically upwards, and towards the valley axes, into tabular red sandstone (facies Db1b) and locally into lacustrine muds (facies Db2). A bird's-eye view of the landscape around Gairloch and Diabaig at this time shows half the basement covered by Torridon Group sediment and the rest bare hills (Fig. 25). The palaeovalley deposits at Diabaig are described in detail on pp. 96–100.

The breccias and tabular sandstones are analogous to facies Ct1 and Ct2 in the Stoer Group and like them are believed to represent alluvial cones derived from basement hills. The only exception to this generalization is the rock-fall breccia involving Stoer Group sandstones at Enard Bay (p. 73). The breccia clasts in the alluvial cones have not come far, for example at Enard Bay no more than 10–100 m from the parent rocks beneath the unconformity (p. 87). Wave ripples and current ripples are common in the tabular sandstones, indicating the persistence of very shallow waters. Trough cross-bedding in sets up to a decimetre thick is rare. So, too, are channels, except at Alligin where they are up to a metre deep. However, there is some evidence for a general drift of material towards the east, presumably due to a weak drainage system fed by sparse local rainfall. For example, sandstone and quartzite pebbles in the Diabaig Formation (facies Db1b) of Inverpolly Forest, and also near Scoraig, are found up to 4 km east of the Stoer Group outcrop which yielded them. Outliers of breccia (Db1a) at A'Mhaighdean [NH 005755] and Fuar Loch More [NH 010765], derive from the Beinn Lair hornblende schist complex about 3 km to the west and SW (Peach *et al.* 1907. p. 315). Trough cross-bedding in the Diabaig Formation at Loch an Eich Dhuibh [NH 018860] indicates palaeocurrents flowing towards the east, whereas at Alligin and Quinag they flowed towards the SE. The Quinag palaeocurrents in the Diabaig Formation are particularly interesting because some of them come *from* the SE, exactly counterposed to those just mentioned. Troughs created by these counterposed palaeocurrents may appear in alternate beds, and even in the same bed. This is most unusual in alluvial sediments (Pettijohn *et al.* 1987, p. 128). Whether the sediments deposited by these counterposed palaeocurrents derive from opposite sides of a palaeovalley, or from sources completely different, might be resolved by examining their mineralogy.

Valley-confined lakes

The red tabular sandstones of facies Db1b typically pass laterally into grey lake shales, facies Db2, best exposed at Diabaig (pp. 97–100) and on Raasay (p. 107). The silt-mud rhythmites that constitute a half of the facies (Fig. 97) contain abundant fragmented and partially carbonized cryptarchs, suggesting seasonal deposition in an oxidizing lake. Intact cryptarchs are preserved in early diagenetic concretionary phosphate at Diabaig (W. L. Diver, pers. comm.) and have also been recovered by Naumova & Pavlovsky (1961) from macerations of correlative shales on Raasay (p. 107). The concretions are formed of francolite (about 20%), calcite (about 10%) and undigested shale (Rodd 1983, table AxI.5; Rodd & Stewart 1992). The lake was shallow, for mud-cracked surfaces are common; there are about 2000 such surfaces in 115 m of shale at Diabaig. Assuming a typical sedimentation rate for lacustrine muds and silts of 500 m/Ma (Håkanson & Jansson 1983, p. 162 & 173; Baltzer 1991) gives an average of one desiccation event per century. The sedimentation rate adopted, however, could be in error by a factor of 2.

The shales are thought to be lacustrine rather than marine because of their low boron content. The boron content of illite separated from brackish or marine shales is in the range 200–500 ppm, whereas illite from six horizons in the grey shale at Diabaig had a boron content between 77 ppm and 113 ppm (Stewart & Parker 1979). The shales lack primary carbonates and are completely devoid of evaporites so that the suggestion by Van de Kamp & Leake (1997) that they were deposited in a hydrologically closed basin is unfounded. The gradual thickening of the shale facies towards the south, shown in Figure 23, suggests that there was a single, large lake that developed over a drainage system as a result of regional warping.

The appearance of turbidites in the upper part of the shales (p. 100), derived from the west like the overlying Applecross Formation, marks the arrival of a major flux of coarse clastic sediment and water. In Raasay, the shales and turbidites are cut by red channel sands up to 8 m thick, also derived from the west, with strongly erosive bases (Fig. 107). The sands are sedimentologically like the lowest beds of the Applecross Formation. The geochemistry of the shales shows that they derive partly from a source quite unconnected with the local basement gneisses (pp. 35–36).

These features suggest that the lake formed a sink for the Applecross alluvial system and gradually filled up with sediment supplied mainly from afar. The turbidites are interpreted as delta-front deposits. Lake levels evidently varied greatly, probably by tens of metres, for there are desiccated horizons between the turbidites, and also phases of deep erosion during which braided sheets of sand spread down the palaeovalleys of Raasay.

Valley-confined rivers

Sheets of gneiss cobble conglomerate up to 14 m thick appear at intervals through the whole of the Diabaig Formation from Inverpolly Forest northwards to the latitude of Canisp and Suilven, a distance of about 10 km perpendicular to the southeasterly palaeocurrent direction. Fine-grained red sandstone cobbles can also be found in the conglomerate, together with rare quartzite, suggesting derivation from the Stoer Group more than 20 km to the west. The conglomerates were deposited by a large river system, though perhaps not synchronously over the whole zone, for they are found in four parallel palaeovalleys. All four, fortuitously exhumed, trend SE parallel to the gneiss foliation in the basement and to the palaeocurrent direction in the conglomerate. They correspond to the valley that now separates the mountains Canisp and Suilven, to the two valleys now occupied by the Cam Loch and Loch Veyatie, respectively, and the valley flanking Cùl Mòr to the SW (Stewart 1972, fig. 7).

River conglomerates of this kind, very like facies Ct4 in the Stoer Group, are found nowhere else in the Diabaig Formation. It is significant that the cobble conglomerate appears as the grey lacustrine siltstones thin to nothing, suggesting not only that the river, or rivers, flowed into the lake, but that the lake margin was stable over time. This in turn implies a hydrologically open lake system.

Unconfined alluvial fans

Upward-fining sandstone units tens or hundreds of metres thick overlie the valley-confined sediments and residual basement hilltops. The best known of these units is the Cape Wrath Member (Applecross Formation), the lower part of which has been shown to

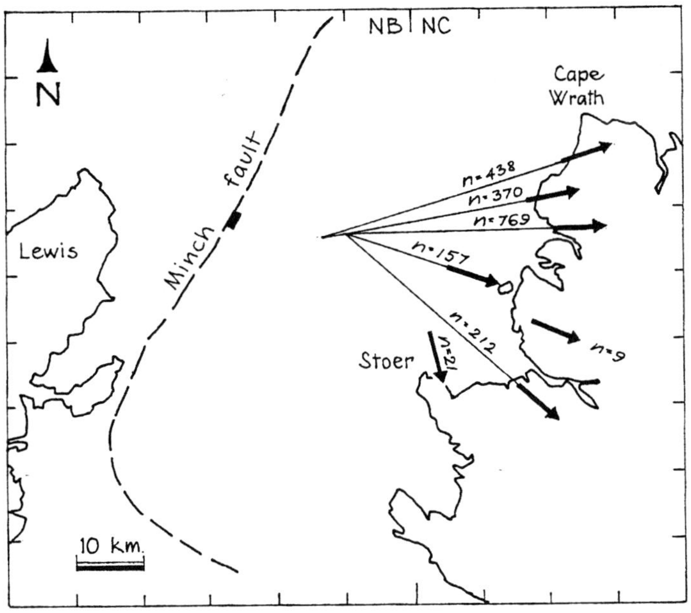

Fig. 26. Palaeocurrent directions in the Cape Wrath Member of the Applecross Formation (Williams 2001, fig. 8). Locations, vector means (measured from grid north) and the number of observations corresponding to the seven arrows are; Cape Wrath, sub-area C, $\theta = 073°$, $n = 438$; Cape Wrath, sub-area B, $\theta = 080°$, $n = 370$; Cape Wrath, sub-area A, $\theta = 089°$, $n = 769$; Handa, $\theta = 110°$, $n = 157$; Ben Dreavie, $\theta = 113°$, $n = 9$; Quinag, $\theta = 133°$, $n = 212$; Rubha Stoer, $\theta = 164°$, $n = 21$. The trace of the Minch fault is from Stein (1988, fig. 6b).

be a giant alluvial fan with a radial palaeocurrent pattern (Williams 1969*a*, 2001). It is about 500 m thick at Cape Wrath and has a radius of about 50 km (Fig. 26), with an apex where the Minch fault intersects the metamorphic basement, about 17 km east of the surface trace (Fig. 22). The sediments are illustrated in Figure 40. Maximum clast size decreases exponentially upwards from a maximum of 120 mm at the base to 30 mm at the top. The largest pebbles reach 180 mm in diameter (Williams 1969*a*, table 1). The conglomeratic beds at the base, however, die out eastwards only 3 km from the coast.

Farther south, the valley-confined sediments are overlain by two overlapping fining-upward units (Fig. 27). Their full extent has not, unfortunately, been tracked, and the available palaeocurrent data have not been collected with sufficient regard to stratigraphic level within the units to determine whether or not they show a radial pattern. Both, however, extend at least 25 km perpendicular to the mean eastward palaeocurrent direction, and probably twice this much. They do not extend south of the Loch Maree fault. According to Nicholson (1993) the coarser part of the upper unit near Achiltibuie (p. 78) contains bars migrating downstream in channels 3–7 m deep, whereas the finer part of the underlying unit represents sheet-flood deposition on a flood plain. These are feasible environments, but Nicholson also claims to be able to tell that the cyclic facies arrangement and the channels arose from purely fluvial processes such as channel switching, rather than source rejuvenation or fan head retreat. His argument is based on observations at a single locality, however, and takes no account of the lateral extent of the units.

The existence of recognizable fans in the mainland Applecross Formation might be because they are close to their source, perhaps a fault scarp. The sediments forming the unconfined bajada stratigraphically above display no fan patterns because the source had abruptly retreated, perhaps to a more distant fault scarp.

The size of alluvial fans has long been known to scale with that of their source area (Bull 1962). The general relationship is

$$F = pAq$$

where F and A are the fan and source areas, respectively, expressed in km^2, while p and q are constants. The constants are known for modern fans in semi-arid areas up to about 200 km^2 in size, but the Cape Wrath fan covers about 1.2×10^3 km^2, necessitating a significant, possibly unwarranted extrapolation of the data. Substituting $p = 0.7$ and $q = 0.8$ (Harvey 1989) gives $A = 11 \times 10^4$ km^2, equivalent to a square with sides about 100 km long. Williams (2001) has suggested that the source area may have been as large as 1.8×10^4 km^2, equivalent to a square with sides of about 130 km. Such an area would coincide with the present position of the Outer Hebrides block. The coarse grain size of the fan suggests an upfaulted block in that area shedding sediment to the east.

Unconfined bajada

The greater part of the Torridon group (Applecross and Aultbea Formations) consists of some 5 km of red sandstones in which significant fining-upward units are absent. No facies subdivision is proposed for these sandstones, apart from the simple field classification based on pebble abundance used in Figure 23. Four of the five facies defined at Cape Wrath by Williams (1969*a*) are

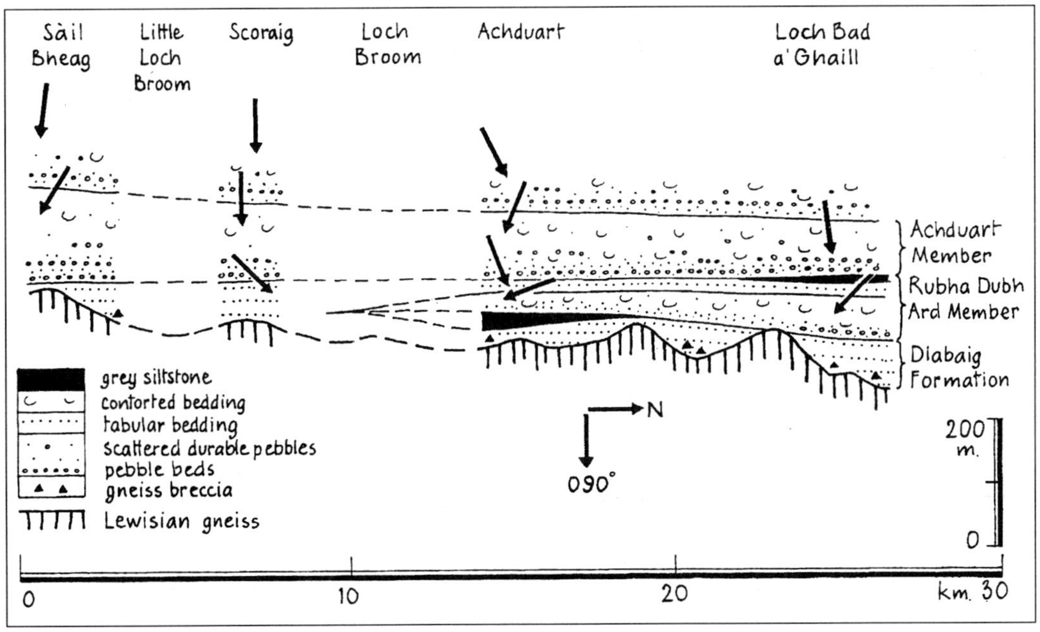

Fig. 27. Diagrammatic section through the unconfined alluvial fan facies of the Applecross Formation from Stac Pollaidh [NC 108106] in the north, to Sàil Bheag [NC 020900] in the south. The section line trends 024°, almost perpendicular to the palaeocurrent vectors shown by the arrows. The arrows are plotted with north to the right, with the point of observation at the arrow head. The datum for the section is the base of the Achduart Member. The palaeocurrent vectors for Sail Bheag and Loch Bad a'Ghaill are from Williams (1969*a*, fig. 12).

from sediments too coarse to be representative of the sediments farther south. Only the finest facies at Cape Wrath (Fig. 40), which is the only one strongly contorted, has anything in common with the Applecross Formation as a whole.

Selley (1965*a*, 1969, 1970) and Williams (1966*c*, 1969*a*, 2001) noted the following environmentally significant features in the pebbly Applecross Formation:

- low palaeocurrent variance;
- medium to coarse grade sediment with virtually no silt or mud;
- common trough cross-bedding in sets 10–100 cm thick, and planar cross-bedding in sets 30–200 cm thick;
- channels 1–2 m deep;
- absence of fining-upward cyclicity typical of meandering rivers.

They concluded that the sands were deposited by braided streams on a braidplain or, at the base of the sequence, on the alluvial fans described in the last section. The best modern analogue would be a deep, perennial sand-bed braided river (Miall 1996). The South Saskatchewan River, with channels up to 600 m wide and an average depth of 3 m is an example. Some even deeper channels have been identified by Nicholson (1993). They contain composite sand bodies up to 9 m thick, formed of downstream-dipping cross-beds, generated by small bedforms that migrated down or across the face of a channel bar.

The average palaeocurrent direction in the Applecross Formation has long been known to be southeastwards. A comprehensive set of measurements from all stratigraphic levels by Nicholson (1993, table 1 & fig. 8) confirms this, giving $\theta = 123^\circ$ for $n = 2802$.

Few sedimentological studies of the Aultbea Formation in its type area have been carried out. It differs from the Applecross in generally lacking pebbles and being medium-grained rather than coarse. Contorted bedding is ubiquitous, making depositional structures difficult to study. The available data from Aultbea, however, do not suggest any significant difference in palaeocurrent direction, or in set thickness (see Fig. 82). Coarsening upward cycles at Aultbea and Toscaig probably represent the advance of flood deposits over floodplain accretions, like the lineage 3 of Bluck (1986). The finer grain-size of the formation probably results from a greater distance from the source, for the zircons it contains are more abraded than those in the Applecross Formation (Rainbird *et al.* 2001). The Aultbea is unlikely to have come from erosion of the Applecross because the source areas of the two formations were distinctly different (Fig. 30).

Some progress has been made by Nicholson (1993) in establishing the palaeohydrological regime of Applecross rivers. The bankfull discharges (Q_b) of 300–3000 $m^3 s^{-1}$ he estimated are, however, difficult to interpret. Precambrian streams rose in barren areas giving almost instantaneous run-off, so that for similar annual rainfall, bankfull discharges were ten to twenty times present-day values (Schumm 1968). Indeed, the resulting intermittent high run-off may be one reason why nearly 90% of all Precambrian fluvial sediments were deposited in braided rather than in meandering channels (Cotter 1978, table 2). Furthermore, the lack of transpiration and evaporation from plant leaves would have roughly doubled the mean annual run-off, compared with a similar climatic regime today (Schumm 1968, fig. 2). In short, the present-day relationship between mean annual run-off and bankfull discharge is not relevant to Precambrian times. Nevertheless, Nicholson proposes figures for mean annual discharge (Q_a) by comparing Applecross rivers with those of the North American plains (Leopold *et al.* 1964, table 7-13) for which $Q_b/Q_a = 4$. For the reasons given above the resulting figures for Q_a are too large by a factor of between 20 and 40, and so too are the sizes of the corresponding drainage basins.

Half to three-quarters of all the beds in the Applecross Formation, and virtually all those in the Aultbea, are strongly contorted. The coarsest four facies of the Cape Wrath Member, however, are not contorted. In other comparable sequences such structures are curiosities rather than commonalities and even though their origin is now well understood, the reason for their frequency in the Torridon Group is not.

The structures have been thoroughly analysed by Owen (1995, 1996*a*, 1996*b*). Some individual beds have been affected by liquefaction shortly after deposition, but more commonly several beds are pierced by fluidized sand diapirs rising from a liquefied buried layer. In addition to these metre-scale structures, detrital iron ore laminae became gravitationally unstable and developed centimetre-scale drops when the sediment was liquefied (Stewart 1963; Selley 1964). The larger structures are predominantly elongated perpendicular to the palaeocurrent direction, and have a tendency to monoclinic symmetry, as if they had undergone simple shear in the down-slope direction (Selley *et al.* 1963, fig. 1).

Seismic shaking has long been known to be capable of liquefying sand. Owen (1995) suggests this as the explanation for the sand diapirs that penetrate several beds, and upwelling groundwater as the cause of the deformation of individual beds. Williams (1970, 2001) claims that the contortions record springs near alluvial fan margins, supplied by infiltration in the upper fan. Such structures are not recorded in model fans (Rachocki 1981, p. 25–110) and are only locally present in natural arid-zone fans (Blissenbach 1954; Bull 1964).

Most of these structures are forms of Rayleigh-Taylor instability and have gravity as their driving force, but there is no evidence that gravity in the late Proterozoic was much greater than now (Stewart 1977, 1980) and so the frequency of the structures remains a mystery.

More recently it has been suggested that some of the Torridon (and Stoer) Group structures result from air forced out of the sediment by rising ground water (McManus & Bajabaa 1998). However, the authors provide neither precise locations nor illustrations of such structures in the Torridonian. The modern example they describe shows small-scale dish-and-pillar structure.

One of the typical features of the Applecross Formation, mentioned above, is the almost complete absence of silt or mud. Only near the base of the formation in the Diabaig–Gairloch area, and between Raasay and Rum, are there significant red shale units. At several localities these shales fill abandoned channels (Fig. 90; Selley 1969, fig. 4), testifying to low stream gradients during early Applecross deposition. Grey shale in the lowermost Aultbea Formation appears to record the invasion of the bajada by extensive lakes. This particular horizon can be followed along strike for at least 20 km, from Dornie to Aultbea, and possibly as far as Toscaig. The abundant microbiota in these shales is the only one from the Torridonian to have been properly described (Zhang Zhongying *et al.* 1981; Zhang Zhongying 1982). Sphaeromorphic cryptarchs present include the new species *Torridoniphycus lepidus*, which occurs as vesicles 10–30 μm in diameter, and vesicle aggregates, thought by Zhang to represent stages in the life-cycle of a single species of coccoidal, endospore-forming blue-green alga. Filaments 3–6 μm wide were identified as *Eomycetopsis crassiusculum* (Horodyski 1980) comb. nov. which also occurs in the Mesoproterozoic Belt Supergroup of Montana. Larger filaments 10–25 μm wide were attributed to the genus *Siphonophycus*. They are very like some in the Cailleach Head Formation (Peach *et al.* 1907, plate LII). They may be discarded sheaths of filamentous oscillatoriacean cyanobacteria. Zhang Zhongying (1982) suggests that the low taxonomic diversity of the biota indicates a restricted aquatic ecosystem, such as a lake or marine embayment.

Cyclothems

The Cailleach Head Formation at the top of the Torridon Group is made up of a series of cyclothems of which the lowest fifteen are perfectly exposed on the cliffs at the type locality. The fifteen cyclothems are described in detail on pp. 80–81 and in Fig. 77. A typical example is shown in Fig. 28, and an interpretation of the facies in Table 11.

The lowest sediments of each cyclothem (facies 1) formed in moderately deep water that was invaded progressively by fluvial braided sediments (facies 2). Facies 2 contains channels up to 0.5 m deep, but the top of the facies is truncated by a flat erosion plane

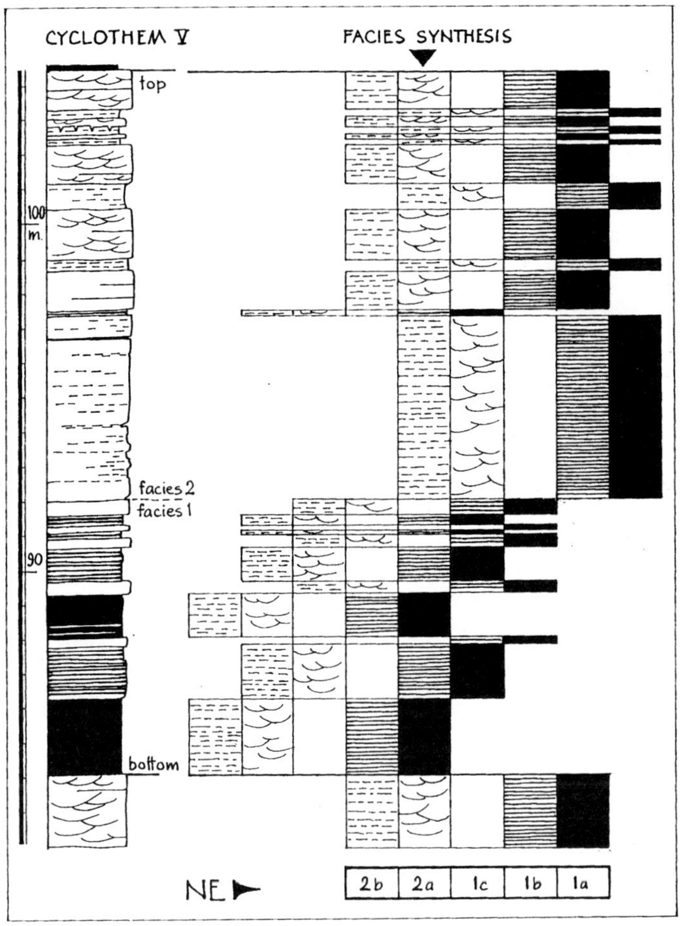

Fig. 28. Cyclothem V in the Cailleach Head Formation. A graphic log of the actual cyclothem is shown in the left-hand column, with a metre scale corresponding to that in Fig. 77. The sediments forming the cyclothem can be generated by the to and fro migration of the facies sequence 2b (proximal) > 2a > 1c > 1a (distal), as shown in the facies synthesis column. The facies are defined in Table 11. Palaeocurrents show that the direction of migration was northeasterly.

suggesting abrupt subsidence beneath wave-base. Facies 1 lacks carbonate, macroscopic pyrite or evaporite minerals, suggesting a well circulated, oxygenated lake, probably, but not certainly, hydrologically open. The minimum water depth is given by the combined thickness of subfacies 1a & 1b, which when decompacted is about 10 m.

The cyclothems may simply reflect asymmetrical oscillations in lake level due to climate change, superimposed on a general subsidence (the sequence is over 600 m thick). Lake levels may have receded *slowly* as evaporation rates increased, allowing an alluvial wedge to advance into the former lake area. As lake levels rose *rapidly* the wedge was inundated and covered by lake sediments. The scenario works best for hydrologically closed lakes which are extremely sensitive to climate change. Several closed lakes in the Mesozoic rifts of eastern North America show long series cyclothems generated by Milankovitch-type climatic cycles (Olsen & Gore 1989). Some of the cyclothems can be traced for over 100 km (Hubert *et al.* 1976). The cyclothems at Cailleach Head are believed to have developed on the distal edge of a bajada with shifting ephemeral channels. So delta abandonment, which played a key role in Carboniferous cyclothem formation, seems an unlikely cause for those at Cailleach Head.

Palaeocurrents in the Cailleach Head Formation flowed northeastwards, at right angles to those in the underlying fluviatile red sandstones of the Applecross and Aultbea Formations. It appears, therefore, that uplift of a well-defined, NNE trending basin margin no longer determined the depositional palaeoslope as it had done previously.

Geochemistry and mineralogy

The Torridon Group is almost entirely composed of arkoses, the chemistry of which has been intensively researched over the last ten years. Attention has been focused particularly on the weathering of locally derived detritus in the Diabaig Formation, the nature of the source rocks, and burial diagenesis. All the sandstones show plagioclase albitization, but the source of the sodium for this transformation remains obscure. Post-depositional potassium metasomatism of the shales is also evident. Detailed consideration of the source rocks for the Applecross and Aultbea Formations is reserved for a later section (p. 40).

Breccias, tabular sandstones and shales in the palaeovalleys

The chemistry of these sediments at the type locality at Diabaig has been examined in detail by Rodd & Stewart (1992) and is shown in Table 12. The breccia matrix and the sandstones are arkoses derived from laterally adjacent gneisses, but weathering has destroyed 70% of the plagioclase and all the amphibole and pyroxene (Tables 13 & 14). The lacustrine shales, however, have quite a different source. The mass balance technique described on pp. 24–25 shows that they contain far too much K and Fe to be derived from the local gneisses

Table 11. *Facies in the Cailleach Head Formation and their origin*

Facies	1a	1b	1c	2a	2b
Lithology	Siltstone to fine sandstone	siltstone to fine sandstone	fine to medium-grained sandstone	medium-grained sandstone (<10% mica)	medium-grained micaceous sandstone
Grain size (mm)	0.05–0.1	0.05–0.1	0.05–0.15	0.1–0.2	0.1–0.2
Bed thickness (cm)	0.01–1	0.01–1	1–300	1–200	1–200
Bed persistency (*p*)	>10 000	>1000	100–500	10–100	10–100
Sedimentary structures	–	linguoid ripples	flat bedding, current lineation, wave ripples, tabular planar cross-bedding, drag marks	trough cross-bedding	trough cross-bedding, channels, siltstone clasts
Colour	grey	yellowish-grey	pale or moderate pink	pale red	greyish red, yellowish green
Sedimentary environment	lake bottom	delta toe	delta top	fluviatile channels	fluviatile channels

The persistency (*p*) of a bed is defined as its lateral extent divided by its maximum thickness.

Table 12. *Chemistry, norms and molecular proportions A–CN–K for gneisses and Diabaig Formation facies at Diabaig*

	A Average gneiss $n=121$	B Breccia $n=3$	C Tabular sandstone $n=3$	D Grey shale $n=2$	E Grey sandstone $n=6$
SiO_2	67.7	71.36	77.41	60.60	76.10
TiO_2	0.4	0.47	0.20	1.09	0.40
Al_2O_3	15.3	13.14	11.72	18.55	9.93
$t.Fe_2O_3$	4.3	4.43	2.14	8.13	4.30
MnO	0.1	0.08	0.03	<0.01	0.11
MgO	2.1	3.10	2.13	2.51	1.91
CaO	3.7	0.76	0.11	0.82	3.07
Na_2O	4.2	3.76	4.33	1.48	2.25
K_2O	2.2	3.29	2.61	3.20	1.22
P_2O_5	–	<0.01	<0.01	0.11	0.06
Total	100.0	100.39	100.68	96.49	99.35
Q	25	35	40	10	52
Plag	45	30	35	11	25
Or	7	15	17	4	4
Ms	4	–	–	–	–
Bi	7	–	–	–	6
Ho	4	–	–	–	–
Cpx	7	–	–	–	–
Chl	–	15	8	10	10
Illite	–	5	–	52	–
Smectite	–	–	–	10	–
A	51.4	54.2	53.6	72.2	56.1
CN	40.6	31.1	33.5	14.3	36.3
K	7.9	14.7	12.9	13.5	7.5

Chemical and mineralogical data are per cent. Column A is an estimate of the average composition of the Lewisian basement at Diabaig, based on 85.2% biotite gneiss (Holland & Lambert 1973, table 4G) and 14.8% Scourie dykes (Holland & Lambert 1973, table 6K). Sediment compositions are XRF analyses from Rodd & Stewart (1992, table 2). A, CN and K are mol percentages of Al_2O_3 (equivalent to the CIA index), CaO and Na_2O, and K_2O, respectively.

Stewart (1995*b*). It appears that micaceous material has been contributed to the lacustrine system by the same rivers that later deposited the overlying Applecross Formation (Rodd & Stewart 1992). The conclusion is not really surprising, for lakes in hydrologically open systems are often fed silt and mud in suspension by a river whereas sand and pebbles tumble in from nearby hills. Nevertheless, it is disappointing, because knowing the composition of the source rocks and the derived sediment, together with the relief, it would have been possible to calculate the ratio of suspended load to dissolved load and reach some useful conclusions about the climate (Stewart 1993*b*).

Alluvial fan and bajada sandstones

The chemistry of these sandstones, which form the bulk of the Torridon Group, has been examined by Stewart (1991*b*), Stewart & Donnellan (1992), Van de Kamp & Leake (1997) and Young (1999*a*). Their modal mineralogy and chemistry are summarized in Figure 29 and Table 15, respectively.

The sandstones are all arkoses, mostly with plagioclase and K-feldspar in equal abundance. Calcium values are uniformly low due to albitization of the plagioclase. Strontium was removed along with the Ca and consequently Rb/Sr for the sandstones is around unity. The averages in Table 15, however, conceal some significant upward trends in composition. Sequential sampling of the lowest kilometre of the Applecross Formation in Coigach, for example, shows a monotonic upward decline in plagioclase and concurrent increase in quartz. The ratio of quartz to feldspar at the base is 0.65 wheras at the top it is 3.3 (Stewart & Donnellan 1992, fig. 4). A similar trend is implied by the chemistry of the fining-upward Cape Wrath Member at Cape Wrath (Williams 2001, table 3) and Quinag (Donnellan 1981). These trends, shown in Figure 31, are probably due to increased weathering of feldspar, particularly plagioclase, rather than any change in source composition. Initially erosion was more rapid than weathering and soils were thin or absent. A progressive reduction of source area relief allowed thicker soils to form with consequently more complete transformation of plagioclase to clay. Increased transport distance alone is unlikely to have changed the ratios of quartz to feldspar or Na_2O/K_2O (Ronov *et al.* 1966, tables 5 & 6; Nesbitt *et al.* 1996).

Progressive changes in source rock composition during deposition of the Torridon Group can be demonstrated from mass-balance calculations like those used on pp. 24–25. The chemistry of the sandstones and shales in Table 16 gives the fence diagrams in Figure 30, which show that in terms of Al and Fe the lowest Applecross came from an iron-rich source (e.g. hornblende–biotite gneiss, or pelitic sediment) whereas the Cailleach Head Formation at the top of the Torridon Group derived from an iron-poor rock (e.g. sandy sediment). Igneous rocks and orthogneisses are an unsatisfactory source for they are incapable of simultaneously providing Al, Fe and K values lying within the fenced areas (Stewart 1995*b*). These inferred changes in source rock have nothing to do with palaeoclimate, weathering or transport distance.

Changes in source area can also be detected by examining the Rb content of detrital feldspars, obtained from double ratio plots of whole rock data (Stewart & Donnellan 1992, figs 5 & 6). K-feldspar in the Applecross Formation in Coigach is relatively rich in Rb (≈420 ppm), but in the Cape Wrath Member at Cape Wrath and Quinag it is significantly less (≈260 ppm). Plagioclase also shows remarkable variations in Rb content despite having been albitized. In the Applecross Formation in Coigach, which is about 2 km thick, plagioclase has less than 5 ppm Rb, whereas in the overlying Aultbea Formation the figure rises to about 140 ppm.

The Sr content of feldspars is notably more uniform than Rb. All whole rock samples of sandstone from Rum up to Cape Wrath have about 250 ppm of Sr in plagioclase, presumably homogenized during albitization (Stewart 1991b, fig. 5*b*; Stewart & Donnellan 1992, fig. 6).

Sodium metasomatism

Sodium metasomatism of clastic sequences is a well-known phenomenon indicating burial temperatures of between 100° and 150°C, a flux of acid formation water and the availability of Na. The transformation of original detrital plagioclase to albite in the Torridon Group sandstones and shales is attested by the very low levels for Ca. What little remains is in epidote. Independent evidence of a pore water flow through the sandstones is given by relatively low $\delta^{18}O$ values in the $<2\,\mu m$ clay faction as compared with similar clays from shale layers (Benner & Hoernes 1994).

There is no evidence for the albitization of K-feldspar, except very locally in the Diabaig Formation. Albitization textures have been reported from the sandstones by many workers, starting with Black & Welsh in 1961, but it was the geochemistry that highlighted its importance. The chemistry suggests the plagioclase has an anorthite content An_4 (Stewart & Donnellan 1992). Optical methods give An_{6-16} (Van de Kamp & Leake 1997).

The amount of Na contributed to the Diabaig Formation during albitization can be seen from Table 13. The sediment now extant (429 g from an original kilogram) contains 12.1 g of Na_2O, compared with 6.8 g before albitization. The source of this extra Na is unknown. As pointed out earlier, there is no evidence whatever that the Diabaig Formation was deposited in an evaporative lacustrine environment as assumed by Van de Kamp & Leake (1997).

The timing of the albitization event has been tentatively fixed at about 985 Ma by whole rock Rb–Sr isochrons obtained from Torridon Group phosphate and shales (Turnbull *et al.* 1996). The

Table 13. *Geochemical budget for sodium in the Diabaig Formation at Diabaig*

Na_2O in a kilogram of source material			Na_2O in a kilogram of sediment and dissolved load		
Gneiss (842 g)			*Breccia & sandstone (250 g)*		
45% Plagioclase (An_{20})	35.4 g		22% Plagioclase (An_{20})	5.2 g	
7% K-feldspar	0.3		16% K-feldspar	0.2	
7% Biotite	0.4				5.4
4% Muscovite	0.2		*Shale (500 g)*		
4% Hornblende	0.3		11% Plagioclase (An_{20})	5.1	
7% Clinopyroxene	0.6		62% Kaolinite & smectite	0.3	
		37.2	4% K-feldspar	0.1	
					5.5
Suspended load (158 g)			*Total dissolved load (250 g)*		
10% Plagioclase (An_{20})	1.5		11.1% Na_2O	27.8	
		1.5			27.8
Total		38.7 g	Total		38.7 g
			Na_2O in a kilogram of sediment now		
			Breccia & sandstone (250 g)		
			22% Albite (detrital)	6.1	
			11% Albite (authigenic)	3.0	
			16% K-feldspar	0.2	
					9.3
			Shale (179 g)		
			11% Albite	2.3	
			52% Illite	0.5	
					2.8
			Total		12.1 g

The assumed proportions of sand, silt and total dissolved load are typical of present-day temperate climates. Sand/silt = 0.5 (Stewart 1993*b*); silt/dissolved load = 2 (Meybeck 1976, 1988 annex). The modal mineralogy for the gneiss is repeated from Table 12A. The plagioclase in the local gneiss is oligoclase (Sutton & Watson 1951, p. 254 & 259). The suspended sediment attributed to the source is supposed to be like Applecross shale (Table 16B), with a mass sufficient to balance K (assumed insoluble) between source and sediment. The present ratio of sandstone and shale is that observed in cross-sections, with the sandstone assumed conserved. The present albite content of the sandstone (22% detrital, 6% authigenic) comes from modal analyses. The dissolved load does not include atmospheric CO_2.

Table 14. *Geochemical budget for calcium in the Diabaig Formation at Diabaig*

CaO in a kilogram of source material			*CaO in a kilogram of sediment and dissolved load*		
Gneiss (842 g)			*Breccia & sandstone (250 g)*		
45% Plagioclase (An_{20})	13.6 g		22% Plagioclase (An_{20})	2.0 g	
7% K-feldspar	0.1		16% K-feldspar	–	
7% Biotite	0.6				2.0
4% Muscovite	–		*Shale (500 g)*		
4% Hornblende	4.0		11% Plagioclase (An_{20})	2.0	
7% Clinopyroxene	11.8		62% Kaolinite & smectite	5.1	
		30.1	4% K-feldspar	–	
					7.1
Suspended load (158 g)			*Total dissolved load (250 g)*		
10% Plagioclase (An_{20})	0.6		8.6% CaO	21.6	
		0.6			21.6
Total		30.7 g	Total		30.7 g

For notes see Table 13.

reliability of this conclusion is discussed in the section dealing with the age of the Torridon Group (p. 45).

Potassium metasomatism

Regional potassium metasomatism may be expressed by the illitization of detrital kaolinite and smectite in shales, and less commonly by the replacement of plagioclase in sandstones by authigenic K-feldspar (Fedo *et al.* 1995; Fedo *et al.* 1997). Illitization has been convincingly demonstrated in the palaeosol beneath the Applecross Formation at Cape Wrath (Retallack & Mindszenty 1994, fig. 11; Young 1999*b*, fig. 6). Retallack & Mindszenty (1994) have suggested that the K entered the palaeosol from the overlying sandstones whereas Young (1999a) thinks that the metasomatism of the palaeosol was part of a regional event that affected all Torridon Group shales.

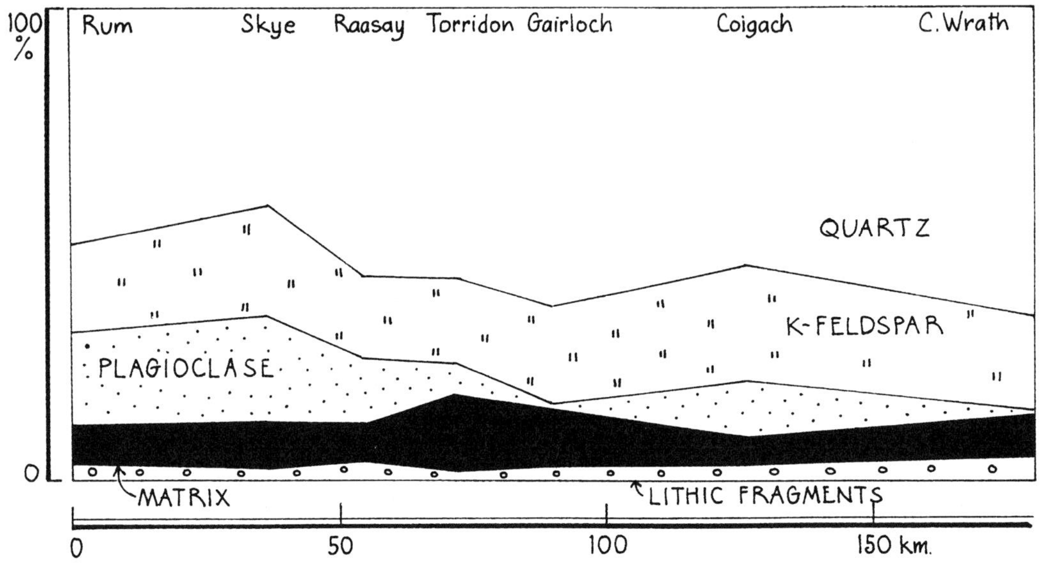

Fig. 29. The modal mineralogy of the Applecross Formation between Cape Wrath and Rum. Data for Rum, Skye and Raasay are from Byres (1972), Torridon from Maycock (1962) and Rodd (1983), Cape Wrath from Williams (1966*a*). The Coigach data are norms.

Metasomatism was inferred by Young (1999*a*) because of the way the shale analyses plot on an A–CN–K diagram (Fig. 31). Shaly sediment normally lies on a track parallel to the A–CN side of the triangle, starting from a point corresponding to the probable source rock – in this case upper continental crust (see p. 40). This track is due to the progressive loss of Ca and Na (but not Al) during plagioclase weathering, while weathering-resistant K-feldspar remains (Nesbitt & Young 1984). Torridon Group shales do not plot along a track like this, but are all displaced towards the K apex. No purely depositional process could have generated such a trend, so it has to be attributed to illitization or authigenic growth of K-feldspar (i.e. metasomatism), or to a more potassic source rock for the clay than for the silt (i.e. mixing). Any of these processes is capable of producing the linear trend shown by shale analyses plotted in A–CN–K space (Young 1999*a*, fig. 5b), or the anomalous clump of points in Figure 31. A possible potassic source for the clay component of the shale is a Palaeoproterozoic shale, most of which have been K metasomatized (Nesbitt 1992).

Table 15. *Chemistry, normative mineralogy and molecular proportions A–CN–K for Applecross and Aultbea sandstones*

	A Rum 3400 m $n=12$	B Skye 1050 m $n=13$	C Raasay 1000 m $n=10$	D Coigach 1000 m $n=59$	E Quinag 600 m $n=27$	F C. Wrath 270 m $n=32$	G Aultbea 600 m $n=25$
SiO_2	79.87	78.24	85.27	82.90	79.08	81.46	82.55
TiO_2	0.37	0.41	0.31	0.32	0.35	0.33	0.29
Al_2O_3	10.71	11.74	8.27	9.01	10.38	10.13	9.85
t.Fe_2O_3	2.00	2.12	1.48	2.27	2.64	2.44	1.80
MnO	0.48	0.68	0.23	1.24	1.40	0.82	0.83
CaO	0.18	0.14	0.26	0.08	0.03	0.02	0.01
Na_2O	2.86	3.12	1.94	1.78	1.34	0.41	1.52
K_2O	3.50	3.58	2.86	3.06	4.65	5.37	4.06
P_2O_5	0.24	0.13	0.09	0.04	0.04	0.03	0.04
Total	100.25	100.20	100.78	100.73	99.94	101.03	100.97
Ba	748	842	613	636	775	861	691
Ce	–	52	–	37	41	45	53
La	–	24	–	19	21	23	25
Ni	<15	<15	<15	4	12	6	2
Rb	91	90	80	79	84	104	92
Sr	96	117	98	60	85	77	61
Th	–	8	–	6	6	7	6.5
Y	–	12	–	8	9	9	12
Zn	9	18	12	10	16	10	9
Zr	195	218	173	172	207	199	206
La/Th	–	3.0	–	3.2	3.5	3.3	3.8
K/Rb	319	313	288	321	460	429	366
A	55.8	56.6	55.8	58.9	58.9	60.9	58.8
CN	24.5	24.7	23.4	19.5	12.5	4.1	15.0
K	19.7	18.7	20.9	21.6	28.6	35.0	26.2

The average composition of Applecross sandstones is shown in columns **A–F**, and Aultbea sandstone (from Rubha Mòr Coigach) in column **G**. The number of samples is *n* from the stratigraphic interval given in metres. Major element data are in per cent and traces in ppm. Analyses are by XRF, columns **A–C** by F. Street (Reading University, Postgraduate Research Institute for Sedimentology), and columns **D–G** by N. Donnellan (Birmingham University, Geology Department). Major and trace element analyses for nine identical Torridonian rock powders done at the two above-mentioned laboratories agree to within 10%, except for Y Ce and La. The data for Y Ce, La and Th in column **B** ($n=8$) were analysed at Birmingham.

Table 16. *Chemistry, norms and molecular proportions A–CN–K for interbedded alluvial sandstones and shales in the Torridon Group*

	A Diabaig		B Applecross		C Applecross		D Applecross		E Aultbea		F Cailleach Hd	
	sst. $n=2$	shale $n=2$	sst. $n=2$	shale $n=1$	sst. $n=2$	shale $n=1$	sst. $n=2$	shale $n=1$	sst. $n=2$	shale $n=2$	sst. $n=1$	shale $n=1$
SiO_2	75.04	60.06	76.67	61.58	81.80	56.00	82.50	52.17	82.88	61.54	73.94	63.64
TiO_2	0.41	1.08	0.48	1.06	0.19	1.18	0.38	1.54	0.53	1.12	0.46	0.35
Al_2O_3	13.34	18.85	10.41	18.37	8.07	20.54	8.55	22.27	8.19	18.50	12.86	20.26
t.Fe_2O_3	3.57	9.97	2.72	7.43	1.02	9.79	2.50	10.89	2.39	6.84	3.30	4.61
MnO	0.05	0.05	0.04	0.09	0.03	0.08	0.01	0.05	0.04	0.08	0.08	0.06
MgO	1.80	3.04	0.95	2.83	0.61	3.80	0.77	4.34	0.92	3.51	1.68	2.84
CaO	0.49	0.22	3.35	0.87	3.82	0.32	0.10	0.07	0.00	0.42	1.28	0.34
Na_2O	2.04	1.05	1.43	1.27	1.20	1.06	1.68	0.59	0.41	1.17	4.17	3.19
K_2O	3.23	5.60	3.90	6.30	3.25	7.01	3.46	7.80	4.64	6.55	2.15	4.60
P_2O_5	0.02	0.07	0.04	0.19	0.01	0.22	0.05	0.27	0.00	0.26	0.07	0.11
Total	99.99	99.99	99.99	99.99	100.00	100.00	100.00	99.99	100.00	99.99	99.99	100.00
Q	42	17.5	47.5	19	56.5	9	57	2	59.5	17.5	37	13
Plag	18	9	11	10	10	9	15	4	2.5	9	36	29
Or	5	5	12	15	14	18	18	18	25	20	10	6
Illite	28	55	17	30	12	33	5.5	56	8	38	7	43
Chl	5	6	2	11	1	17	3	13	4	12	8	8
Mu	–	–	4	11	–	11	–	–	–	–	–	–
A	63.4	70.2	55.6	64.6	53.9	68.6	56.3	70.0	59.0	66.8	53.1	65.7
CN	20.0	7.3	21.8	11.3	23.3	6.0	19.0	3.5	4.8	7.6	37.3	18.2
K	16.6	22.5	22.6	24.0	22.8	25.3	24.7	26.5	36.2	25.6	9.6	16.1

Formation names head each column. Number of samples analysed is *n*. The sources of the samples and analyses follow. Column **A**: Upper Diabaig (Stewart 1995*b*, table 2), Columns **B** & **C**. Big Sand, near Gairloch (Stewart 1995*b*, table 3), column **D**. Sheigra, near Cape Wrath (Williams 2001, table 3 analyses 2 & 7 from facies FA2), column **E**: Dornie, near Achiltibuie (Stewart 1995*b*, table 4), column **F**: Cailleach Head (Donnellan 1981, samples CH1 & CH2).

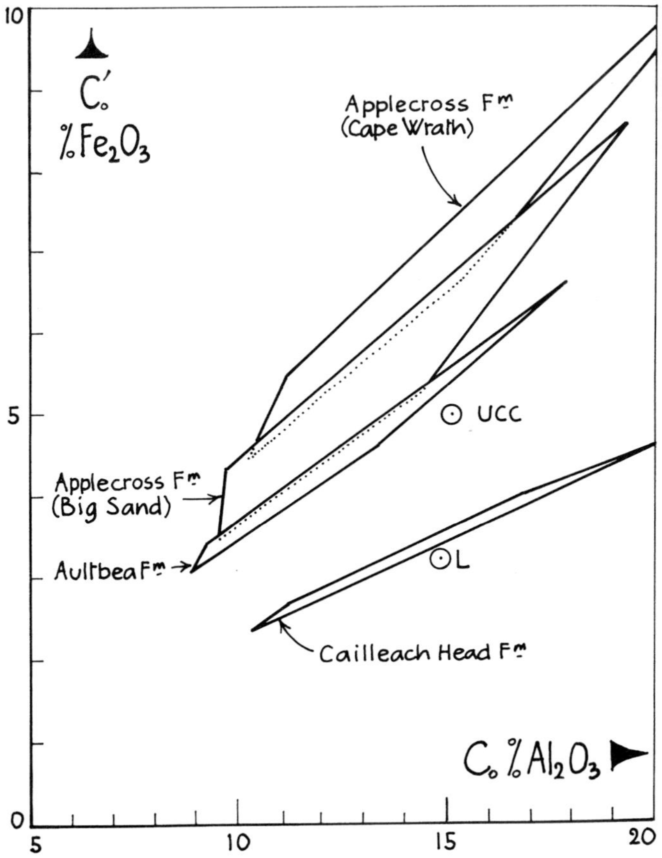

Fig. 30. Fence diagrams showing the composition of Torridon Group source rocks in terms of Al_2O_3 and t.Fe_2O_3. The diagrams are based on the analyses of interbedded sandstones and shales from the Applecross, Aultbea and Cailleach Head Formations, given in Table 16. Also shown are the average Laxfordian (L) from Table 1C, and average upper continental crust (UCC) from Taylor & McLennan (1985, table 2.15).

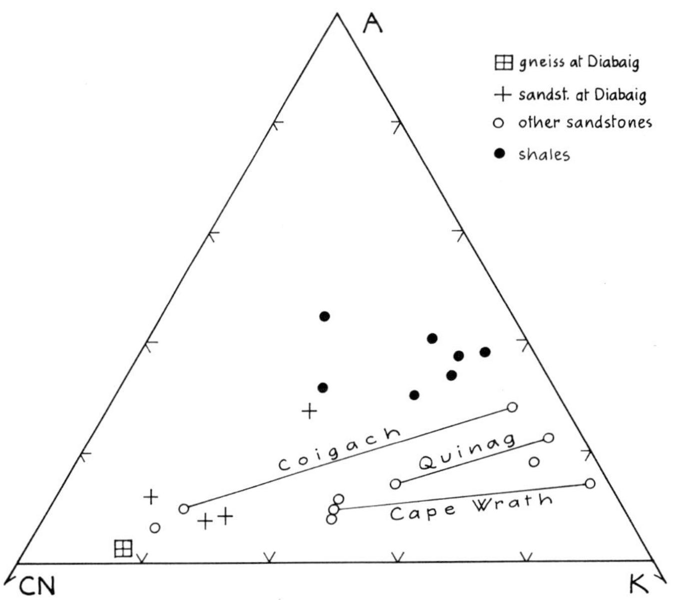

Fig. 31. The composition of Torridon Group sandstones and shales in A–CN–K space. The tie lines join sandstones at the top and bottom of major fining-upward sequences: Cape Wrath (Williams 2001, table 3, anal. 1 & 6); Quinag (Donnellan 1981, anal. Q14 & Q34); Coigach (Donnellan 1981, anal. C3 & C64). The data for other sandstones, the shales and the gneiss come from Tables 12 & 16. Note that the diagram is truncated at the plagioclase–K-feldspar join, i.e. CN = K = 50.

The nature and location of the source rocks

The location of the source terrain that yielded the quartz and feldspar forming the bulk of the Torridon Group is important in identifying the kind of sedimentary basin in which it formed. Some authors have proposed that Torridon Group sediment has travelled 1000–2000 km (Williams 1969*b*; Rogers *et al.* 1990; Nicholson *in* Anderton *et al.* 1992; Nicholson 1993; Rainbird *et al.* 2001), whereas others have argued that it was no more than 250 km (Williams 2001), or even less (Watson 1977; Stewart 1982). Thousands of kilometres of transport suggests a continental sag basin, but a local source terrain points to a rift or molasse basin.

The age of the detritus may also be relevant in identifying the type of basin. If all the detritus is similar in age to the sediment then the basin is probably a molasse that formed adjacent to an eroding orogenic belt. Detrital ages much older than the sediment are ambiguous. They may, and probably do, come from durable detrital grains or pebbles recycled repeatedly through earlier sediments, metasediments or paragneisses, with little or no bearing on either the age or location of the terrain that yielded the sediment. Another possibility is that relatively old basement has contributed sediment to a rift basin which transects it.

Selley (1966) concluded that the Applecross Formation derived directly from 'high-grade quartzo-feldspathic metamorphic rocks', and subsequent work has tended to follow the same line of thought. The palaeocurrent directions, and the way they converge on the Outer Hebrides block (Fig. 26), have been particularly influential in directing attention to the Lewisian gneiss complex as a source. However, the erodable continental surface is not usually made of gneiss. Geological maps of the continents (Blatt & Jones 1975) and mass balance calculations for large river systems (Stewart 1993*b*) show that crystalline basement rocks form only 36% of the present continental surface. The rest is sediment, of which shale is by far the most important lithology, forming 36% of the total area. There is no reason to think that the Proterozoic eon was any different. The vast expanses of Precambrian basement exposed today are simply due to the erosion of contemporaneous sediment and its incorporation in younger rocks (Gregor 1985; Taylor & McLennan 1985, p. 105–109; Garrels 1988). In other words, in the absence of definite indications to the contrary, a predominantly sedimentary source area should be assumed for any clastic sediment, even where it overlies a gneissic basement. If the source area for the Torridon Group was as described, the sands would have been derived mainly from the weathering of older sandstones and basement rocks, and almost half the suspended load would come from the weathering of older shales. Erosion generates roughly seven times more shale than sand. Consequently the chemistry of shales is much more important than that of sandstones in assessing source rock chemistry. Indeed, to a first approximation, the composition of the shale is that of the source rock.

Some useful conclusions about the origin of Torridon Group sediment can be had from the ratios La/Th and Th/Sc in shales, both of which tend to reflect source lithology rather than subsequent weathering or sedimentary differentiation (Taylor & McLennan 1985, p. 47 & 103), and also from the rare earth patterns. The ratio $La/Th \approx 4.1$ for Torridon Group shales (Young 1999*a*, Table 1) is similar to that for Taylor & McLennan's estimate of average upper continental crust ($La/Th = 2.8$), or the Laxfordian ($La/Th \approx 4$: Table 1), but much lower than the ratio in the lower crust ($La/Th = 10$), or the Scourian ($La/Th = 52$) in which Th has been depleted. The ratio in Torridon Group sandstones has a weighted average of 3.3 (Table 14; Williams 2001, table 3), even closer to the upper crustal estimate.

The ratio Th/Sc is about 0.75 for the shales (Young 1999*a*, table 1) and 0.97 for upper crust. There are no Sc data for the Lewisian. Th/Sc is near unity for post-Archaean shales, but only around 0.4 for Archaean ones (Taylor & McLennan 1985, fig. 5.3), suggesting that some Torridon Group silts and clays may have originally come from the Archaean (but not the Scourian – see below).

The chondrite normalized rare-earth pattern for Torridon Group shales (Fig. 32) is similar to that for average post-Archaean

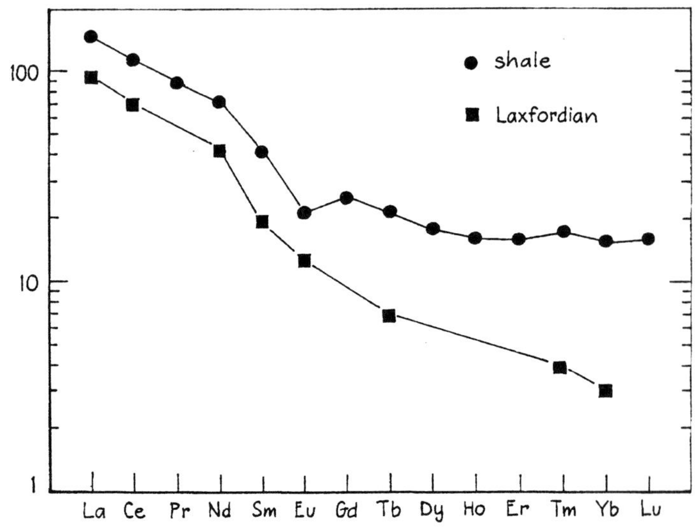

Fig. 32. Chondrite-normalized rare-earth patterns for Torridon Group shales (Young 1999*a*, fig. 13b) and Laxfordian gneisses (Weaver & Tarney 1981). Note that the shales have a negative europium anomaly lacking from the Laxfordian.

upper continental crust except that the abundances are greater in the shale. Patterns for the Laxfordian and Scourian basement rocks do not compare well with the shale because they both lack the negative Eu anomaly which the shale has, and both show heavy rare earth depletion. These rare-earth element patterns deal a mortal blow to the hypothesis of a Lewisian source area for the Torridon Group espoused by Stewart & Donnellan (1992) and Williams (2001).

In brief, the source area for the greater part of the Torridon Group (Applecross and Aultbea Formations) was post-Archaean upper continental crust, but not the Lewisian gneiss complex. Geochemical mass-balance calculations show that the source terrain was largely sedimentary or metasedimentary and became progressively less basic during deposition of the group (Stewart 1995*b*). These conclusions are essentially in accord with those of Van de Kamp & Leake (1997) and Stone *et al.* (1999).

Pebbles, microcline, zircon and mica in the Applecross Formation

The Applecross Formation in contact with the Lewisian commonly contains large gneiss clasts, but away from the unconformity the pebbles are small and siliceous. Hicks (1878) was the first to wonder what the source of these siliceous pebbles was. Later workers have invested much time examining them in the hope of identifying a specific source area, but without seriously considering that they might have been recycled through earlier sediments. The identification of the source terrain is better approached by examining first cycle detrital zircon and mica grains.

The most extensive account of the pebbles is by Williams (1969b) who examined thin sections of 325 of them, extracted from the Applecross Formation near Cape Wrath. Additional data are given in Peach *et al.* (1907, p. 278–284), Anderton (1980), and for Rum by Black & Welsh (1961). The pebbles can be divided into five main suites, the relative abundance of which is shown in Table 17. The zircon grains in the Applecross and Aultbea Formations have been studied in detail by Rainbird *et al.* (2001). There is, as yet, relatively little published data on mica ages.

Quartz is usually white vein quartz, only very rarely the blue type characteristic of granulites. Nearly all the grains show signs of strain and many have deformation lamellae. Two sections of pebbles from Cape Wrath contain tourmaline blades in only slightly strained vein quartz. One of these pebbles has given an $^{40}Ar/^{39}Ar$ age of about 1950 Ma (Allen *et al.* 1974).

Quartzite pebbles are mostly strongly deformed fine-grained metaquartzites. Undeformed orthoquartzite is four times less

Table 17. *Regional variations in the pebble suites of the Applecross Formation in NW Scotland*

	Raasay $n=219$	Torridon $n=268$	Gairloch $n=876$	Coigach $n=100$	Stoer $n=300$	Handa $n=400$	Cape Wrath $n=3927$
Quartz	47	35	36	39	59	52	63
Quartzite	12	33	26	13	18	32	24
Chert, jasper	4	5	7	12	7	9	8
Igneous	36	21	26	33	11	5	3
Schist, gneiss	1	6	5	3	5	2	2

The number of pebbles counted is *n* and the data in per cent. Data sources: Raasay & Scalpay: Selley (1966); Torridon: Maycock (1962, table III); Gairloch: weighted average from Lawson (1970, table 9-4) and Peach *et al.* (1907, p. 333); Coigach: Williams (1969*b*, table I-18); Stoer: Williams (1969*b*, table I-17); Handa: Williams (1969*b*, table I-15); Cape Wrath: Williams (1969*b*, tables I-17 to I-14). Quartzite includes metaquartzite, quartzite and sandstone.

abundant than metaquartzite. The orthoquartzites often show well-rounded grains covered by a ferruginous pellicle. MacKie (1899) describes such a pebble from the Applecross Formation at Kinlochewe within which is embedded a clast of an even earlier quartzite, likewise with a ferruginous pellicle around its grains. Banded iron formation is a rare type of quartzite recorded both by Teall (in Peach *et al.* 1907, p. 284) and Williams (1969*b*).

Up to 0.6% of the pebbles at Cape Wrath are tourmalinized quartz and quartzites (Williams 1969*b*, table I). Similar pebbles are recorded by Teall. A survey of sixteen thin sections of tourmalinized pebbles collected by P. Allen from the Applecross Formation in Assynt and Applecross shows that the host rock is usually undeformed orthoquartzite, less commonly metaquartzite, vein quartz or chert. The tourmaline is black, with pleochroism typically green (rarely yellow or blue) to colourless. It shows a variety of habits; blades randomly arranged, radiating, or aligned along bedding, stumpy crystals formed on detrital grains of quartz, needles in cross-cutting quartz veins, and irregular masses. Several specimens show two generations of tourmaline growth with different habits, including one of those for which Allen *et al.* (1974) reported a Laxfordian age. Out of nine tourmaline-bearing pebbles dated by $^{40}Ar/^{39}Ar$, two can be classed as Scourian, four Laxfordian and two Grenvillian (Allen *et al.* 1974; Allen 1991). The dispersion of ages in each group is wider than for the zircon grains described below, probably due to variable argon loss.

Red sandstone pebbles of medium grain-size form up to 1.5% of pebbles at Cape Wrath (MacKie 1899; Peach *et al.* 1907, p. 279; Williams 1969*b*). Williams suggests they may be derived from the Stoer Group, but those he describes differ from the Stoer sandstones now exposed by being feldspathic sandstone rather than arkose, and in having more K-feldspar than plagioclase.

Chert and jasper: The cherts can have almost any colour, including black, so that in the field they are sometimes hard to distinguish from fine-grained tourmalinized quartzites. Under the microscope a wide range of oolitic, spherulitic and botryoidal structures are apparent, though many specimens are structureless. Rhomb-shaped pseudomorphs of quartz or iron-ore are common. As Williams (1969*b*) points out, these structures indicate silicification of carbonates. Identical pebbles in the Jura Quartzite have been described in detail by Anderton (1980) who noted their similarity to the ferruginous oolitic cherts known from banded iron formations in eastern Canada, and suggested that they might have come from the Applecross Formation. Similar silicified oolite pebbles occur in the Carboniferous Millstone Grit of England (Gilligan 1919). Muir & Sutton (1970) claimed to have found microfossils in two of the chert pebbles from the Applecross Formation at Cape Wrath, but Peat (1984) has shown them to be fungal contaminants or, in some cases, inorganic structures.

Igneous pebbles are porphyry with phenocrysts of quartz, orthoclase or acid plagioclase. The phenocrysts are about a millimetre in size, set in a reddish ground mass which is often banded. The rocks are rhyolites are rhyodacites. An Rb–Sr isochron for K-feldspar phenocrysts extracted from such pebbles gives them a Laxfordian age of 1578 ± 50 Ma (Moorbath *et al.* 1967, table 2) almost identical to that for the microcline pebbles described in the next paragraph.

Microcline pebbles up to 2 cm in size from the basal Applecross Formation at Cape Wrath and Achiltibuie give an Rb–Sr isochron age of 1569 ± 40 Ma (Moorbath *et al.* 1967). The high initial ratio $Sr^{87}/Sr^{86} = 0.728$ suggests that the crystals existed long before 1569 Ma and that the isochron dates a heating event. The microcline pebbles, however, have Rb and Sr contents that set them apart from the sand-size K-feldspar in the Applecross Formation (Stewart & Donnellan 1992). It is to be hoped that sand-sized K-feldpar and plagioclase from the Torridon Group will one day be dated, for it constitutes the only fraction of the sediment truly representative of the source rocks.

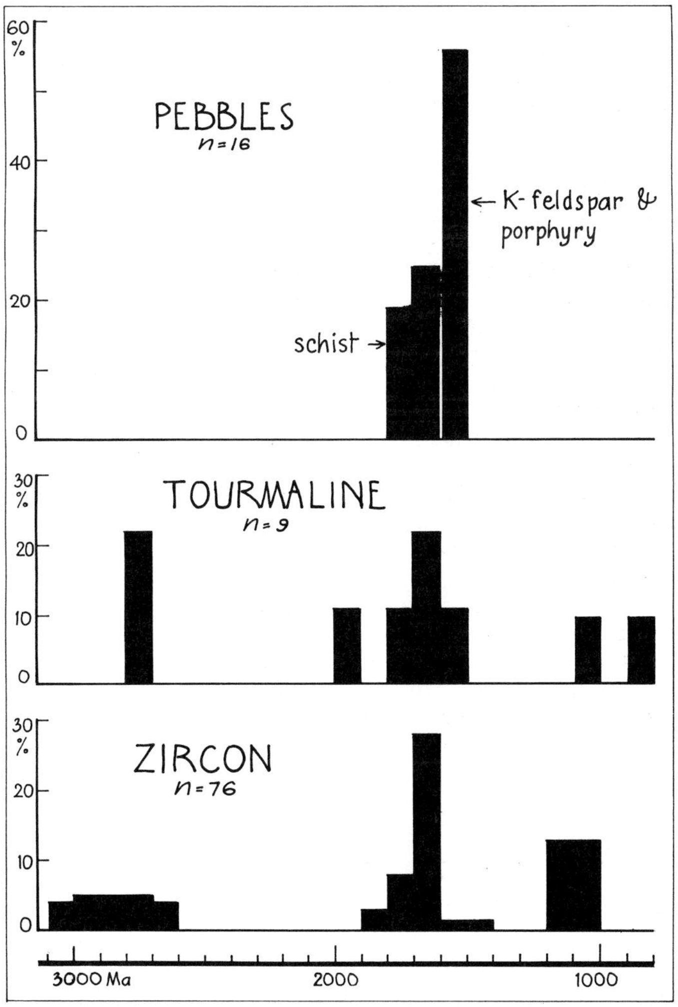

Fig. 33. Age histograms for the Torridon Group. Zircon ages are from Rainbird *et al.* (2001, fig. 7), tourmaline in quartzose pebbles from Allen (1991), and pebbles of muscovite schist, porphyry and K-feldspar from Moorbath *et al.* (1967). Note that the main zircon peak on a probability density plot is actually double, with one mode at 1650 Ma and the other at 1800 Ma.

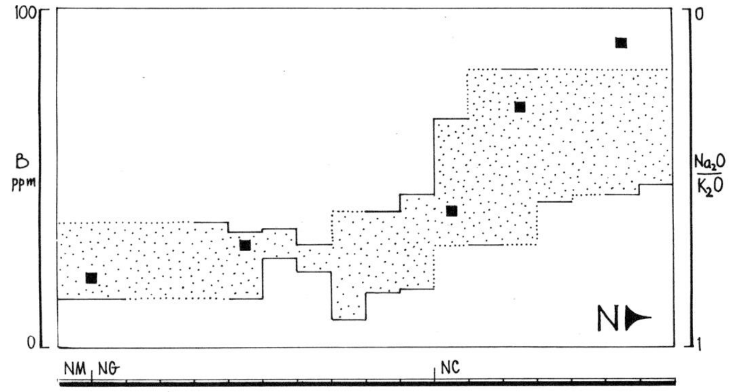

Fig. 34. Concentration of boron in stream sediment over the Torridon Group, from Rum to Cape Wrath (British Geological Survey unpublished data). The ratio Na_2O/K_2O in the underlying sandstones (from Table 15) is shown by black boxes. The boron data are averages for the 10 km grid squares shown at the foot of the diagram. The maximum values (solid lines at upper edge of stipple) are in mountainous areas and the minimum values (lower edge of stipple) along the coast where boulder clay has diluted the concentration.

Metamorphic pebbles of gneiss and schist, though sparse, are important in having the same mineralogy as the sandstones of the Applecross and Aultbea Formations. They are thus more likely than the other pebbles to represent the main lithology of the source area. Quartz–muscovite schists with grains <2 mm predominate. The quartz forms elongate, strained grains. Many specimens show cataclastic textures and some are mylonitized. Common accessories are microcline, perthite, sodic plagioclase and epidote. The muscovite gives a Laxfordian age by K–Ar (Moorbath *et al.* 1967, table 4).

Zircon grains ($n = 76$) have been separated from the Applecross and Aultbea Formations and dated singly by the U–Pb concordia technique (Rainbird *et al.* 2001). The results (Fig. 33) largely confirm earlier work on an Applecross sample of 22 zircons by Rogers *et al.* (1990), published only in abstract. The main difference is that Rogers *et al.* failed to find any ages in the range 1720–1860 Ma. Roughly 20% of the grains from the Applecross and Aultbea Formations are Archaean and another 20% Grenvillian.

Micas with Grenville ages appear to be common in the Torridon Group and also in the Lewisian gneiss complex. Muscovite in the Diabaig Formation dated by Rb–Sr and K–Ar gives ages of about 1150 Ma (Moorbath *et al.* 1967, table 5). Detrital muscovite in the Applecross and Aultbea Formations gives Grenvillian ages (Prave *et al.* 2001 & pers. comm.). An even younger detrital mica age of 997 Ma has been reported by Dempster *et al.* (*in* Bluck *et al.* 1997), but no further details of this sample are available.

The regional distribution of pebble suites, summarized in Table 17, shows a definite increase in quartzose types at the expense of the igneous suite towards Cape Wrath, but not the abrupt change in pebble type between Coigach and Gairloch claimed by Williams (1969*a*). There is no evidence of any trend going stratigraphically upwards through the Applecross Formation, nor in the palaeocurrent direction from Rubha Stoer to Ben More Assynt. A northward increase in the boron content of present-day stream sediment, mainly due to tourmaline, correlates well with the ratio Na_2O/K_2O (Fig. 34) but only roughly with the proportion of siliceous pebbles. Since Zr in stream sediment mimics B it seems that all resistate minerals are more highly concentrated in the north.

The source area

All available age data are summarized in Figure 33. The easiest ages to interpret are the youngest, which are Grenvillian. Both Rogers (pers. comm.) and Rainbird *et al.* (2001) found that about 20% of the zircon grains they examined fell into this group. The degree of abrading is variable, suggesting that the zircon grains were imprisoned in pebbles, some for a longer time than others. However, no Grenville age pebbles have so far been identified in the Torridon Group. The Grenville micas present in the sediment are particularly significant for they are most unlikely to have been recycled.

The Grenville belt lay just to the SW of the Torridon Group as shown in Figure 39, and has left an imprint in the Lewisian gneiss complex. For example, high-grade metamorphic biotites from some Outer Hebridean gneisses have an Rb–Sr age of 1091 ± 70 Ma, due to regional reheating (Cliff & Rex 1989). Similar biotites from basement shear zones on the mainland of Scotland give K–Ar ages of 1168 ± 30 Ma and Rb–Sr ages of 1190 ± 40 Ma (Moorbath & Park 1972). The Grenville micas in the lowermost Torridon Group (Diabaig Formation) almost certainly came from the local Lewisian gneiss complex. The micas and sharply faceted zircons with Grenville ages at higher levels probably came directly from Grenville metamorphic and igneous rocks.

The next oldest group of ages, which includes those from the microcline grains and igneous pebbles described earlier, was formerly regarded as late Laxfordian. Mantle-derived rocks of this age from Rockall Bank have been suggested as a suitable source for some Applecross zircons and tourmalines (Morton & Taylor 1991). Rainbird *et al.* (2001), however, have shown that the age histogram has a double mode, with peaks at 1650 Ma and 1800 Ma. They point to possible sources in the trans-Labrador granitoid belt ($\approx$1650 Ma), the Ketilidian of Greenland (1790–1860 Ma) and the trans-Scandinavian igneous belt in Baltica, whence the zircons could have been brought by major rivers flowing towards Scotland, parallel to the Grenville orogenic belt. They also suggest that the oldest zircons, dating from the Archaean, may have been reworked from the Stoer Group, or have come from Greenland.

Greenland, Labrador and Baltica are unlikely source areas for the zircons because they were the scene of extensive anorogenic magmatism between 1270 Ma and 1510 Ma (Gower *et al.* 1990) of which there is little trace in the zircon suite. These areas are also an unlikely source for the sediment, as distinct from zircons, because the alluvial fans forming the lowest part of the Applecross Formation (that yielded the zircon data) almost certainly had a source area over the Minch and Outer Hebrides, thousands of kilometres from Greenland, Labrador or Baltica (see p. 33).

The only explanation for the multiplicity of detrital ages concentrated in a small area near Scotland is that most of the Torridon Group was derived from earlier, unmetamorphosed sediments of diverse ages overlying the Lewisian gneiss complex. As mentioned earlier, the continental surface is, and always has been, covered mainly by sediments and a source area of this kind is *a priori* preferable to any other. The source sediments would need to have been deposited before 1000 Ma by large rivers draining the Laurentian shield (Scottish continental shelf and Rockall Bank) and the mountains of the Grenville orogen. These sediments are not preserved beneath the Torridon Group, and they cannot be identified with the Stoer Group which, unlike the Torridon Group, had its source in the local Lewisian and is relatively rich in plagioclase. Only a small proportion of Torridon Group sediment came directly from Grenville metamorphic and igneous rocks.

The hypothesis now offered recalls that of Nicholson (1993) and Rainbird *et al.* (2001) except that they regard the Torridon detritus as having come directly, rather than indirectly, from the shield and the Grenville belt. Evidence for indirect derivation, i.e. recycling, comes from the presence of red sandstone pebbles in the Applecross (see above) and quartzites pebbles within quartzite pebbles (MacKie 1899; Williams 1969b). The heavy minerals (including zircon) recorded as present in the quartzite and sandstone pebbles are obviously also second cycle (MacKie 1899, 1926).

Palaeomagnetism

Palaeomagnetism can contribute to understanding of both palaeoclimate and the age of the sediments. The palaeomagnetism in the Torridon Group is mainly due to hematite, much of which has martite texture indicating that it was once magnetite (Stewart & Irving 1974). Little magnetite now remains in most Torridon Group sandstones for the ratio Fe_2O_3/FeO is in the range 5–20, similar to the Stoer Group (Van de Kamp & Leake 1997, table 2; Williams & Schmidt 1997, table 1). Magnetite has, however, been detected thermomagnetically in heavy mineral concentrates from the Torridon Group (Irving 1957; Parry 1957; Schwarz 1971; Smith *et al.* 1983). The transformation of magnetite to martite starts in soils and near-source sediments but takes many millions of years to complete (Van Houten 1968). The successful fold tests reported by Irving & Runcorn (1957) and Irving (1964, fig. 5.10) were obtained from samples that had not been thermally cleaned. The magnetization of these samples may therefore stem from residual detrital magnetite as well as from diagenetic hematite.

Several lines of evidence suggest that the magnetization of the Torridon Group occurred during the transformation of magnetite to hematite. In Rum both magnetite and hematite contribute to the remanence, with opposite polarities, indicating that the transformation spanned at least one polarity reversal of the Earth's field (Robinson & McClelland 1987). The fact that palaeomagnetic reversals at Loch Torridon are not parallel to facies boundaries even over quite short distances (see Fig. 100) also suggests that the magnetization of the hematite is indeed secondary, reflecting progressive burial and heating while the Earth's field reversed (Turner 1979). The palaeosol at Cape Wrath gives further evidence of secondary magnetization. The ratio Fe_2O_3/FeO is about unity in the bed rock but rises through the palaeosol to reach a value of about 10 in the most altered, clayey parts (Retallack & Mindszenty 1994, table 7). The magnetization is due to magnetite in the bed rock but hematite in both the palaeosol and the overlying Torridon Group, with similar mean directions for palaeosol and sediment (Williams & Schmidt 1997). The simplest explanation for these observations is that the palaeosol and the sediment were magnetized together as the magnetite was oxidized, and that no field reversals occurred in the meanwhile.

As in the Stoer Group there is substantial variation in D and I from rocks at the same stratigraphic level (cf. Smith *et al.* 1983, table 1), perhaps due to varying degrees of hematitization of magnetite. If this is correct then it follows that the mean D and I for the Torridon Group, and derived palaeopole, are averages for an indeterminate time period.

The best estimate of the magnetization vector is $D = 129°$, $I = +61°$, which gives unit weight to twelve sampled sections of the Torridon Group scattered over the entire outcrop (Buchan *et al.* 2000, table 3). The inclination implies a palaeolatitude of 42°S, with a likely uncertainty of $\pm 10°$. The corresponding pole position is 19°S, 222°E for $A_{95} = 23°$.

Weathering and palaeoclimate

Clues to the palaeoclimate may be had from the palaeosol beneath the basal unconformity, described earlier, the mineralogy of the Torridon Group sandstones, the absence of primary carbonates, and the palaeolatitude deduced from palaeomagnetism.

The palaeosol at Sheigra, near Cape Wrath, contains minute carbonate micronodules with ferruginous growth banding interpreted by Retallack & Mindszenty (1994) as typical of soils forming in subhumid climates, with seasonal rainfall amounting to an 600–1000 mm annually (i.e. Köppen type Cs or Cw). In such soils calcium leached during degradation of feldspars and mafic minerals is carried down through the soil and concentrated by evaporation at a wetting front, whereas in more humid climates it is totally removed from the soil profile in solution (Retallack 1990, p. 161).

The palaeoclimate during deposition of the Applecross Formation can also be estimated from the modal feldspar and quartz content of the sandstones, assuming that they are predominantly first cycle and were derived from calk-alkaline precursors (Van de Kamp & Leake 1997). The data, plotted on Figure 35, indicate a temperate climate (Köppen type Cf and Df). However, the mineralogy of sand-grade material from the Diabaig Formation provides a better basis for estimating the palaeoclimate because it is entirely of local derivation. It also suggests a temperate climate, but warmer and drier than for the Applecross, with a summer drought like that of the Mediterranean basin today (Köppen type Cs).

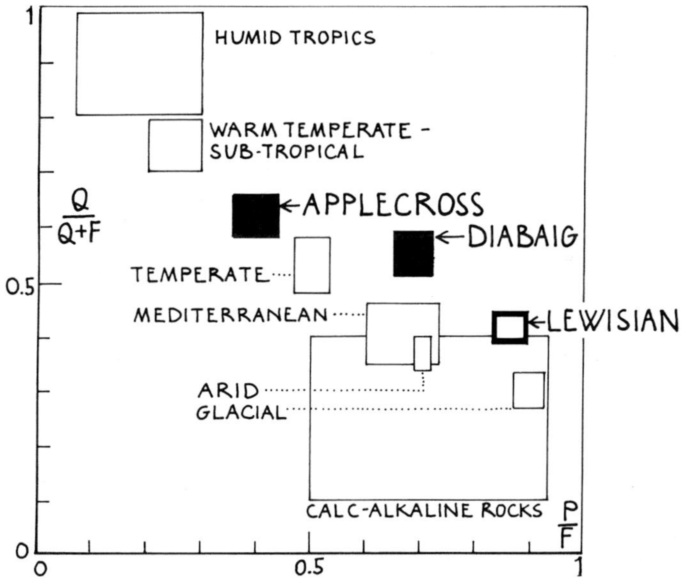

Fig. 35. The modal mineralogy of Torridon Group sandstones (Diabaig and Applecross Formations), compared with first cycle recent sands derived from calk-alkaline basement (Van de Kamp & Helmold 1991). Minerals used are quartz (Q), plagioclase (P) and total feldspar (F). Sandstone mineralogy is from Van de Kamp & Leake (1997, table 1). The modal mineralogy of the Lewisian gneiss complex at Diabaig is also shown.

The palaeoclimate during deposition of the Diabaig Formation cannot be determined from the concentrations of Na_2O and CaO in the dissolved load, shown in Tables 13 & 14, because of the uncertainty in the relative masses of sand, shale and dissolved load. The relative masses used in the Tables are averages for large rivers in temperate climates today, that vary by a factor of 5 depending mainly on the relief (Meybeck 1977, 1988, annex). Source rock also plays a role in determining the dissolved loads. The lack of carbonates in the basement near Diabaig must have contributed to the low value for CaO in solution relative to Na_2O.

The absence of lacustrine carbonates in the Diabaig Formation, and the absence of pedogenic carbonates from even the slowly deposited parts of the Torridon Group (e.g. the lower Applecross Formation) are significant. It will be recalled that pedogenic carbonates are present in the Stoer Group (p. 8 & Fig. 47). They are also found at the same palaeolatitude in the alluvial red beds of the almost contemporaneous Keweenawan (Kalliokoski 1986). Their absence from the Torridon Group suggests a climate that was neither arid nor semi-arid, but on the contrary, rather wet.

The climatic zones deduced above are broadly consistent with the palaeolatitude of 30°–50° derived from the palaeomagnetism, though bearing in mind that the sediments were deposited near the middle of a huge continent (Rodinia) a drier climate, more like that of central Asia today, might have been expected.

Basin analysis

Key facts already described in this volume that need to be embraced by any tectonic model for the Torridon Group are listed below.

- A conformable sequence of coarse fluviatile sediments at least 6 kilometres thick in the Torridon Group on the mainland (Fig. 23), and, with the Sleat Group, over 4.5 km in Skye (Fig. 20).
- An outcrop now bounded by two major fractures about 80 km apart (Fig. 1).

- Crustal thickness of 25–30 km beneath NW Scotland (e.g. Blundell *et al.* 1985); slightly thinner than normal continental crust.
- Tripartite facies succession; local breccia (Db1) → lake deposit (Db2) → fluvial sands fining upwards (Ax, Ab, CH), suggesting a decelerating rate of subsidence (cf. p. 27).
- Arkosic detritus in the Applecross and Aultbea Formations derived from relatively old continental crust and sediment, with only about 20% from the nearly contemporaneous Grenville orogen (pp. 41–42).
- Palaeocurrents orthogonal to the boundary faults. Most come from the west, with a mean direction of 123° (p. 34), almost exactly perpendicular to the Moine thrust and the Minch fault, both of which strike at 030°. Near the base of the Torridon Group at Stattic Point, Quinag and Ben More Assynt (see pp. 56 & 86) some palaeocurrents come from the east and SE.
- A source area for the lowermost Applecross Formation located over the Outer Hebrides and adjacent continental shelf (p. 33).
- A source area that retreated northwestwards (p. 33).
- A western boundary fault, near the present Minch fault (Fig. 26), from beyond which came a supply of pebbly material throughout deposition of the Applecross Formation, suggesting recurrent uplift of the source terrain (Watson 1977).
- Tectonic activity during deposition of the Diabaig Formation at Stattic Point, recorded by soft-sediment brecciation and overfolding on a palaeoslope towards the NW, probably related to movement on the Coigach fault, and by large-scale mass flow deposits moving SW at Diabaig.
- Ubiquitous multilayer contortions in the Applecross and Aultbea Formations, due to seismic shaking during sedimentation (p. 34).
- Pre-lithification tensional stress, roughly perpendicular to the boundary faults, shown by dilatational sandstone dykes striking NE, near the base of the Torridon Group at Rubha Stoer, Stattic Point, Gairloch and Diabaig.

The dimensions of the Torridon Group allow no more than three alternative depositional settings: a foreland basin, a thermal relaxation basin, or a rift. The hypothesis of a foreland basin related to the Grenville orogenic belt was first suggested by Sutton (1963), and recently resuscitated by Prave *et al.* (2000) and Rainbird *et al.* (2001). Foreland basins develop next to orogenic belts due to lithospheric loading by outward spreading thrust sheets (Allen & Allen 1990, p. 245; Leeder 1999, p. 524–8). The Grenville belt lay to the SW of the basin (Fig. 39), and is of similar age to the Torridon Group. By comparison with Alpine and Himalayan foreland basins, the sediments in a Grenville-related foreland basin should have been deposited by palaeocurrents coming from the SE, bearing pebbly Grenville-age detritus. Since the proximal edge of a foreland basin is overthrust by the adjacent orogen the source area should be advancing basinwards, in this case towards the NW. These features are the exact opposite of those in the Torridon Group, and the foreland basin model is consequently untenable.

A thermal relaxation (failed rift) basin has been proposed by Nicholson (1993) as a suitable site for Torridon Group deposition. The basin is thought to have followed rifting during which the 3.5 km thick Sleat Group was deposited. The main evidence for a basin much larger than the present outcrop of the Torridon Group is the discovery of deep channels in the Applecross Formation, interpreted as a record of major perennial rivers. As explained elsewhere (p. 34) this interpretation of the channels is almost certainly wrong. Further objections concern the thickness and nature of the post-rift sediments. Accepting Nicholson's suggested basin stretching factor of $\beta = 2$ the post-rift, Torridon Group sediments should be less than half the thickness of the syn-rift Sleat Group (Cochrane 1983), whereas in fact they are twice as thick. In addition, post-rift sediments are normally marine clastics quite different from the Torridon Group. This hypothesis is therefore also rejected.

An origin by rifting better explains the facts listed above, especially the boundary fault spacing. The slight crustal thinning is consistent with this interpretation (Fairhead 1986). A symmetrical arrangement of horsts and grabens was suggested by the writer (Stewart 1982) but the faults that cut the Torridon Group and bound it to the NW and SE almost all dip eastwards, suggesting that a half-graben is a better model. Of the faults shown in Figure 22, however, few have any connection with rifting. Most behaved as normal faults during Mesozoic extension (Judd 1878, p. 669–673; Roberts & Holdsworth 1999). Some of those active during the Mesozoic era also behaved as normal faults in the late Palaeozoic era (Stewart 1993*a*; Potts *et al.* 1995). Faults that date back to Proterozoic times include the Outer Hebrides fault zone, that now bounds the basin to the west, initiated as a thrust about 1700 Ma ago (Lailey *et al.* 1989; Fettes *et al.* 1992, p. 151, 170; Butler *et al.* 1995). On the east side of the basin Precambrian eastward-dipping normal faults cutting the Torridon Group, originally proposed by Soper & England (1995), have been located in the Moine thrust zone by Butler (1997).

Some faults were definitely active during deposition of the Torridon Group. The exponential increase in pebble size in the basal Applecross Formation at Cape Wrath (p. 67) and the convergence of palaeocurrent directions on the Minch fault (Fig. 26) strongly suggests that the fault was then active. Movement on the Coigach fault during deposition of the Diabaig Formation at Stattic Point has already been mentioned. Intermittent uplift of the Outer Hebrides block along the Minch fault can be inferred from the supply of durable pebbles in the Applecross Formation.

The proposed Torridon rift in NW Scotland is shown diagrammatically in Figure 36. Slight uplift west of the Minch fault initiated the supply of sediment and water that created the Diabaig lake and formed the lowermost sparsely pebbly Applecross Formation units. Stream gradients were low and silty overbank deposits and channel plugs common. Extensive grey shale units in

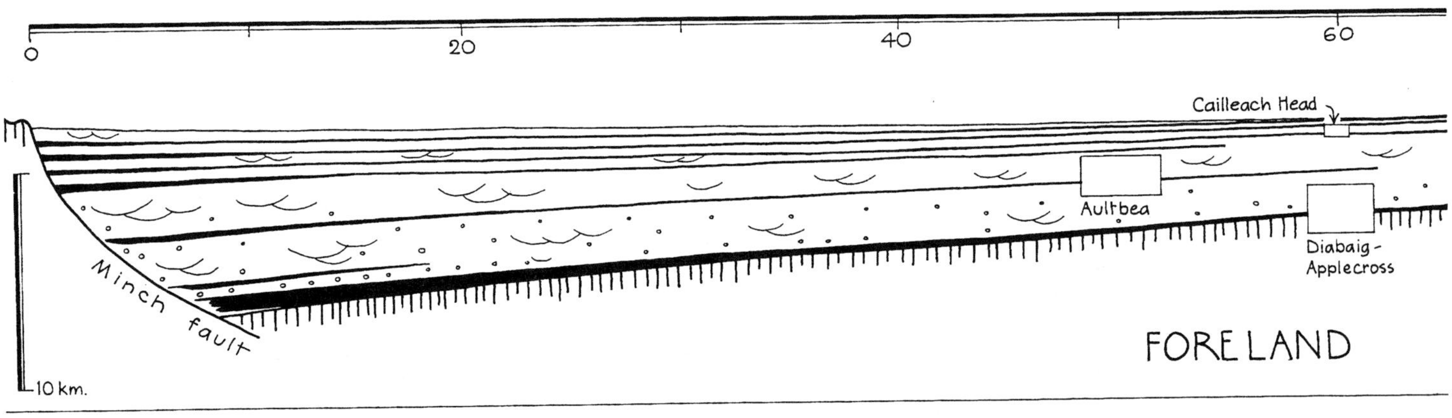

Fig. 36. A speculative true-scale section of the basin which received the Sleat and Torridon Groups. It shows the basin in the latitude of Skye before late Proterozoic warping and the development of the Caledonian Kishorn and Moine thrusts. The section satisfies what is known of the stratigraphy, but the parts excised by Caledonian thrusting, and the boundary faults, are necessarily hypothetical.

the lowermost Applecross Formation from Raasay southwards may represent the fringes of the Diabaig lake. Vigorous uplift of the western source area then started, pouring pebbly Applecross sediment over the subsiding rift floor and overwhelming the Diabaig lake. The coarsest sediments (facies FA1) are seen at Cape Wrath and Handa only because those areas are nearest to the source. According to Williams (2001) rifting occurred first in the Cape Wrath sector and only later to the south. However, there is no evidence that the Cape Wrath Member is older than the rest of the Applecross. In the absence of well-defined time planes it could just as well be contemporaneous, or even younger than the finer grained sandstones generally found at the base of the Applecross Formation, as it is at Quinag (p. 55).

The supply of pebbly sediment ceased when the fall line retreated abruptly. This may denote the end of the rifting event and the start of post-rift lithospheric flexure, which allowed onlap of the sediments onto the flanking basement – generating the so-called steer's head geometry (Cochrane 1983). The sediment deposited during this phase, the Aultbea and Cailleach Head Formations, had travelled further and is consequently finer than that forming the Applecross.

Age and correlation

A maximum age for the Torridon Group is given by a detrital zircon in the Aultbea Formation, dated by the single grain U–Pb method at 1046 ± 26 Ma (Rainbird *et al.* 2001). A minimum age comes from the late Lower Cambrian fossils in the unconformably overlying Eriboll and An t-Sròn Formations (Cowie & McNamara 1978; Huselbee & Thomas 1998), dated at about 520 Ma (Bowring & Erwin 1998).

The time of sedimentation has been studied most recently by Turnbull *et al.* (1996). The best estimate is given by the early diagenetic phosphate concretions in the Diabaig Formation that yielded a whole rock Rb–Sr isochron age of 994 ± 48 (2σ) Ma and a Pb–Pb isochron age of 951 ± 120 Ma.

The whole rock Rb–Sr isochron age of 977 ± 39 Ma obtained from shales in the Applecross Formation is probably unreliable. It was based on the assumption that the Sr in the shales was in plagioclase, and had been homogenized during albitization. But some of the Sr in the shales is in fact in K-feldspar (Table 14). A whole-rock X-ray diffractogram of one of the dated shales (Turnbull *et al.* 1996, fig. 6a) has a well-defined orthoclase peak indicating the presence of about 15% K-feldpar. If this K-feldspar is like that in the adjacent sandstones (Stewart & Donnellan 1992, fig. 6) it contains about 20% of the total Sr in the shale. If the K-feldspar is detrital, as seems probable, then the date obtained is partly inherited from the source rock. An attempt to fix the time of diagenesis by Rb–Sr dating of the finest grained illite failed – all size fractions <2 μm gave an age of 667 ± 66 (2σ) Ma of which the significance is uncertain.

The Rb–Sr isochron age of 994 ± 48 Ma obtained from phosphate concretions is adopted as the best estimate for the depositional age of the Torridon Group, remarkably close to the estimate of 1040 ma based on a comparison of Torridon palaeomagnetic pole positions with those for Laurentia (Smith *et al.* 1983).

Precambrian inter-regional correlation can be attempted using lithostratigraphical, geochemical and mineralogical methods, but usually reduces to time equivalence, for which some kind of objective criterion is useful. Consider two rock units with mean isotopic ages t_1, t_2 and associated standard deviations σ_1 and σ_2 (where $2\sigma_1 \leq 2\sigma_2 \leq 50$ Ma). It is suggested that the criterion

$$|t_1 - t_2| \leq 2\sigma_1$$

should be satisfied for the rock units to be worth considering as time correlatives.

Early attempts at Torridonian correlation were based on the lithological similarity of the Torridon Group and the physically adjacent psammitic rocks of the Moine Supergroup (Geikie 1895; Peach & Horne 1930, p. 199; Kennedy 1951; Sutton 1963). The ages of the Torridon Group and the Moine Supergroup are, in fact, comparable using the criterion given above. Single grain U–Pb concordia ages for detrital zircons of igneous origin in the Moine of Sutherland are as young as 1005 ± 18 (1σ) Ma, ranging up to 1850 Ma (Kinny *et al.* 1999). Very few zircons give Archaean ages. Derivation of both Torridon Group and Moine from upper continental crust has been deduced from their chemistry by Stone *et al.* (1999), and is consistent with their similar oxygen isotope patterns (Benner & Hoernes 1994). A rift setting has been proposed for the Torridon Group (p. 44) and also for the Moine (Soper & Anderton 1984; Soper *et al.* 1998). The two units, however, may nevertheless have evolved far from each other (see Chapter 5), and had mineralogically different source areas. The latter is suggested by the diverse colours of their detrital zircon suites (MacKie 1923), and by their very different tourmaline contents – reflected in the boron content of present day stream sediment (Plant 1984). Boron concentrations are 2–5 times higher over the Torridon Group than over the Lewisian or Moine. Prave *et al.* (2001) claim that the ages of zircons in the Moine psammites of Morar and the Applecross Formation are similar, spanning the whole interval from the Neoproterozoic to the Archaean, but the significance of this is debatable. To be certain that the similarity is due to a common source, the colour, shape, zonation and rare-earth content of the zircons need to be compared, not just the ages. And unless it can be shown that the sediments are definitely first cycle, a common source for the zircons could be interpreted as meaning that the Applecross derived by erosion from the Morar Moines, or *vice versa*.

More is to be gained by looking at the eastern margin of the Laurentian plate, of which NW Scotland formed a part before the opening of the Iapetus ocean, to find regionally related rifting events contemporaneous with the Torridon Group. Three rifting events which have been suggested as correlative with the Torridon

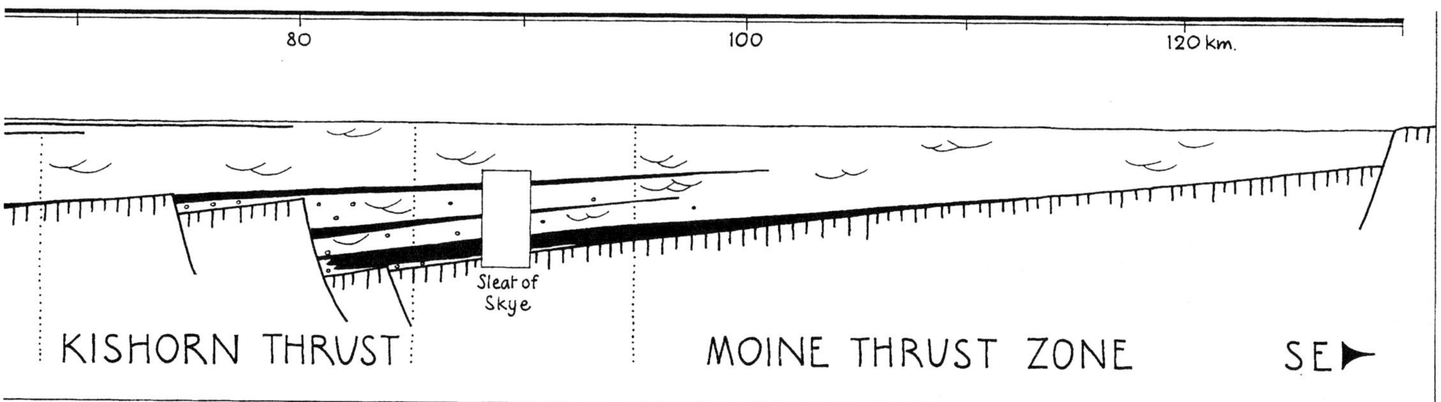

Group are the Wakeham Supergroup, the Double Mer Formation and the Keweenawan. In addition, there are several thick, non-rift sequences in the North Atlantic area that are, or may be, partly correlative.

The Wakeham Supergroup in northeastern Quebec is contained in a 200 km long basin tectonically overlying Grenvillian basement, and was suggested as a correlative of the Torridon Group by Turnbull *et al.* (1996). It consists of slightly metamorphosed clastic sediments, rhyolites, basalts and gabbros, intruded by post-tectonic granites dated at about 990 Ma. Rhyolite in the lower part of the supergroup has been dated at about 1270 Ma (Martignole *et al.* 1994). The supergroup is supposed to have formed during gravitational collapse of the Grenville orogen so that the correlation with the Torridon Group can be based only on the ages, which barely satisfy the criterion given above.

Of the many late Neoproterozoic rifts that cut the eastern margin of Laurentia (Hoffman 1989), the Lake Melville rift in Labrador was only about 1000 km from Scotland on pre-Iapetus reconstructions. The Double Mer Formation, which fills the rift, consists of unmetamorphosed, cross-bedded reddish-brown sandstones and conglomerates, up to 5 km thick, once considered as possibly correlative with the Torridon Group (Gower 1988). The sandstones contain about 50% feldspar, of which half is plagioclase. The overall similarity to the Torridon Group is clear but the Double Mer unconformably overlies Grenville basement and so postdates orogenic uplift at about 960 Ma. Furthermore, the palaeomagnetic pole position of the Double Mer sediments is like that of the nearby Long Range dykes (13.0°S, 178.6°E) dated at about 600 Ma (Murthy *et al.* 1992). Thus the Double Mer is far too young to correlate with the Torridon Group – it is, rather, to be regarded as a result of early Iapetus rifting (Windley 1995, p. 231–2).

The Keweenawan Supergroup is a 30 km thick sequence of unmetamorphosed basalts and continental sandstones deposited in the Mid-continent Rift that cuts the Laurentian shield and is truncated by the Grenville orogen (Hinze *et al.* 1997; Ojakangas *et al.* 2001). The rift extends over 2000 km from the Grenville front at Lake Erie, through Lake Superior to southern Kansas, and has been attributed to extensional stress generated by the continental collision during the Grenville orogeny (Dalziel *et al.* 2000). The lavas that form most of the lowest 20 km of the Keweenawan formed between 1109 and 1087 Ma, mainly as a result of fissure eruptions. The fluviatile sediments that dominate the upper 10 km of the Keweenawan were deposited between 1087 and about 1000 Ma, and are generally regarded as post-rift. They become more mature upwards, with lithic sands derived from Keweenawan volcanics in the Copper Harbor Formation (Oronto Group) at the base, to predominantly quartz-rich sands in the Orienta Sandstone (Bayfield Group) at the top. The sandstones show fining-upward sequences deposited by high sinuosity rivers. The Keweenawan sequence is not entirely conformable. For example, the Jacobsville Sandstone at the top rests on a well-developed palaeosol in the underlying Freda Sandstone. The Keweenawan is slightly older than the Grenville orogen, which cuts it at a high angle about 3400 km from Scotland on a pre-drift reconstruction. The main difference from the Torridonian lies in the abundance of mafic igneous rocks, which are present throughout the entire length of the Mid-continent rift. The sandstones at the top of the Keweenawan, however, are plausible correlatives of the Torridon Group in terms of age (they just fulfil the contemporaneity criterion given above), tectonic setting and, in general terms, sedimentary environment.

Of the non-rift successions the most interesting is the Krummedal sequence in East Greenland, deposited in the interval 960–1050 Ma (Watt *et al.* 2000; Kalsbeek *et al.* 2000), and therefore almost an exact time correlative of the Torridon Group. It consists of over 3500 m of clastic sediments, with limestones and metavolcanics, all metamorphosed to amphibolite grade shortly after deposition, and again during the Caledonian (Higgins 1988). The detrital zircon population is dominated by Mesoproterozoic grains, whereas Archaean grains are rare. The lithologies, zircon age spectrum and depositional age for the Krummedal sequence have been used as grounds for lithostratigraphic correlation with the Moine Supergroup in Scotland (Winchester 1988; Kalsbeek *et al.* 2000). However, the presence of significant limestone units shows that the Krummedal sequence is probably marine, unlike the Torridon Group.

Other possible correlatives are to be found in the Thule Supergroup, the Hecla Hoek succession and the Eleonore Bay Supergroup. The Thule Supergroup occupies an intracratonic basin about 200 km across, stretching from west Greenland, across northern Baffin Bay to Ellesmere Land. The unmetamorphosed clastic sediments filling the basin were deposited between 1300 Ma and 650 Ma (Dawes 1997, figs 2 4 & 120). The metsedimentary and metavolcanic Hecla Hoek succession in Svalbard contains rocks dated at about 970 Ma in Ny Friesland and Nordaustlandet (Harland 1997, p. 103, 114 & 121–125). The oldest part of the Eleonore Bay Supergroup in east Greenland also dates from about 950 Ma, for example the clastic sediments of the Nathorst Land Group (Sønderholm & Tirsgaard 1993, fig. 29). The Rivierdal Group of east Greenland, tentatively suggested as a correlative of the Moine Supergroup and the Torridonian by Higgins *et al.* (2001), is probably too young to correlate with either. In other words, there are insufficient data to apply the contemporaneity criterion given earlier (p. 45) to the Rivierdal Group, or any other of the thick Neoproterozoic sedimentary successions preserved in the Caledonides north of Britain.

Chapter 5
Overview

General aspects of the Torridonian briefly reviewed in the following paragraphs include the significance of the similarity in depositional style shown by the three component groups, their burial history, palaeocontinental setting and regional correlation.

Depositional style

The Stoer, Sleat and Torridon Groups are thick fluvial successions, each of which tends to become finer upwards. The last two groups both show an upward progression from locally derived basal breccias into lake deposits, followed by a thick fluvial sequence, as if they were deposited in a regime of decelerating subsidence. In addition, both were derived from progressively more acid source rocks, as the source terrain expanded to embrace not just local basic gneisses, but a wider range of rock types, including sediments. These features are consistent with deposition in an extensional basin subjected to two distinct stretching events. The Stoer Group must represent a much earlier stretching event, for it was lithified and deeply eroded before the Torridon Group was deposited upon it. The upward facies progression in the Stoer Group is like that in the other groups, except that the lacustrine phase (facies Ct3, see Fig. 5 & p. 9) is underdeveloped.

The orientation of the basin is suggested by the palaeocurrent directions. Stoer Group directions are bimodal; 77% of the currents flowed westwards ($\theta = 270°$, $n = 282$) and the remainder eastwards ($\theta = 069°$, $n = 84$). In the Sleat Group 65% flowed eastwards. In the Torridon Group virtually all the currents flowed ESE ($\theta = 123°$), almost exactly perpendicular to the Minch fault and the Moine thrust which both strike NNE (030°). The palaeocurrent directions suggest a single fault-bounded sedimentary basin striking roughly NNE and receiving sediment from the flanks. Evidence of tectonic instability and fault activity is evident from the time the earliest Stoer and Torridon Group sediments were deposited (pp. 20 & 44), even though the soft-sediment contortions which formed during sedimentation and thought to be seismically induced, are noticeably absent from the lowest units (Clachtoll and Diabaig Formations). Contortions are also absent from the basal Sleat Group (Rubha Guail and Loch na Dal Formations). The reason for the lack of contorted bedding in the lowermost formations and their abundance at other stratigraphic levels remains a mystery.

The tectonic instability recorded in the sediments, taken together with the palaeocurrent directions and evidence for decelerating subsidence, suggests rifting. The rift-bounding faults are surmised to have dipped mainly eastwards, so that some of them were transformed into thrusts by Caledonian compression (Brewer & Smythe 1984; Blundell *et al.* 1985). A similar conversion of normal faults into thrusts occurred when the Mid-continent rift was compressed during the Grenville orogeny in the Lake Superior region (Cannon 1994). The rift faults in NW Scotland may, indeed, have controlled the orientation of the much later Caledonian orogenic margin, as suggested by Stein (1988). Similar control is seen in east Greenland (Higgins *et al.* 2001).

Later extension of the Torridonian basin reactivated some of the thrusts as normal faults (Brewer & Smythe 1984) so that the western part again subsided and received some 4 km of mainly continental sediment during the Triassic period (Steel & Wilson 1975; Stein 1992; Hitchen *et al.* 1995).

Burial history

The total thickness of the Stoer Group exposed today is only 2 km, but albitization of the entire sequence indicates that the highest

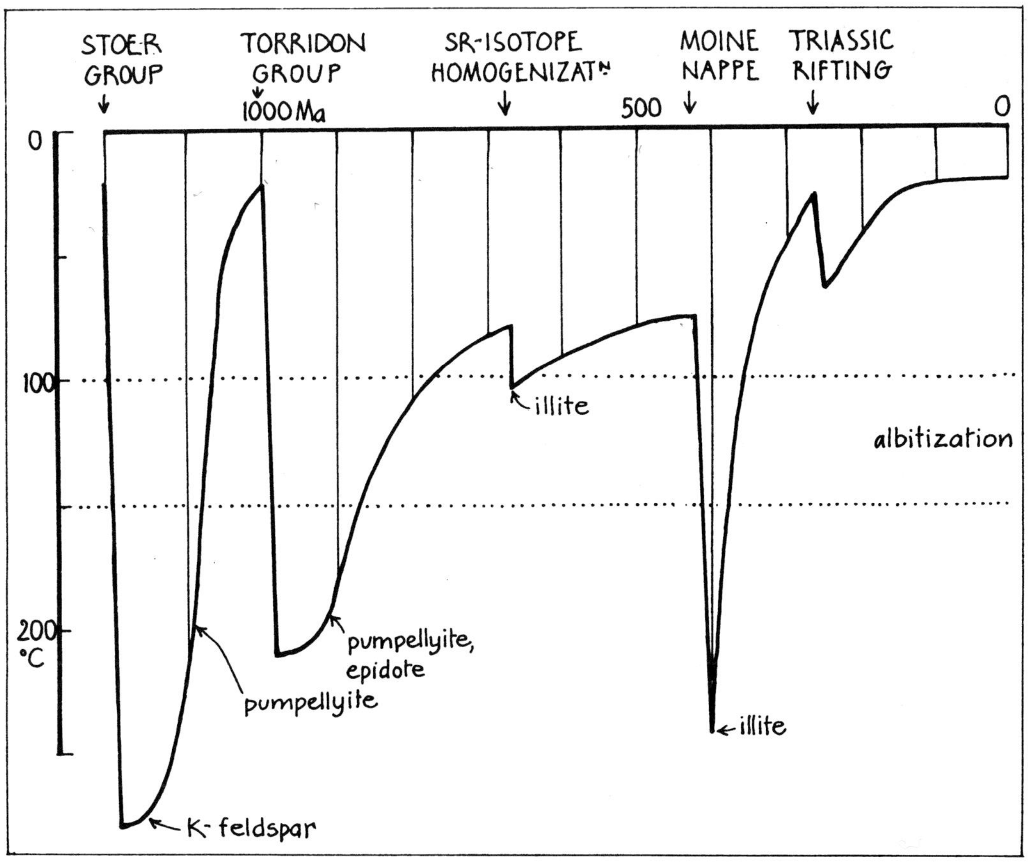

Fig. 37. Torridonian thermal history. The graph attempts to summarize the main depositional episodes, thermal events and mineral crystallization phases recorded by a sample of sediment at the base of the Stoer Group at Stoer.

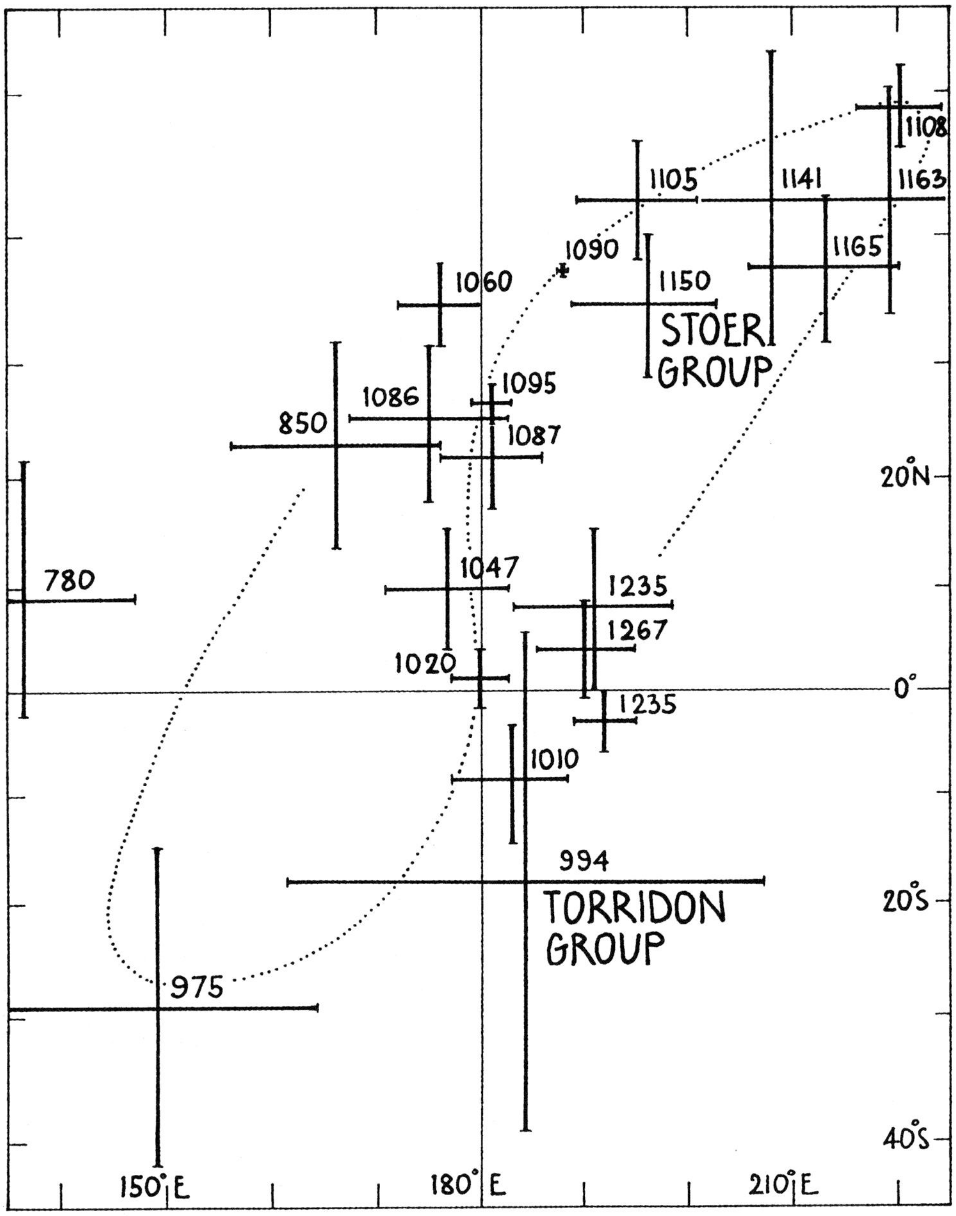

Fig. 38. Selected palaeomagnetic poles for Laurentia, detailed in Table 18. Mercator projection. The ages are in millions of years (Ma) while the bars estimate the uncertainty in the pole position at the 95% probability level (A_{95}). The Gardar and Tugtutôq poles from Greenland have been rotated −12.2° about an Euler pole at 66.6°N, 240.5°E (Roest & Srivastava 1989). The Stoer and Torridon Group poles have been rotated −38° about an Euler pole at 88.5°N, 27.7°E (Bullard *et al.* 1965). The apparent polar wandering path defined by the poles is dotted. The Keweenawan track starts at 1108 Ma (top right) and ends at 975 Ma (bottom left).

Table 18. *Selected Proterozoic palaeomagnetic poles from Laurentia*

Rock unit	Age (Ma)	Pole lat. (deg. N)	Pole long. (deg. E)	A_{95} (deg.)	Reference
Western N. America intrusions	780	09	136	12	Buchan *et al.* (2000)
Grenville B overprint	850 ± 50	23	166	10	Buchan *et al.* (2000)
Grenville A overprint	975 ± 50	−29	149	15	Buchan *et al.* (2000)
Torridon Group	994 ± 48	−19	222	23	Buchan *et al.* (2000)
Jacobsville Sandstone	1010 ± 40	−09	183	6	Weil *et al.* (1998)
Freda Sandstone	1020 ± 40	01	180	3	Weil *et al.* (1998)
Nonesuch Shale	1047 ± 40	10	177	6	Weil *et al.* (1998)
Copper Harbor Conglomerate	1060 ± 30	35	176	4	Weil *et al.* (1998)
Michipicoten volcanics	1086 ± 2	25	175	8	Weil *et al.* (1998)
Lake Shore traps	1087 ± 2	22	181	5	Buchan *et al.* (2000)
Mamainse Pt. volcanics	1090 ± 7	38	188	1	Weil *et al.* (1998)
Portage Lake volcanics	1095 ± 3	27	181	2	Buchan *et al.* (2000)
Upper Osler lavas	1105 ± 2	43	195	6	Buchan *et al.* (2000)
Logan sills	1108 ± 1	49	220	4	Buchan *et al.* (2000)
Abitibi dykes	1141 ± 1	43	208	14	Buchan *et al.* (2000)
Stoer Group	1150 ± 50	35	234	7	This volume
Tugtutôq	1163 ± 2	42	226	11	Buchan *et al.* (2001)
Late Gardar dykes	1165 ± 15	37	222	7	Buchan *et al.* (2000)
Sudbury dykes	1235 ± 5	−03	192	3	Buchan *et al.* (2000)
Middle Gardar dykes	1235 ± 15	05	202	8	Buchan *et al.* (2000)
Mackenzie dykes	1267 ± 2	04	190	5	Buchan *et al.* (2000)

beds were once buried to a depth of 3–4 km, sufficient to raise the temperature by the required 100°C. The growth of pumpellyite at the base of the Stoer Group indicates temperatures of at least 125°C, possibly as much as 230°C (Hay *et al.* 1988), corresponding to over 3 km of burial (p. 20). Another estimate of the maximum temperature at the base of the group may be given by the fluid inclusion homogenization temperatures of 279 ± 35°C in the K-feldspar + calcite vein suite (S. J. Hay, A. E. Fallick, P. J. Hamilton, pers. comm.) described on p. 20. These temperatures would have been quite sufficient to have albitized the entire sequence shortly after deposition, as shown in Figure 37. Another possibility, however, is that the Stoer Group was albitized at the same time as the Torridon Group, perhaps at about 675 Ma when both of them experienced strontium isotopic homogenization at temperatures under 150°C, recorded by illite grains <2 μm (Turnbull *et al.* 1996). Turnbull *et al.* rejected this hypothesis because of the long interval between deposition and albitization, but such intervals are by no means rare (e.g. Fedo *et al.* 1997).

Epidote has been reported from the basal beds of the Torridon Group at Upper Loch Torridon and Raasay (*q.v.*), in gneiss just below the sediments near Canisp and Cùl Mòr (Peach *et al.* 1907, p. 306–7), and in hematite-quartz veins cutting hornblende schist next to the unconformity at Slattadale and Beinn Lair (Peach *et al.* 1907, p. 244 & 315). Much or all of this 'epidote' may be pumpellyite, indicating temperatures of 125–230°C following renewed depression of the Lewisian to depths of over 3 km during deposition of the Torridon Group. This would have been sufficient to have caused the observed albitization of the Torridon Group and any permeable sandstones remaining beneath it.

A much later thermal event is predicted when the Moine thrust moved into position during the Silurian. The crystallinity of illite in Torridon Group sediments, conodont coloration in the Cambrian, and thermal modelling suggest that temperatures of 275°C were attained in the footwall (p. 31).

Palaeomagnetism and palaeogeography

The palaeomagnetic poles for the Torridonian are near to contemporaneous poles from Laurentia (Fig. 38 & Table 18), leaving no doubt that NW Scotland was an integral part of that shield when the Torridonian sediments were deposited. The Stoer and Torridon Group poles correspond to the beginning and end, respectively, of the so-called Keweenawan apparent polar wandering track. The track is by far the best defined part of the Proterozoic polar wandering path because it is based on sections from the Keweenawan Supergroup, many of which have been dated accurately using the U–Pb system on zircon and baddeleyite, whereas those that are less well dated can nevertheless be placed in stratigraphic sequence (Weil *et al.* 1998; Buchan *et al.* 2000). Stoer Group magnetization should date from about 1150 Ma as Laurentian poles of this age are closest to that for the Stoer (Fig. 38). The magnetization is most unlikely to be older than 1200 Ma because the polar wandering track doubles back on itself at that time, so that poles from Laurentia at 1250 Ma are near the equator (Fig. 38). The Torridon Group pole has a similar position to that for the uppermost Keweenawan (1010 Ma), and scarcely overlaps that yielded by Grenville cooling remanences dated at 950–1000 Ma. It is unlikely to be any younger, for the polar wandering track then performs another loop, with poles dating from around 800 Ma much farther north than those for the Torridon Group. The dates deduced from the magnetization of the Stoer and Torridon Groups are in close accord with the respective isotopic ages of 1100–1200 Ma (p. 22) and 946–1042 Ma (p. 45).

The positions of the Stoer and Torridon Group poles in Figure 38 have been corrected for Mesozoic–Cenozoic opening of the North Atlantic by applying the rotation of Bullard *et al.* (1965), which was originally obtained by fitting the continental shelf edges on the two sides of the ocean. The Gardar poles from Greenland were positioned on Figure 38 by using the rotation axis obtained from sea-floor spreading patterns in the Labrador Sea (Roest & Srivistava 1989). The resulting fit of Scotland NW of the Moine thrust zone (the Hebridean terrane) with Laurentia during the early Neoproterozoic can be regarded as well established. The Cambro-Ordovician Durness sequence, unconformably overlying the Torridon Group, contains fossils belonging to the American faunal province, confirming that the Hebridean terrane was still part of the Laurentian shield as late as the Arenig (Salter 1859; Huselbee & Thomas 1998). The Hebridean terrane is shown, together with the palaeomagnetically derived palaeolatitudes, in Figure 39. On tectonic grounds Baltica is generally believed to have lain just south of Greenland but the lack of reliably dated palaeopoles for Baltica over the period 800–1200 Ma mean that its precise position and orientation are poorly constrained.

The maps in Figure 39 show that the Stoer and Torridon Groups formed in quite different geographic positions. The Stoer Group was deposited on the passive margin of Laurentia, only a few hundred kilometres from an ocean, while the Torridon Group formed in the heart of the continent Rodinia, close to the Grenville orogenic belt. The subtropical steppe palaeoclimate deduced from Stoer Group sediments is consistent with its palaeolatitude and marginal continental position, but the Torridon Group palaeoclimate is problematic. The available data (p. 43) indicate a climate at least as wet as the present-day Mediterranean. Considering that the area was in the heart of a large continent, in latitudes between 30° and 50°, a much drier climate, like that of present-day central Asia, might have been anticipated.

The Torridon Group rift is cut orthogonally by the Grenville orogen (Fig. 39) in just the same way as the Mid-continent rift (Davidson 1995, fig. 2). Both rifts are essentially contemporaneous with the climactic phase of the orogeny. Indeed, the latter has been claimed to be a direct consequence of the Grenville collision (Gordon & Hempton 1986). Neither, however, show much evidence of Grenville derived detritus. The same story is repeated in the Rhine graben, the sediments of which, though contemporaneous with the Alpine foreland basin nearby, have little detritus derived from the Alps. Perhaps the external forebulge due to the downward flexure of the lithosphere beneath the weight of advancing nappes (Allen & Allen 1990, fig. 6.12) was sufficient to keep the foreland and rift basins separate in each case.

The position of the Torridon Group near the edge of the later Iapetus ocean has been taken to mean that it records an early stage in the break-up of Laurentia–Baltica (Stewart 1982). But the first oceanic crust dates from about 600 Ma (Windley 1995, p. 231–2; Svenningsen 2001), which makes the Torridon Group (994 Ma) rather too old to have any direct connection. The similarity in age between the Torridonian and Keweenawan suggests that they record rifting events earlier in Laurentian history, more nearly contemporary with the Grenville orogenic belt. With the exception of the Krummedal sequence in east Greenland the other Proterozoic supracrustal successions preserved in northern Canada, east Greenland and Svalbard (p. 46), are much too poorly dated to satisfy the contemporaneity criterion given on p. 45. In any case, none of them are old enough to correlate with the Stoer Group. It will, no doubt, become possible to construct tectonic maps showing sedimentary basins of Torridonian age in Laurentia when more single grain U–Pb dates become available.

In Figure 39 Scotland is shown as a single entity, as if it had remained a part of Laurentia when the Iapetus ocean started to open at about 600 Ma. This is the usually adopted configuration (e.g. Torsvik & Rehnström 2001, fig. 8; Cocks 2001, fig. 1) but there is no palaeomagnetic evidence to prove it, and the geology is ambiguous. Inevitably, the idea of Scottish tectonic integrity has constrained ideas on the origin of the rocks. For example, Gregory (1915) was convinced that Moine schist could be detected amongst pebbles in the Applecross Formation, MacKie (1928) was sure that the Moine zircon suite could be identified in the Torridon Group, and Anderton (1980) suggested that certain Applecross pebbles had found their way into the Dalradian Jura Quartzite. The possibility of Moine–Torridonian correlation has been the subject of lively discussion for over a century. More recently Prave (1999) has suggested that the Dalradian was eroded from the Moine Supergroup

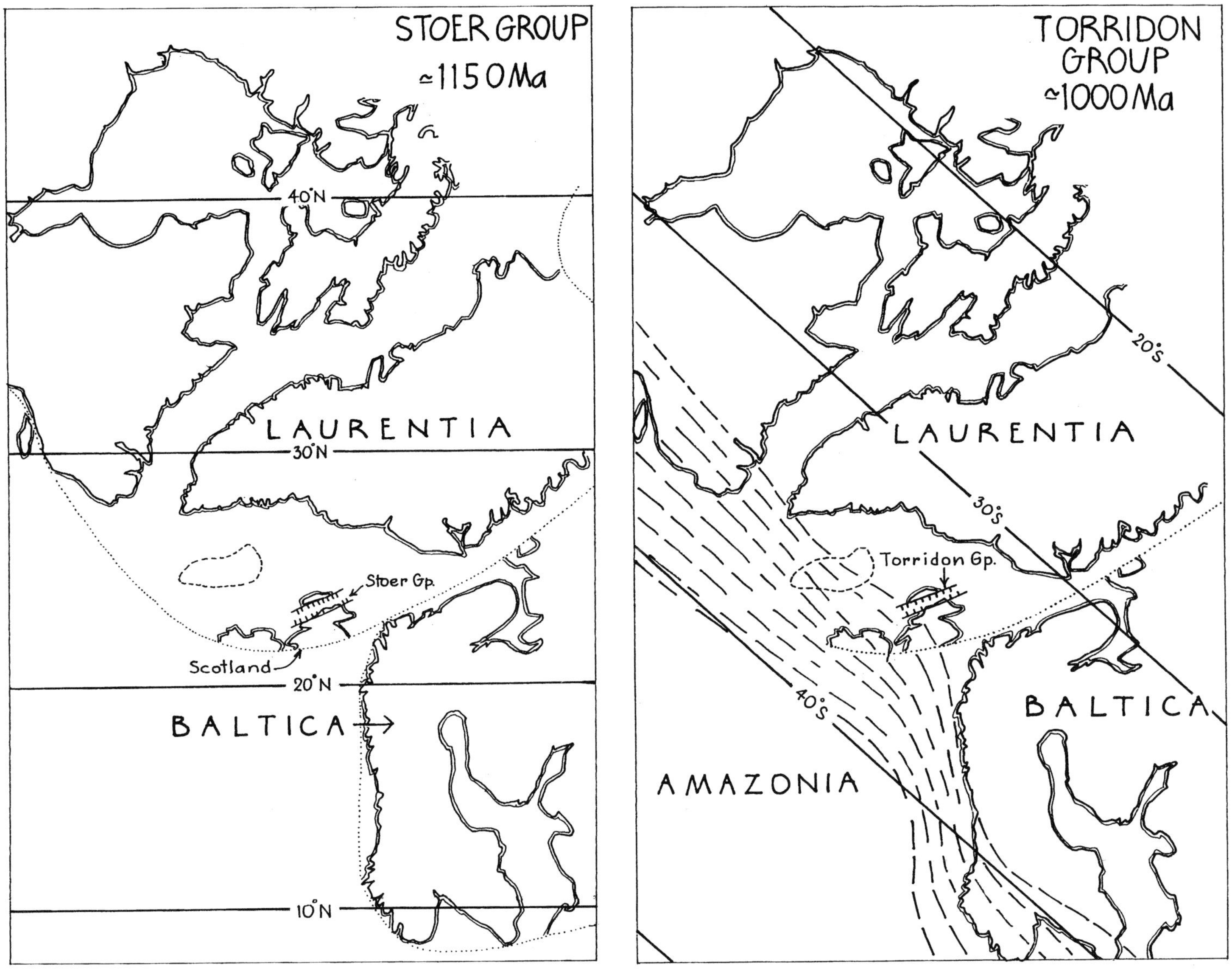

Fig. 39. The palaeocontinents during Stoer and Torridon Group deposition (Mercator projection). The left-hand map is based on Gower *et al.* (1990, fig. 1) and Buchan *et al.* (2000, 2001). The Rockall plateau is shown pecked and the margins of the continental blocks dotted. The blank area south of Scotland is ocean. The right-hand map shows the continent Rodinia, formed by closure of the Grenville orogeny (dashed) at about 1000 Ma, based on Dalziel (1997, fig. 11), Weil *et al.* (1998) and Buchan *et al.* (2000). The palaeolatitudes (from palaeomagnetism) are only approximate.

during the latest Neoproterozoic, again on the assumption of presumed proximity.

Scotland can be divided into at least five NE trending terranes (Stone *et al.* 1999) bounded by major dislocations, with differing crustal velocity and density structure (Rollin 1994) and differing geological histories. The first of these is the Hebridean terrane, composed of Archaean and early Proterozoic gneisses overlain by unmetamorphosed Torridonian sediments, the subject of this memoir. The Northern Highlands terrane, covered by the Moine Supergroup, was deposited at about 1000 Ma and metamorphosed to amphibolite grade, with the production of garnets (Vance *et al.* 1998) and pegmatites, at about 800 Ma (Rogers *et al.* 1998). The Hebridean and Northern Highlands terranes are separated by the Moine thrust, which dips gently southeastwards for at least 100 km but fails to penetrate the lower crust. The Grampian terrane, covered by the Dalradian Supergroup, is separated from the Northern Highlands by the steeply dipping Great Glen fault. Neodinium isotopic data show that the fault cuts through the continental lithosphere and separates geochemically distinct mantle domains (Canning *et al.* 1998), suggesting large-scale transcurrent movement along the fault.

Metamorphism of the Laurentian margin of the Iapetus ocean during the middle Ordovician (Soper *et al.* 1999) brought the rocks of the Northern Highlands and Grampian terranes to amphibolite grade. Closure of the Iapetus ocean in the Silurian reheated and deformed the Northern Highlands and Grampian terranes, and generated the Moine thrust (Dallmeyer *et al.* 2001). All these terrains, together with the Highland Border Complex, the Midland Valley terrane and the Southern Upland terrane to the SE, were essentially in their present relative positions by the late Ordovician (Hutchison & Oliver 1998), but where were they before, in particular while the Torridonian was forming?

If the Moine Supergroup was involved in a late Proterozoic, post-Grenville, orogeny, then it left no trace (e.g. deformation, isotopic ages, molasse) in the Hebridean terrane. If the metamorphism arose from extension, as claimed by Dalziel & Soper (2001), there are no isotopic ages to prove it in the Hebridean terrane. For these and other reasons it has been proposed that the Northern Highlands and Hebridean terranes formed far apart and were juxtaposed by strike-slip movement, not just by Caledonian thrusting (Bluck *et al.* 1997; Bluck 2001). Although Bluck *et al.* imply that most of this strike-slip movement was Caledonian, it might have been earlier, perhaps associated with transform faulting along the Laurentian margin when the Iapetus ocean started to develop in the Vendian. The Iapetus fracture changed azimuth by about 60° near the British islands, perhaps accommodated by a triple junction (Soper 1994), but otherwise by oblique-slip or transform motion along either azimuth. If the reconstruction of Rodinia preferred by Dalziel & Soper (2001, fig. 8b) is adopted, such motion could have left slices of Amazonia along the Laurentian margin. The hypothesis seems marginally more plausible than the alternative, and if it is true for the Northern Highlands terrane it is probably also true for the Grampian.

Any proposals to the link the Torridonian with the Moine and Dalradian Supergroups should be viewed with reserve until the time when the structural integrity of Scotland can be demonstrated, rather than assumed. On geochronological and structural grounds the most promising correlative for the Torridonian remains the Keweenawan (p. 46).

Chapter 6
Directory

The following chapter is intended to provide a comprehensive field description of the Torridonian, replacing that given in the Geological Survey's NW Highlands memoir of 1907, and citing all relevant literature. Stratotypes and palaeocurrents are described, along with the section lines used to construct the regional stratigraphic sections (Figs 4 & 23), but detailed consideration of topics such as geochemistry, diagenesis, sediment source areas, palaeomagnetism and basin tectonics is contained in Chapters 2–5. For convenience the rocks are described under thirty-three compact subareas, most of which are shown on Plate 2. This plate also locates all figured maps and sections. The Directory starts with Cape Wrath and continues with sub-areas progressively farther south.

Cape Wrath

The Geological Survey divided the Cape Wrath succession into three stratigraphic units, without definitely correlating any of them with the formations of the Torridon Group that they had established farther south (Peach *et al.* 1907, p. 292). They described the lowest unit (30 m thick) as coarse pebbly sandstone or basal conglomerate; the middle unit (300 m thick) as coarse red sandstone with pebble bands, and the uppermost unit (75 m thick) as fine-grained mottled purple and yellow sandstones.

Remapping by Williams (1966*a*, 1969*a*, 2001) enabled him to subdivide the sequence stratigraphically into five main lithological units, which he called facies associations (FA) even though they always appear in the same order (cf. the definition of facies given on p. 2). Each successive facies is finer than the one below, as shown by the representative logs in Figure 40. The two lowest (FA1 & FA2) correlate roughly with the Geological Survey's lowest unit, the next two (FA3 & FA4) correlate with the Geological Survey's middle unit, and the last (FA5) correlates with the Geological Survey's uppermost unit. The Geological Survey and Williams agree that the whole succession at Cape Wrath has a total thickness of about 430 m. Williams' mapping has been integrated into 1:50 000 sheet 113 published by the British Geological Survey in 1998, except that the original tripartite division of the Applecross Formation has been retained. A map showing all five facies has recently been published by Williams (2001, fig. 3).

The Geological Survey and Williams assumed that the Cape Wrath sequence belonged to the Applecross Formation because of its lithology and distinctive suite of durable pebbles. The palaeomagnetism of 18 samples from facies FA3, FA4 and FA5 (Smith *et al.* 1983 & supplementary publication 24015) and a further 29 samples from facies FA1 and FA3 (Williams & Schmidt 1997), supports this view. The whole sequence at Cape Wrath was formally designated the Cape Wrath Member of the Applecross Formation by Stewart & Donnellan (1992). The base was defined on the coast at Droman, 4.5 km NW of Kinlochbervie [NC 18475930]. The top is arbitrarily placed at the highest stratigraphic level exposed in facies FA5, at An Grianan [NC 265627] (Williams 2001). The Cape Wrath Member has also been identified by Williams farther south, at Handa, Ben Dreavie, Quinag and Stoer. Correlation with the sequences on Handa island and Ben Dreavie (*q.v.*) appears reasonable in view of the small distances involved, lithological similarity and unconformable contact with basement. At Quinag (*q.v.*) the greater part of the Torridon Group forms an upward-fining sequence that is lithologically very like that at Cape Wrath. There are also geochemical similarities with the Cape Wrath Member, providing good grounds for correlation (Stewart & Donnellan 1992). At Stoer the situation is less clear. The Applecross Formation unconformably overlying the Stoer Group is not nearly so coarse as farther north and moreover lies in the same latitude as Quinag where the Cape Wrath Member is underlain by about 70 m of undifferentiated Applecross Formation. It is thus uncertain which part, if any, of the Stoer sequence can be correlated with the Cape Wrath Member.

The succession at Cape Wrath fines upwards from a flat unconformity that cuts the Lewisian gneiss complex. Maximum clast diameter declines logarithmically upwards from 120 mm at the base of the succession to 30 mm at the top. Palaeocurrents deduced from trough cross-bedding axes in FA1–FA3 in the Cape Wrath area, and from stratigraphically equivalent beds at Handa, Ben Dreavie, Quinag and Stoer, have a fan-like arrangement, intersecting at a point about 30 km west of the present coastline (Fig. 26).

The two lowest facies thin eastwards. Facies FA1, indeed, wedges out completely within 3 km of the coast. Coarsening-upward cycles 6–8 m thick are a feature of facies 2 and can be seen clearly in Figure 40. They represent prograding fan lobes, similar to those described from the Devonian sediments of Norway by Steel *et al.* (1977) and Gloppen & Steel (1981). The palaeocurrents in facies 1–3 systematically change azimuth with latitude whereas those in facies 4 & 5 are virtually constant, suggesting that the former formed on a giant alluvial fan and the latter on a braid plain (Williams 2001).

Palaeocurrent data from the Applecross Formation south of Cape Wrath, at Stoer, Reiff, Stac Polly, Achiltibuie, between Sàil Bheag and Sàil Mhór (immediately south of Little Loch Broom) and at Torridon, also have a fan-like arrangement, that Williams (1969a) interpreted as a second alluvial fan with its median line through Loch Broom. However, the difficulty of identifying the Cape Wrath Member outside the type area has induced Nicholson (1993) to question the existence of the southern fan. The palaeocurrents used to establish it were drawn, in fact, from a variety of levels within the Applecross Formation. At Reiff the section sampled lies near the top of the formation. At Loch Lurgainn (Inverpolly Forest), Horse Sound (Achiltibuie) and the area between Sàil Bheag and Sàil Mhór palaeocurrents were measured in the lower Applecross Formation just above the Rubha Dubh Aird Member. At Torridon they come from the whole formation. There is no stratigraphic correlation between these sections and it follows that the southern fan is very probably spurious.

Palaeosols along the basal unconformity, attributed by Williams (1968) to pre-Applecross weathering, were identified by him at five localities: on the cliffs at Cape Wrath [NC 251731], Poll a'Mhurain [NC 185618], Sheigra [NC 182607] and Droman [NC 185594], and at the foot of a waterfall near Rhiconich [NC 279546]. Unfortunately the locality at Sheigra, which was by far the best, is now covered by a rock fall. The palaeosols were re-examined by Retallack & Mindszenty (1994) who confirmed their Precambrian age, but concluded that they had been heated to no more than 120°C and were compacted to only 80% of their original thickness. This is quite surprising in view of the thickness of the overlying Torridon Group (at least 5 km) and the superposition of the 12 km thick Moine nappe during the Silurian (Stewart 1995*a*). Nevertheless Williams & Schmidt (1997) have shown that the directions of magnetization of the weathered gneiss, the palaeosols and the overlying Torridon Group are all the same. Perhaps the original Precambrian palaeosol was subjected to subsequent hydrothermal alteration. The weathered zone at the Cape Wrath locality mentioned above, which is generally about a metre thick, increases to 8 m where it is cut by a fault. Potassium metasomatism of the palaeosol at Sheigra has been demonstrated by Young (1999*b*).

Inland exposures are sufficient to map the unconformity and establish that it is generally flat over some 200 km^2, but the actual surface is rarely seen. However, it is well exposed at An Socach [NC 264585] and again 1.5 km to the east, where it forms a ridge about 150 m high. The gneiss in contact with the Applecross at An Socach is unweathered and covered by a selvage of gneiss breccia.

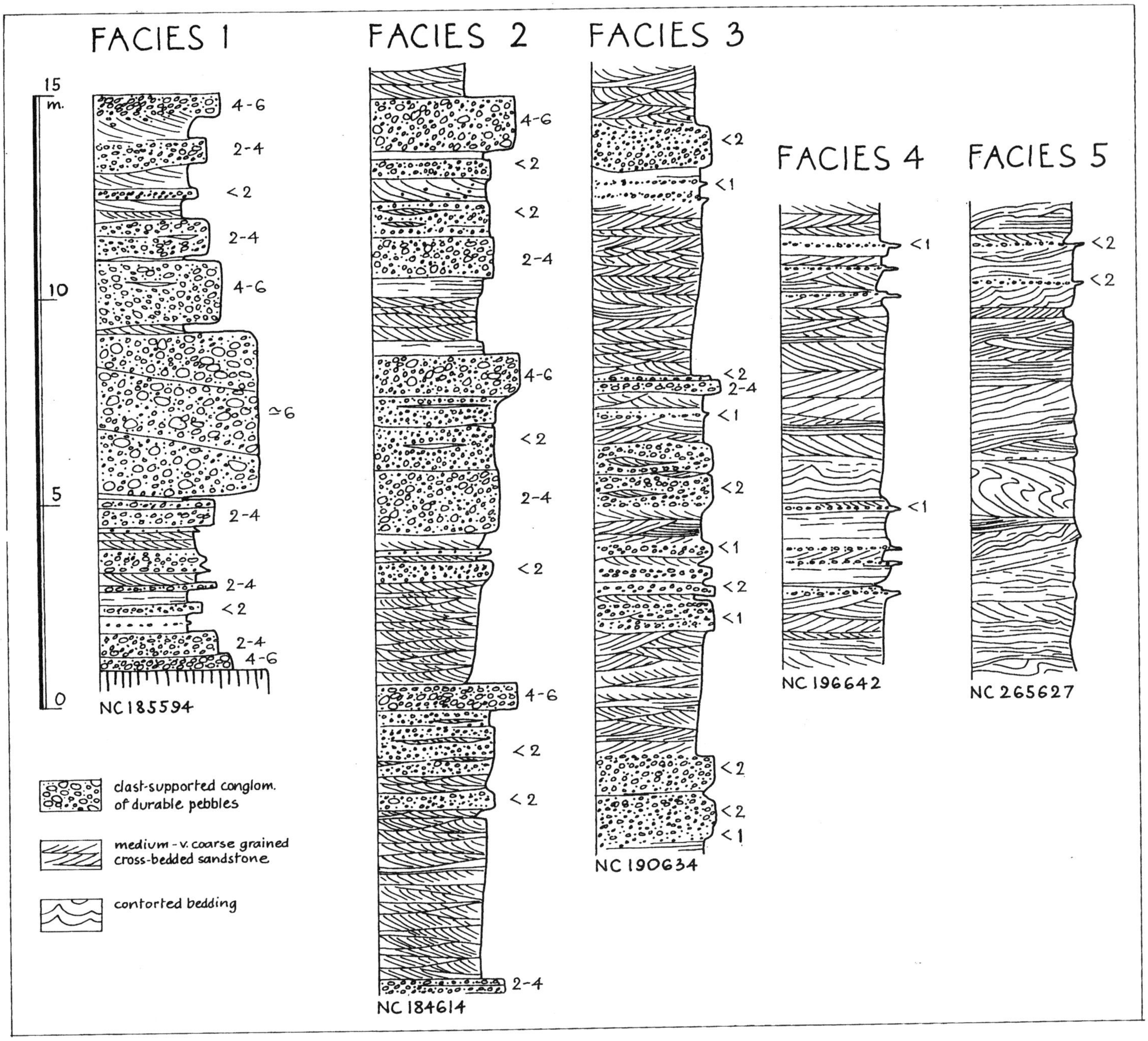

Fig. 40. Facies of the Applecross Formation at Cape Wrath (based on Williams 1969*a*, Fig. 14, ©1969 The University of Chicago). Sedimentary structures are shown as seen. The average pebble size is given in centimetres.

Handa

Handa is a cliff-bound island only 12 km from outcrops of the Cape Wrath Member near Kinlochbervie. No 1: 50 000 geological map is presently available for this area. According to Williams (1966*a*, p. 73) the island is formed of units FA1–FA4 of the Cape Wrath sequence, with a total thickness of about 350 m. The lowest beds, exposed on the northeastern cliffs, are coarse sandstones containing pebbles with a maximum diameter of 120 mm, like the coarsest seen at Cape Wrath; pebble types are similar. These same beds are exposed again in small islands a short distance to the NE where they appear to be only 40 m above the Lewisian basement. The fault that separates the island from the Lewisian basement on the mainland has only a small throw for it fails to displace SW dipping granite sheets exposed a few kilometres NE of the island (Stewart 1993*a*). Further evidence of limited displacement is given by the existence of a small outcrop of gneiss breccia, probably basal Torridon Group, on the mainland about 5 km east of Handa [NC 187461; J. D. Crossley, pers. comm.]. It seems probable, therefore, that the rocks of Handa immediately overlie the basement and belong to the Cape Wrath Member.

The vector mean palaeocurrent direction from trough cross-bedding is about 100° ($n = 228$), according to Williams (2001, table 5), and 110° ($n = 325$) according to Nicholson (1993, table 1).

Ben Dreavie

Ben Dreavie lies about 3 km SW of Loch Stack and about 13 km from the nearest Applecross Formation exposures at Cape Wrath and Quinag. According to Williams (1966*a*, p. 74) some 15 m of pebbly sandstones belonging to the cross-bedded cobbly facies (FA1) form the summit of the mountain, at about 480 m above sea level. Pebbles frequently reach 10 cm maximum diameter. The unconformity with the underlying gneiss is not exposed. The mean palaeocurrent direction is 110° ($n = 9$) according to Williams (2001, table 5).

Quinag

The Torridon Group outlier at Quinag is shown in Figure 41. No up to date one-inch or 1:50 000 geological map is currently available for this area; the Torridon Group is undivided on the one-inch to the mile geological map of Assynt (British Geological Survey 1923).

The main element in the succession is a pebbly, upward-fining member closely resembling the Cape Wrath Member (Applecross Formation) in its type area, except that here it is underlain by a thin sequence of non-pebbly Applecross sediments. It was originally detected at Quinag by Peach *et al.* (1907, p. 299) and correlated with the Cape Wrath sequence by Williams (1969*a*) on the basis of its palaeocurrents. A similar conclusion can be drawn by using double ratio plots to model the distribution of rubidium (in K-feldspar and plagioclase) and yttrium (in apatite and zircon) in the sediments. In both pebbly sequences the K-feldspar is poor in Rb and the plagioclase relatively rich, whereas the lowermost, pebble-free Applecross Formation at Quinag, and all the Applecross Formation in Coigach, is quite different (Stewart & Donnellan 1992). The average yttrium content of apatite in the Cape Wrath sequence and the pebbly Applecross sediments at Quinag is about 8300 ppm, whereas in the lower, non-pebbly Applecross sediments at Quinag it is only 5500 ppm. In Coigach virtually all the yttrium is in zircon. There is also a marked similarity between the ratio Zn/MgO (in chlorite) between the upper, pebbly part of the Quinag sequence and the Cape Wrath Member in its type area, but a dissimilarity with the Coigach Applecross sediments. In short, the source of the pebbly, fining-upward sequence at Quinag was like that of the type Cape Wrath Member, but distinctly different to that of the Applecross Formation underlying it at Quinag, and to the south in Inverpolly Forest and Coigach generally. For these reasons only the pebbly, fining-upward sequence at Quinag can be correlated with the Cape Wrath Member.

Basement relief at Quinag in Applecross times reached 400 m so that all the strata up to that level in the stratigraphic column (Fig. 42) are truncated against it. This is a marked contrast to the essentially flat unconformity underlying the Cape Wrath Member at Cape Wrath.

The lowest beds at Quinag are like the tabular sandstones of the Diabaig Formation in having moderate lateral persistency ($p \approx 100$), occasional rippled surfaces and a lack of contorted bedding. The only well-developed basal breccia is exposed about 1.5 km SSW of Spidean Còinich [NC 20212641] where gneiss blocks reach 3 m in size. Diabaig sandstones are well exposed by the main road to Lochinver, unconformably overlying gneiss. The sandstones contain some seams of red siltstone. There is no non-Lewisian detritus at this locality. Most of the pebbles are angular to sub-rounded in shape. Some have been interpreted as wind-faceted (Peach & Horne 1914, p. 11; MacGregor & Phemister 1972, p. 27). Both the sediment and the underlying gneiss have been deeply weathered, possibly during the late Cenozoic. Hay *et al.* (1988) record pumpellyite veins cutting both fresh and weathered gneiss, but not the Torridon Group.

Stratigraphically equivalent Diabaig beds outcrop in the NE of the outlier and can be conveniently studied by the road from Skiag Bridge to Kylesku Ferry [NC 239300]. These sandstones show small-scale trough cross-bedding with sets up to a decimetre, and contain both porphyry and Lewisian pebbles. Decimetre-sized gneiss clasts occur near the basal unconformity. Palaeocurrents generally flowed

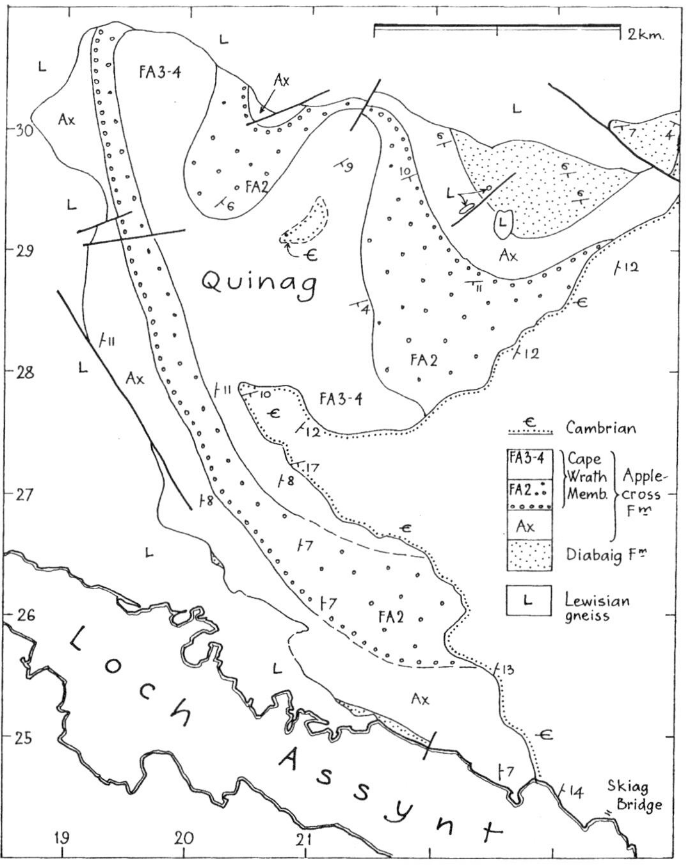

Fig. 41. Outline geological map of Quinag based on the Geological Survey and Williams (1966*a*), revised by the author. See Plate 2 for location.

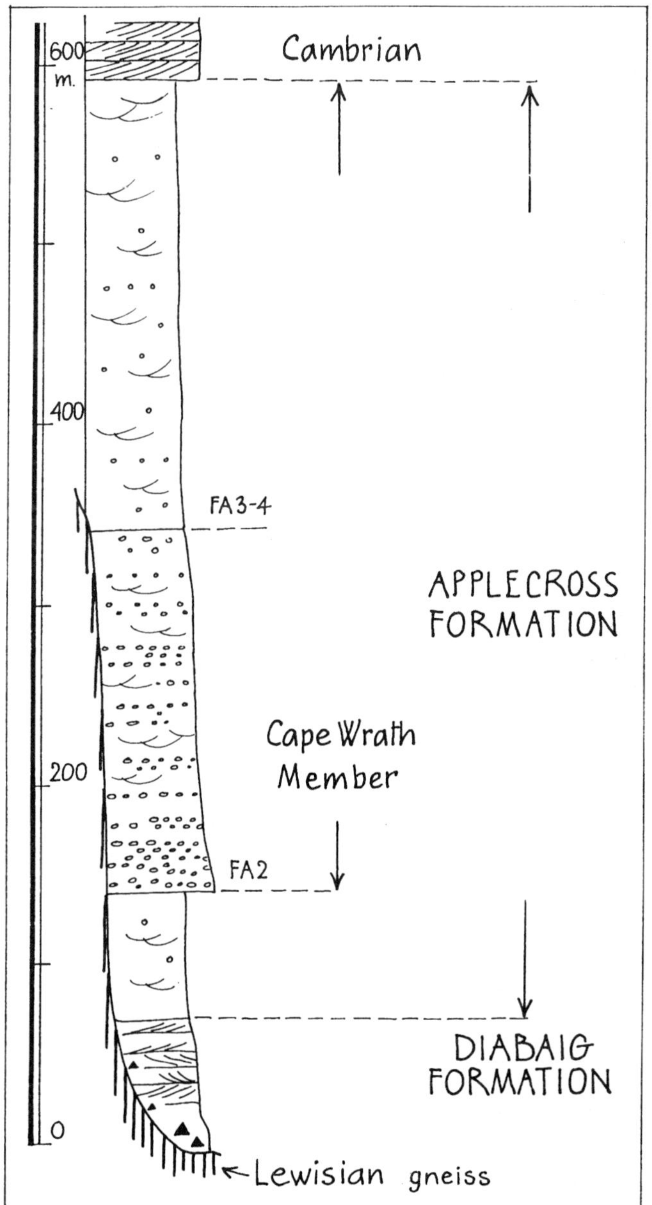

Fig. 42. Graphic log of the Torridon Group succession at Quinag.

towards the SE, as in higher strata at Quinag (Williams 2001, fig. 8), but in many beds the palaeocurrents are counterposed. Such counterposed palaeocurrents can be found all over the outcrop, sometimes producing a kind of herringbone cross-bedding, for example just east of the small gneiss inlier [NC 227291]. Palaeocurrents derived from the east or SE are also a feature of the pebbly sandstones forming the Ben More nappe, 10 km to the SE [NC 307197]. The tabular sandstones just described at Quinag grade upwards into typical coarse-grained sandstones of the Applecross Formation that form a distinct unit about 80 m thick beneath the Cape Wrath Member. These sandstones also show anomalous palaeocurrent directions, with $\theta = 247°$, $n = 11$ [NC 206302]. According to Williams (2001, table 2) they belong to facies FA3.

The upward-fining Cape Wrath Member of the Applecross Formation is best studied on the accessible southern slopes of Spidean Còinich [NC 206263 to NC 210270]. Outcrops farther north are precipitous. The base of the Member is marked by a flood of exotic pebbles. These pebbles are 2–3 cm in diameter in the south of the outlier and 3–4 cm in the north. They appear over a stratigraphic interval of 1–10 m with little obvious change in the sandstone matrix. This conglomeratic development (200 m thick) has been referred by Williams (1969*a*) to his tabular pebbly facies FA2 and is shown as such in Figure 42. Pebble abundance diminishes upwards and a boundary marking the level at which less than 10% of the rock is pebbly has been drawn on the map (Fig. 41) and stratigraphic column (Fig. 42). This stratigraphic level is roughly equivalent to the bottom of Williams' cross-bedded gravelly facies FA3. Sandstone grain size is medium to coarse in this upper part of the Member, but noticeably finer above it, corresponding to Williams' facies FA4. Palaeocurrents obtained by Nicholson (1993, table 1) from the Cape Wrath Member at Quinag flowed towards the SE ($\theta = 129°$, $n = 123$). Williams (2001, table 5) gives a similar result.

Nicol (1857*a*, p. 25) records copper carbonate from the Torridonian of Quinag, but without specifying a precise location. Heddle (1901, vol. 1, p. 146) records azurite from the south cliff (i.e. Applecross Formation) at Quinag. Copper mineralization has also been seen along the unconformity exposed by the main road along Loch Assynt (R. E. Holdsworth, pers. comm.).

Rubha Stoer

This area includes the greater part of the peninsula of Stoer, west of the Coigach fault. The rocks are all red sandstones with the characteristic pebble suite of the Applecross Formation and were mapped as such by the Geological Survey in 1885. A revised geological map by the author has been integrated into 1:50 000 sheet 107W, published by the British Geological Survey in 2002.

The structure is dominated by a roll-over anticline on the hanging wall of the Coigach fault (Stewart 1993*a*, fig. 4). Dips change from about 15° west near Stoer lighthouse to about 20° east, near Raffin. The axis of the anticline trends NE to intersect the Coigach fault near Rubh' an Dunain. Consequently the general dip along the northern coast is westerly, with the lowest strata seen on the north side of Bay of Culkein. The total stratigraphic thickness exposed is about 1100 m making no allowance for fault displacements. Palaeomagnetic samples from the coastal section west of Meall Dubh (Irving & Runcorn 1957, sites B69–B71) show polarity reversals suggesting that the rocks here, about 700 m above the base, are high in the Applecross Formation, possibly near the top.

The lowest bed seen immediately west of the Coigach fault, 300 m SSW of Culkein jetty [NC 03843300], is a conglomerate unit at least 4 m thick, formed of tightly packed durable pebbles about a centimetre in size. Very coarse, pebbly sandstones about 35 m thick, are also seen at Rubh' an Dunain, in a fault isolated block. About 20% of the beds at Rubh' an Dunain are coarse enough to be described as gravel. The pebbles, however, are no more than 2 cm across, all durable types. Interbeds of ripple-bedded red or green siltstone and fine sandstone, up to a decimetre thick, are common.

West of Rubh' an Dunain, as far as Loch an Achaidh (250 m stratigraphically higher), the beds are coarse and still contain silty interbeds up to 20 cm thick. Pebbles are generally less than 3 cm across, either scattered throughout the sandstone or arranged in sheets up to a decimetre thick. About 20% of the section is contorted. The rest of the sequence is similar, except for the absence of silty interbeds and a greater proportion of contorted beds. For example, at Cluas Deas near the lighthouse, and at Balchladich Bay about 35% of the sequence is contorted.

Palaeocurrents measured by Williams (1969*a*, fig. 12) at Rubh' an Dunain and farther west have a vector mean 076° ($n = 52$), while Nicholson (1993) got $\theta = 095°$ ($n = 172$).

Evidence of tensional stress after deposition is provided by a dilatational sandstone dyke 1.5 m wide that cuts the Applecross

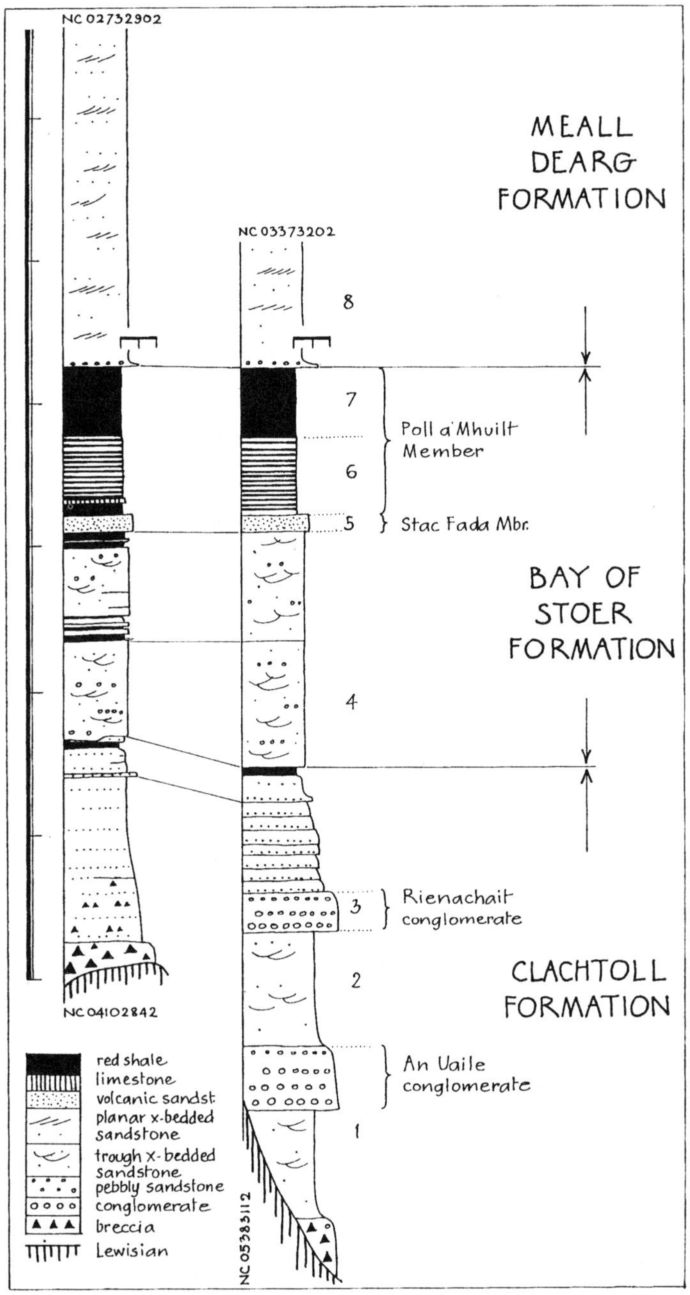

Fig. 43. Stratigraphic logs of the Stoer Group at Stoer. The left-hand log extends from Stoer old graveyard westwards along the north side of Bay of Stoer. This is the type section for the formations and members of the Stoer Group. The right-hand log is an equivalent section about 3 km to the north, showing the Geological Survey's subdivisions (1–8) of the Stoer Group (Peach *et al.* 1907, p. 301) together with the current nomenclature (Stewart 1991*a*). The thickness scale is in units of 100 m, while the grain size scale spans −4 to +4 ϕ units (16–0.06 mm).

Formation on the shore 250 m west of the Coigach fault at Balchladich Bay [NC 02353054]. The dyke strikes parallel to the Coigach fault and dips 80° west.

Stoer, Clachtoll and Clashnessie

The area includes all the Torridonian rocks east of the Coigach fault at Stoer, most of which belong to the Stoer Group. The type area for the Stoer Group was established here because of the almost perfect exposure and simple structure (Stewart 1969). A geological map by the author, scale 1:10 000, has been deposited with the British Geological Survey in Edinburgh and is integrated into 1:50 000 sheet 107W (published by the British Geological Survey in 2002). Figure 43 shows the type section and equivalent strata a few kilometres to the north. The coastal exposures constituting the type section have been designated as a Geological Conservation Review site (Mendum *et al.* 2003).

The rocks were mapped by the Geological Survey in 1887 and divided into nine lithostratigraphic units, shown in Figure 43. When in 1893 the Torridonian strata were formally subdivided into formations the rocks equivalent to the Stoer Group at Stoer were placed erroneously in the Diabaig Formation (described as being 'peculiarly developed' by Peach *et al.* 1907, p. 301); the Applecross Formation was correctly identified. The surveyors had failed to appreciate the significance of the contact between the Stoer and Torridon Groups at Bay of Culkein. It was identified by Williams (1966*b*) as an angular unconformity.

The type section

The type section established by the writer for the constituent formations of the Stoer Group (Stewart 1969, 1991*a*) has its base about 33 m east of the old graveyard at Stoer [NC 04102842] and continues westwards along the low cliffs on the north side of Bay of Stoer, to Meall Dearg [NC 02732902]. Three formations are present: the Clachtoll Formation (Ct) at the base, followed by the Bay of Stoer Formation (BS) and the Meall Dearg Formation (MD). The formations and lithofacies in the type section are shown in Figure 43. Individual lithofacies that recur throughout the formations are numbered as described on p. 2. The formations and lithofacies are described in the following paragraphs, starting at the base.

The Clachtoll Formation at Clachtoll

The formation is characterized by containing only detritus derived from the local basement gneisses. Around Clachtoll it occupies a palaeovalley about 150 m deep, as shown in Figure 5. Massive breccio-conglomerate (facies Ct1) derived from the immediately underlying Scourian gneisses at the base of the section (Fig. 44) passes upwards into tabular pebbly sandstone (facies Ct2) shown in Figure 45. Imbrication of small pebbles in both facies shows current movement towards the west. The boundary between Ct1 and Ct2 is defined by 50% sandstone content. The breccio-conglomerate and tabular pebbly sandstone facies (Ct1 & Ct2) are about 50 m thick near the centre of the palaeovalley [NC 043273].

Breccio-conglomerate on the south side of the palaeovalley is particularly interesting because the source of the clasts can be located accurately. About 350 m east of A'Clach Thuill a tongue of massive breccio-conglomerate (Ct1), resting on gneiss, dips westwards into the sea [NC 04102664]. The lithologies of the gneiss blocks reflect those of the basement within 50 m north and east of

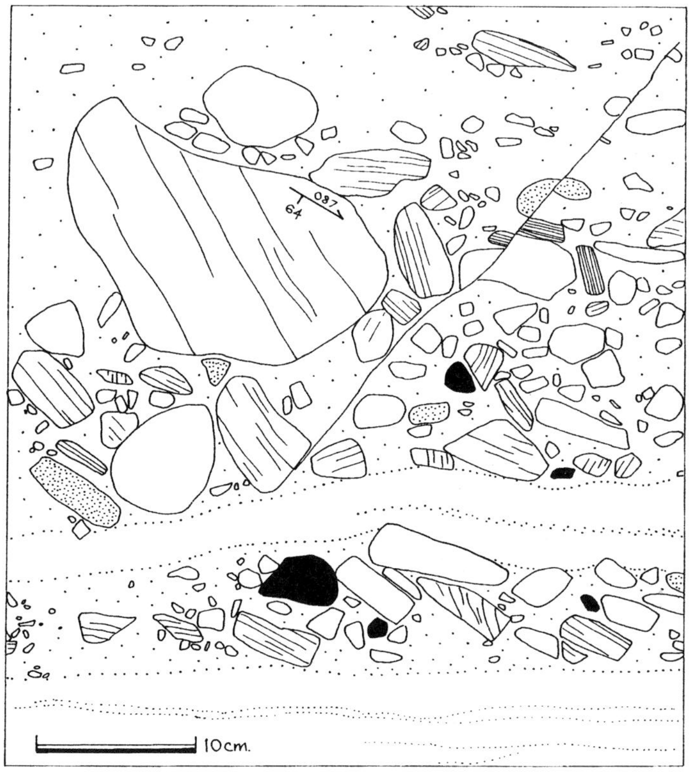

Fig. 44. The breccio-conglomerate facies Ct1 at Clachtoll [NC 04352730], traced from a rock-surface. All clasts over 2.5 mm are shown. Ultrabasic clasts are black, basic ones stippled, acid gneiss unornamented. Unornamented bands are medium to fine grained sandstone. Note the imbrication at the foot of the figure.

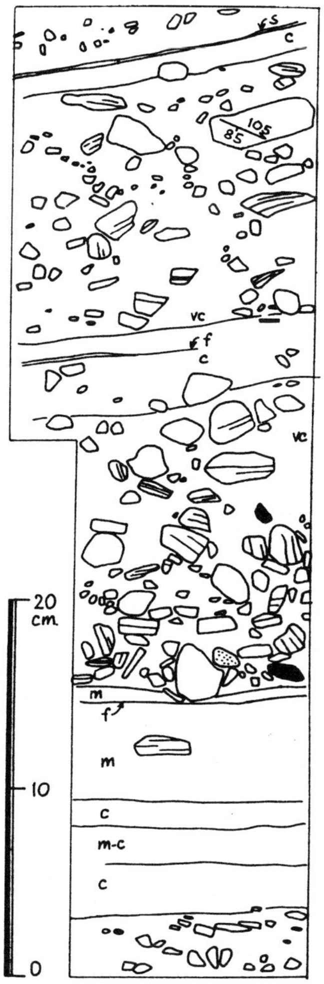

Fig. 45. The tabular sandstone facies Ct2 at Clachtoll [NC 04332729], traced from a rock surface. All clasts over 2.5 mm are shown. The rest is sand and silt. Grain size is indicated thus; s = silt, f = fine sand, etc. Ultrabasic clasts are black, basic ones stippled, acid gneiss unornamented. Note the imbricate structure at the foot of the illustration.

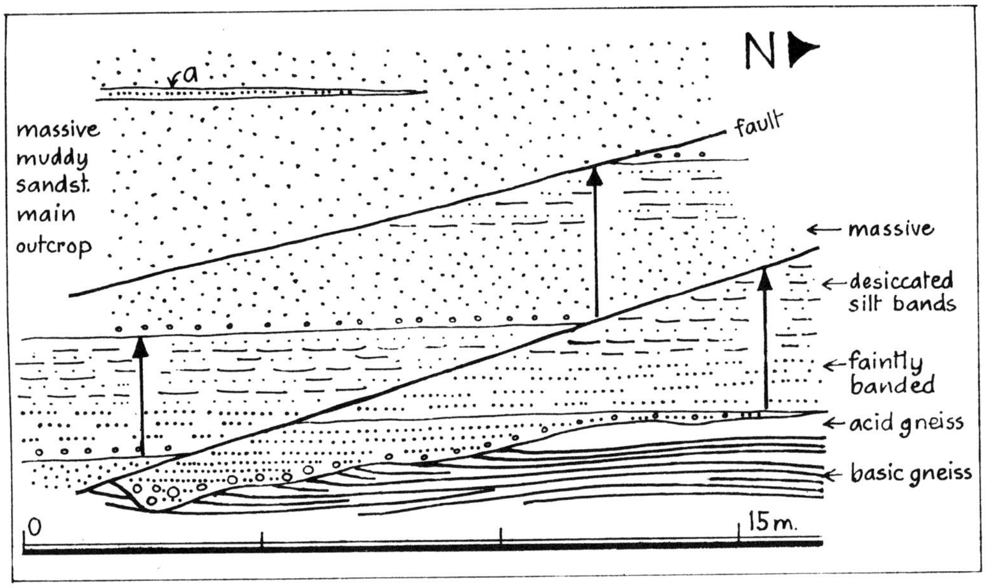

Fig. 46. Stratigraphic profile of muddy sandstones forming the lowermost Clachtoll Formation and their contact with gneiss, at Clachtoll [NC 04052669]. Fining-upward units are indicated by arrows. The whole section is exposed below half tide, after a storm has swept the bay clean of sand. The coarse grey sandstone band (**a**), however, above high water mark on the west side of the bay, is always visible, together with an unpredictable thickness of the underlying beds.

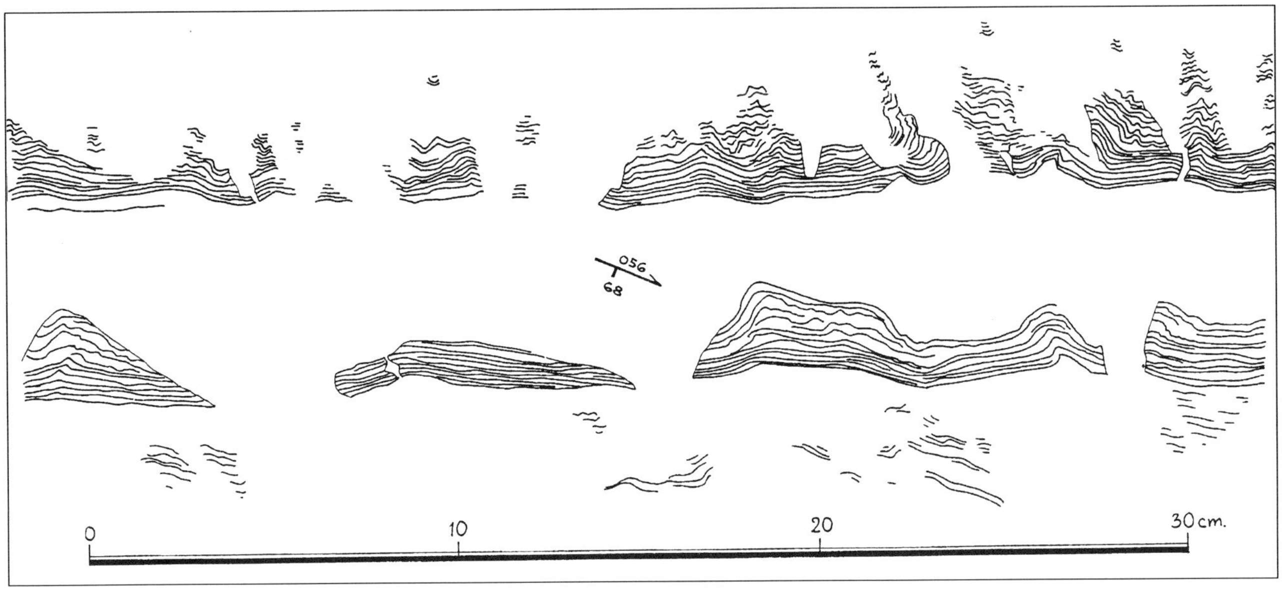

Fig. 47. Desiccated limestone sheets in the muddy sandstone facies Ct7 at Clachtoll. The tracing was made about 50 m north of A'Chlach Thuill [NC 03812678].

Fig. 48. Photograph of intersecting desiccation cracks exposed on a bedding surface in muddy sandstone (Ct7) at Clachtoll [NC 03772727]. The stratigraphic position of the bed is shown in Figure 49. The ruler is 20 cm long.

the outcrop, viz. altered picrite, basic and intermediate gneiss, and quartz. Rounded to well-rounded picritic clasts predominate in the lowest metre of the breccia exposed along the northern edge of the tongue. The breccia matrix at this level is ill-sorted sand, rich in picritic debris. Rare dilatational sedimentary veins a few centimetres wide insinuate themselves along foliation planes in the gneiss for up to a metre below the unconformity. The breccia is about 5 m thick and, apart from the picritic base, mainly composed of acid and intermediate gneiss clasts. Clast size diminishes stratigraphically upwards from 20–30 cm at the base to about 4 cm at the top where the breccia is interbedded with red sandstone. The rounding of the acid to intermediate gneiss clasts also improves upwards, from subangular at the base to subrounded at the top. Down dip (and down current) the breccia passes into red sandstones, exposed at low water springs.

According to Davison & Hambrey (1996, 1997) the breccio-conglomerate is a diamictite, thought by them to be a lodgement tillite formed beneath a glacier. The upward grading belies this, as does the striking difference in shape between acid intermediate and ultra-basic gneiss clasts at the base of the deposit. In any case, the deposit is clast supported, not matrix supported as in a diamictite (Flint *et al.* 1960).

The basal breccia and tabular sandstone is overlain at Clachtoll by 200 m of reddish-brown muddy sandstone (facies Ct7), petrographically a ferruginous greywacke. The lowest 130 m of the facies are completely massive apart from a few fining-upward cycles at the base (Fig. 46). What appear to be bedding planes, for example in the road side quarry [NC 04232727], are probably joints. Joints with the same orientation are seen in the bedded muddy sandstones stratigraphically higher in Clachtoll, where they make a small angle with bedding.

The uppermost 70 m of the muddy sandstone is divided into decimetre-thick beds by calcareous sheets a centimetre or two in thickness (Fig. 47). Millimetre-sized holes in the sandstones are perhaps casts of former evaporite needles. The sandstone beds are cut by diffuse, intersecting sand veins (Fig. 48) that appear to be relics of desiccation cracks that have closed and been deformed during repeated episodes of rewetting. Similar veins are present throughout the muddy sandstone facies, though less sharply defined than those in Figure 48, suggesting that the fine-grained matrix of the sandstone was originally a swelling clay. A graphic log of the beds is shown in Figure 49. Similar beds separated by carbonate sheets only millimetres thick are seen near the northern margin of the palaeovalley, on the type section.

If the two lowest facies at Clachtoll, Ct1 and Ct2 represent small alluvial cones, the origin of the muddy sandstone is more problematic. Critical evidence is provided by the rapid lateral passage from the tabular sandstone into the muddy sandstone, clearly seen at Gruinard Bay and on Horse Island (Achiltibuie) though not at Clachtoll. This equivalence, together with the carbonate sheets in the upper part of the muddy sandstone, suggest the former presence of an ephemeral lake. Rapid sedimentation at its inception is recorded by the above noted metre-thick graded beds of muddy sandstone (Fig. 46). As the palaeovalley filled the rate of sedimentation diminished so that for long periods little clastic material was contributed to the lake and the carbonate sheets developed.

The banded muddy sandstone contains a 2 m bed of very coarse sandstone (Ct2) which can be traced right across the Clachtoll palaeovalley and north to the coast at Port Cam, a distance of 6 km overall (Fig. 49). Small-scale trough cross-bedding near the top of the bed shows palaeocurrents flowing towards the west ($\theta = 260°$, $n = 18$). A few metres above the sandstone bed on the north shore of Bay of Stoer [NC 03782831] a shale sequence (facies Ct3) about 5 m thick is exposed, grey for the first half metre but otherwise red (Fig. 50). Ripple-bedded sandstones appear in the upper part of the shale with palaeocurrents directed to the NE. The uppermost of these sandstones contain pebbles of gneiss and quartz (but no quartzite) up to a centimetre across. They are erosively overlain by red sandstones containing rounded pebbles of gneiss and quartzite (the Bay of Stoer Formation) that form an impassable coastal promontory (Fig. 51) at high tide [NC 03712834].

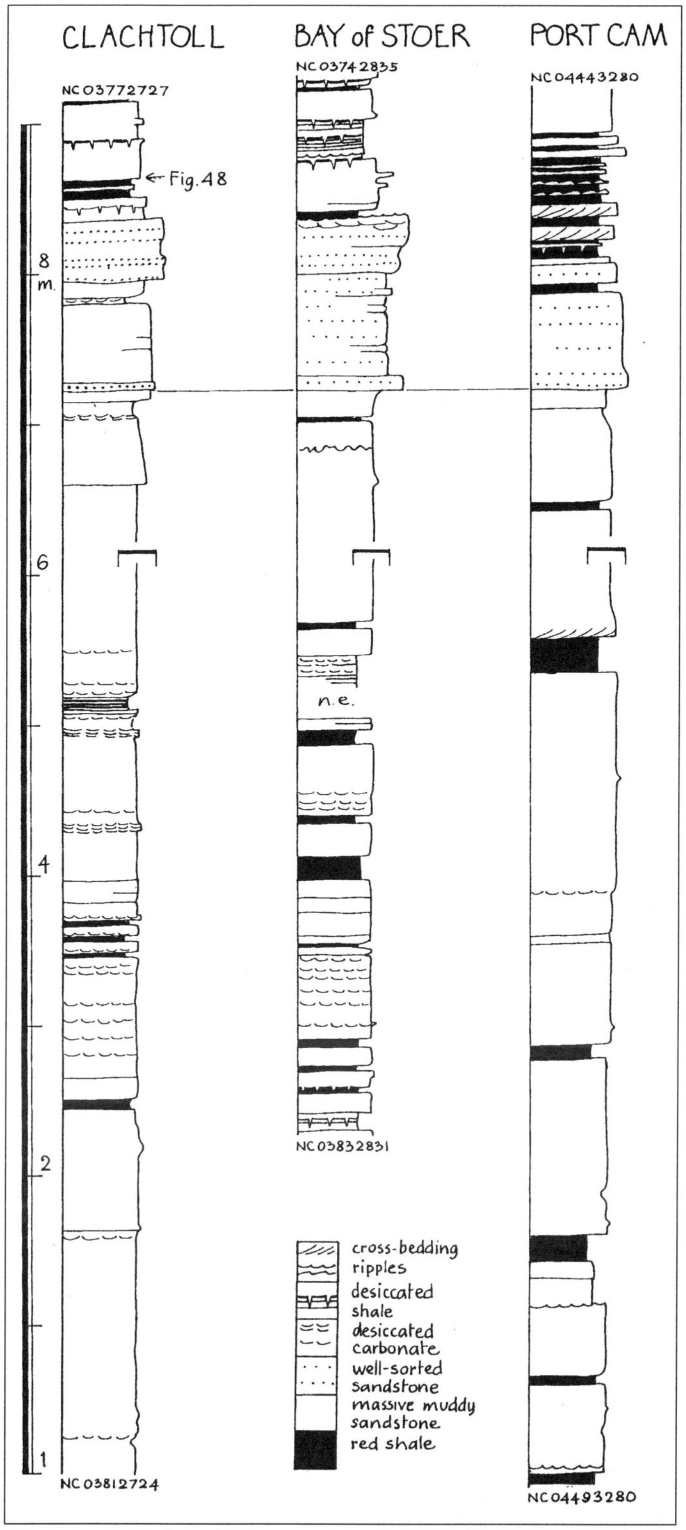

Fig. 49. Graphic logs of three correlative sections in the Clachtoll Formation. Most of the sediments are muddy sandstones (facies Ct7), but the well-sorted very coarse sandstone which appears near the top of the logs is a horizon belonging to facies Ct2 which can be mapped for about 6 km from Clachtoll in the south to Port Cam on the northern coast of Stoer peninsula. The grain size scale spans 0–4 ϕ units (1–0.06 mm).

Towards Bay of Clachtoll the basement rises rapidly so that the massive muddy sandstone (Ct7) is either directly in contact with gneiss, or separated from it by a few metres of breccia (Ct1), or by tabular sandstone with gneiss clasts (Ct2). Small patches of facies Ct1 and Ct2 cling to the low gneiss cliff that rises from the sand on the SE side of the bay [NC 040269], demonstrating that it is a freshly exhumed Proterozoic hillside.

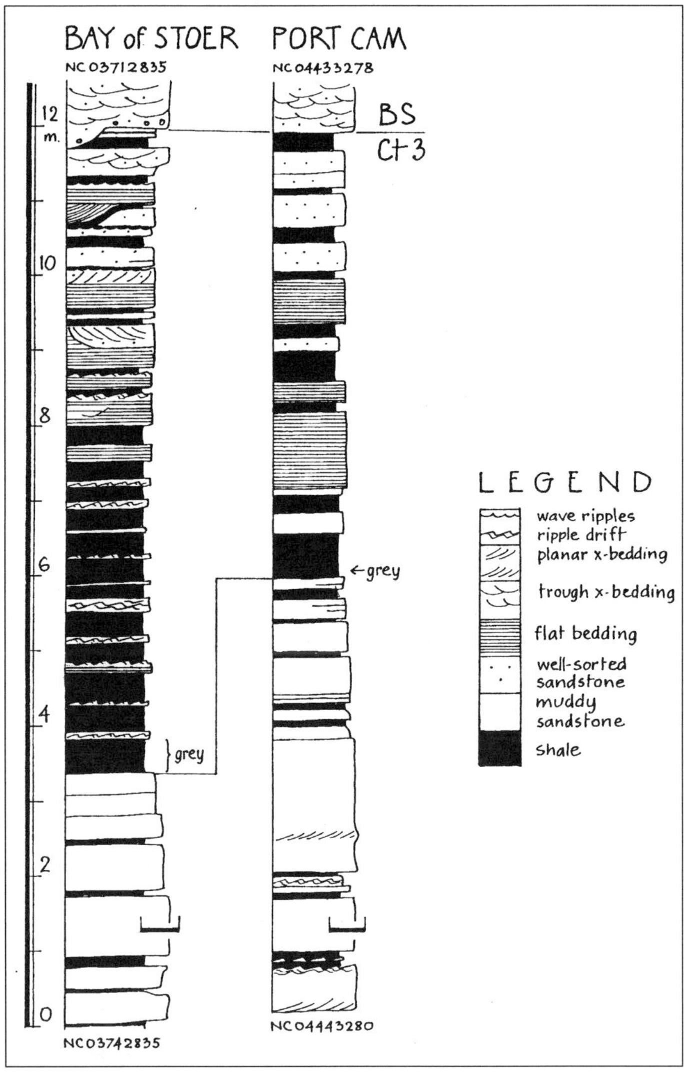

Fig. 50. Graphic logs of the shale facies Ct3, immediately beneath the Bay of Stoer Formation (BS). The left-hand log shows the type section at Bay of Stoer. The right-hand log is the equivalent section at Port Cam, 4 km to the north. Muddy sandstones below the level of the thin grey shale belong to facies Ct7. The grain size scale spans 0–4 ϕ units (1–0.06 mm).

Across Bay of Clachtoll the stratigraphy changes substantially. The uppermost 40 m of the Clachtoll Formation, consisting of bedded muddy sandstone (Ct7), the very coarse sandstone bed (Ct2) and the red shale (Ct3), are all absent, and the sandstones tentatively correlated with the Bay of Stoer Formation appear to interdigitate with the Clachtoll Formation (Fig. 52). The Bay of Stoer Formation also differs significantly across the bay. Possible explanations for these changes are proposed in the section dealing with the Bay of Stoer Formation (p. 64).

Sandstone dykes at Clachtoll

A gneiss hill 350 m east of A'Clach Thuill [NC 041267], recently exhumed from beneath the muddy sandstones, is traversed by dilatational veins filled by fine red sediment (Fig. 53) seen passing into (but not cutting) the overlying muddy sandstone (Fig. 46). Some veins follow the gneiss foliation, but most transect it, with no preferred orientation discernible. The sediment in some veins contains segregations of quartz, feldspar, and calcite that appear to fill cavities of unknown origin. Quartz and feldspar coat the margins of the cavities, and the calcite, which formed last, fills the centre. The quartz is colourless and forms prisms up to a centimetre long. The pink and white feldspars crystals, less than a millimetre in length, have Cape Finisterre habit, and almost all are twinned on the Carlsbad law. A similar mineral assemblage that occupies cracks in the underlying gneiss is associated with a green mineral, either epidote or pumpellyite, and is seen to be cut by silt-filled veins. Nearby there is a southward-facing joint plane about 4 m above high water mark [NC 04082670] coated with the same green mineral, that cuts across a network of silt-filled fractures. It appears, therefore, that the pumpellyite formed both before and after formation of the sediment-filled veins. Potassium feldspar with Cape Finisterre habit is not a low temperature form (Smith & Brown 1988, p. 515) indicating that it crystallized when the veins were deeply buried.

Where the gneiss hill descends to the beach [NC 04052670] dilatational fractures for a few metres below the unconformity have hematite coated walls and are filled by interlaminated silt and fine sand dipping westwards like the overlying muddy sandstone (Fig. 46). This shows that the veins were formed before regional tilting of the Stoer Group, and long before deposition of the Torridon Group.

Elsewhere on the gneiss hill zones of intense brecciation containing only slightly misaligned gneiss blocks can be identified.

Fig. 51. The contact between the Clachtoll and Bay of Stoer Formations on the type section at Stoer. The rocks form a promontory on the coast 300 m SSW of Stoer House [NC 03712834]. The observer looks west. The top of the map case marks the contact. The hammer is 50 cm long.

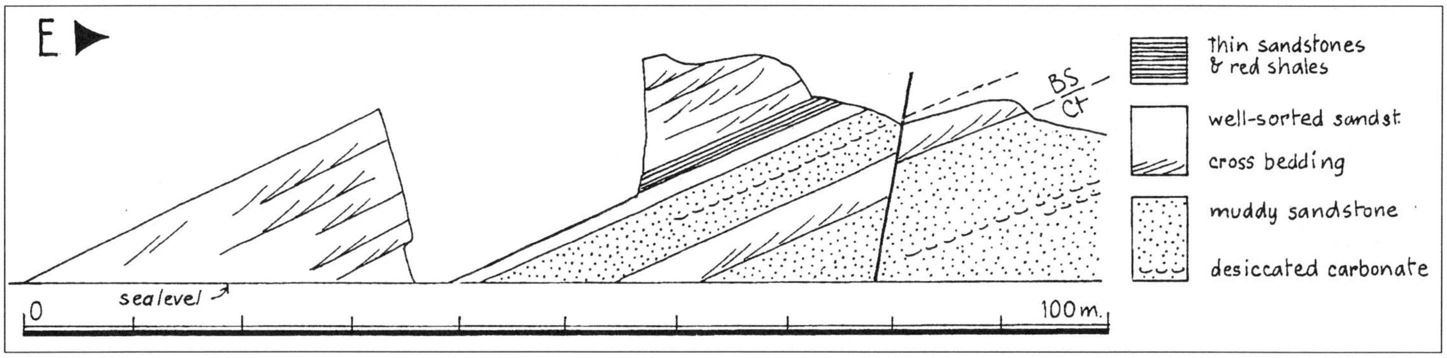

Fig. 52. True-scale section showing the boundary between the Clachtoll Formation (Ct) and the Bay of Stoer Formation (BS) at A'Chlach Thuill [NC 038267]. The name means 'the rock with a hole through it'. The hole doubtless originated by erosion of the interval of thinly bedded, rippled sandstones and desiccated shales shown on the section. The top of the hole eventually collapsed.

They strike roughly NE, parallel to the foliation. The thickest is about 5 m wide and coincides with a major fault that borders the hill to the SE [NC 04102670]. The fault juxtaposes veined and shattered acid gneiss on the NW with a picrite dyke 100 m wide on the other side, about 100 m from high water mark (Barber *et al.* 1978, fig. 12). Slickenfibres in various directions cut the matrix of the breccia. Another NE striking breccia zone, 0.5–1.0 m wide, can be traced up the hill past the 'Clachtoll fault zone' of Beacom *et al.* (1999, Fig. 5b).

Fractured gneiss can also be found SE of Clachtoll, for example on the point of Rubha Leumair [NC 042261]. Fractures in gneiss coated with hematite and a green mineral (probably pumpellyite)

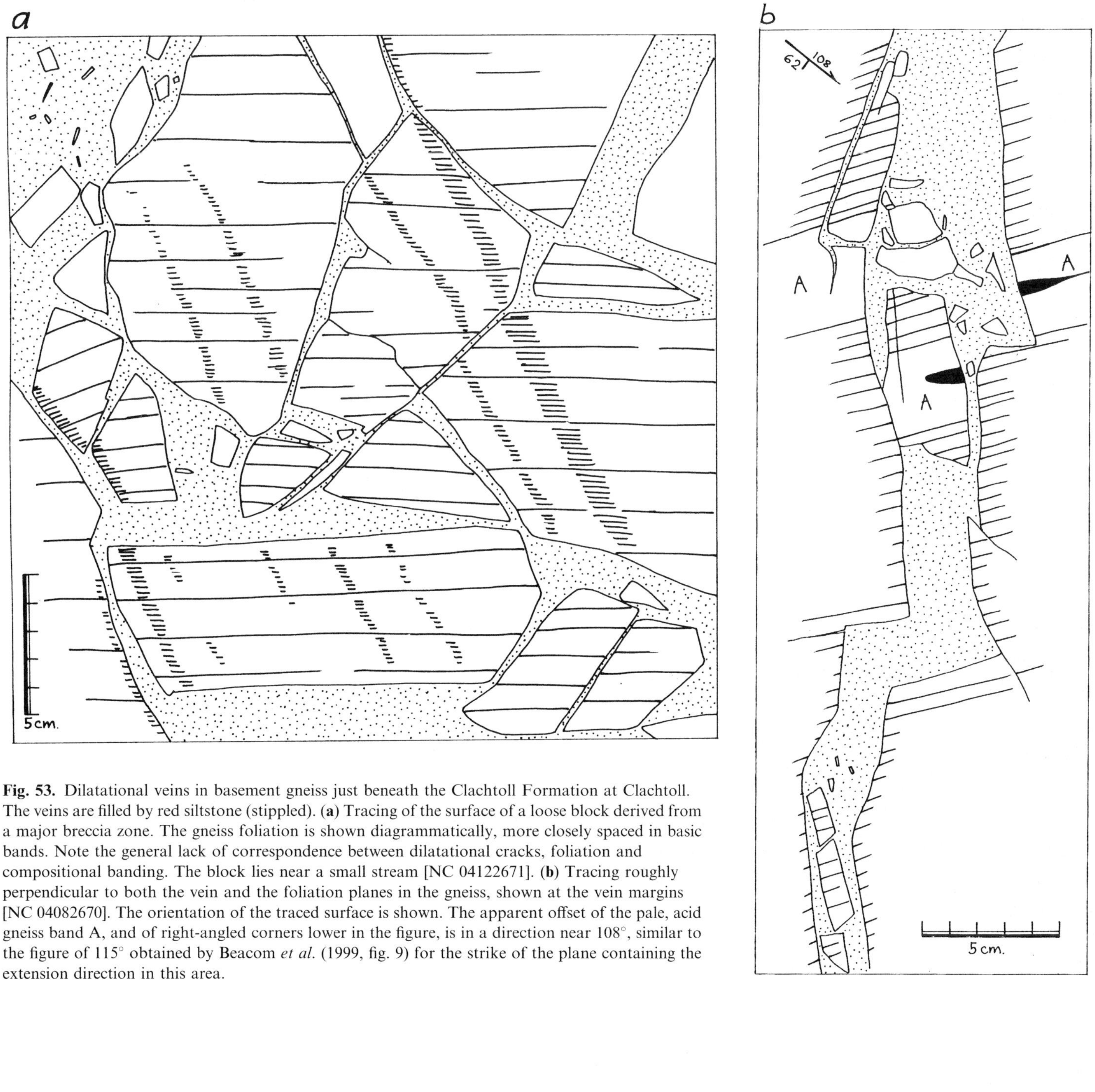

Fig. 53. Dilatational veins in basement gneiss just beneath the Clachtoll Formation at Clachtoll. The veins are filled by red siltstone (stippled). (**a**) Tracing of the surface of a loose block derived from a major breccia zone. The gneiss foliation is shown diagrammatically, more closely spaced in basic bands. Note the general lack of correspondence between dilatational cracks, foliation and compositional banding. The block lies near a small stream [NC 04122671]. (**b**) Tracing roughly perpendicular to both the vein and the foliation planes in the gneiss, shown at the vein margins [NC 04082670]. The orientation of the traced surface is shown. The apparent offset of the pale, acid gneiss band A, and of right-angled corners lower in the figure, is in a direction near 108°, similar to the figure of 115° obtained by Beacom *et al.* (1999, fig. 9) for the strike of the plane containing the extension direction in this area.

are seen again a kilometre east of Rubha Leumair, at localities 17 & 18 of Barber *et al.* (1978, p. 47) [NC 050262], and at locality 14 of Barber *et al.* (1978, p. 43) [NC 057255]. Hematite-coated veins have also been described and figured by Beacom *et al.* (1999, fig. 7) from a point 150 m SE of the Clachtoll exposures.

Hay *et al.* (1988, Fig. 5c) identified some extensional veins in the gneiss at Clachtoll containing K-feldspar and calcite as belonging to the oldest of the four post-Laxfordian crack-sealing phases identified by them in the Lewisian basement of Assynt, predating hematite-filled microcracks that they associate with the Stoer Group. Fluid inclusion homogenization temperatures for the K-feldspar–calcite assemblage in the microcracks are in the range 200–320°C (S. J. Hay, pers. comm.) suggesting a burial depth of about 7 km for a geothermal gradient of 40°C km^{-1}, but it is not certain that these minerals formed at the same time as those in the sediment-filled veins.

Davison & Hambrey (1996, 1997) have suggested that the sediment-filled veins at Clachtoll are due to hydrofracturing by subglacial meltwater during deposition of the Stoer Group, but the greatest principal compressive stress was evidently not vertical as required by the glacial model. The brecciated gneiss they identify from this locality (Davison & Hambrey 1996, fig. 3e), and attribute to shearing by moving ice, has recently been claimed by Beacom *et al.* (1999) to belong to their Clachtoll fault zone.

Beacom *et al.* (1999) show that the Clachtoll fault is normal, strikes NE at 060° and dips NW at 53°. The orientation suggests that it belongs to a suite of pre-Torridon Group fractures identified by them cutting the basement at Gairloch and Canisp, that resulted from WNW to ESE regional extension. The fault, however, is not seen to cut the Stoer Group at Clachtoll. Beacom *et al.* claim that two vein forming events can be identified near the Clachtoll fault zone, the older one identified by being passively filled by relatively fine-grained, dark red, laminated clastic sediment. This older vein set is supposed to have formed when shear planes sub-parallel to major rift faults were held open during extension. The second set, which may contain fragments of the older vein fill (Beacom *et al.* 1999, fig. 5a), was produced by the forceful injection of fluidized sediment. However, field evidence for these two vein filling events is hard to see. Beacom *et al.* believe that the sediment-filled veins, particularly the latter set, are contemporaneous with the main phase of movement on the Clachtoll fault, and consequently with regional extension.

The most significant features of the veins appear to be:

- The absence of any preferred orientation (Beacom *et al.* 1999, figs 2 & 9).
- Their extension in a WNW to ESE direction (Beacom *et al.* 1999, fig. 9).
- The presence of laminated sediment in the veins.
- Their abundance beneath the Stoer Group at the only locality where basement gneiss forming a palaeotopographic high is draped by impermeable sediment. For example, only 80 m to the south [NC 04102664] there are scarcely any veins in the basement beneath the breccia facies Ct1.
- Their formation at a depth of 3–4 km (see p. 20).

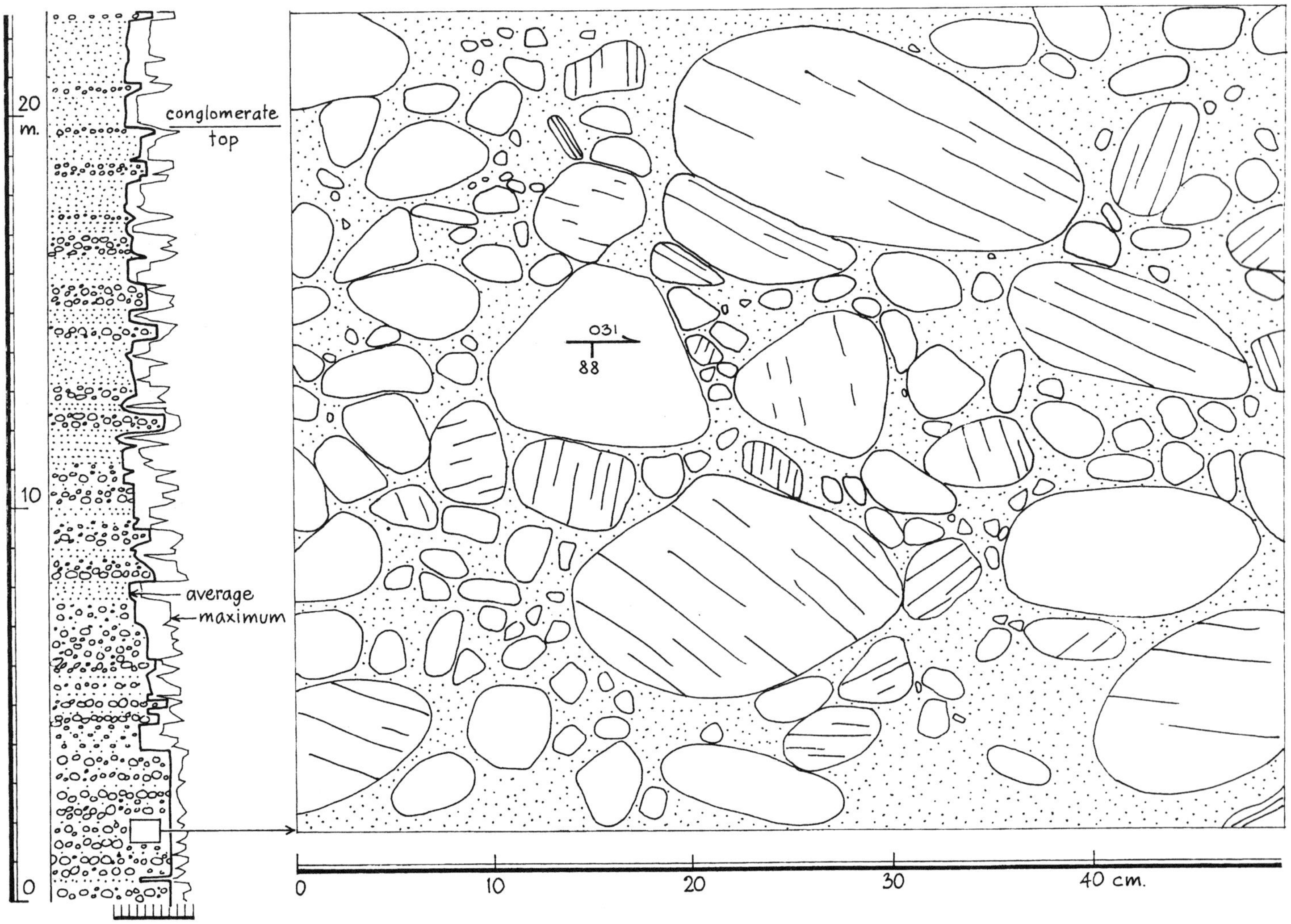

Fig. 54. The Rienachait conglomerate, facies Ct4, at Rienachait [NC 04423020], excluding the lowest 2 m which are not exposed. The graphic log shows average and maximum grain size, measured every 10 cm. The maximum was that found within 0.5 m laterally from the tape. The grain size scale spans +2 to $-8\ \phi$ units (0.25–250 mm). The top of the conglomerate, as mapped, is shown on the log. The tracing of cobble outlines in the lower part of the conglomerate shows all clasts over 2.5 mm.

The laminated sediment indicates that the veins filled up with water and sediment whilst they were held open by high fluid pressure. Tensile stress alone could not have achieved this at a depth of 3–4 km. This suggests that they formed after deposition of the Stoer Group when lateral compression forced fluids upwards out of the gneiss, creating effective hydrostatic tension beneath the impermeable cover at Clachtoll. Where the gneiss was overlain by breccias and sands the fluids leaked through them and fractured the shales above. The least principal stress was horizontal near the unconformity but vertical at higher stratigraphic levels. Consequently, calcite filled dilatational fractures roughly parallel to bedding, formed by the crack–seal mechanism (Ramsay & Huber 1983, p. 248), are abundant in the shales of the Clachtoll Formation and Poll a' Mhuilt Member, and are particularly well displayed on the smooth walls of the western sea cave [NC 03242863]. Locally, the veins pass laterally into patches of calcite cemented breccia up to several cubic metres in volume. The breccias were formed by normal faults that intersect bedding parallel to the strike and cut the rock into a mosaic of lozenges 0.5–1 cm in size.

The Clachtoll Formation at Clashnessie

North of Stoer village the Clachtoll Formation thickens into another palaeovalley (Fig. 5), but the sediments filling it are lithologically quite different to those seen at Clachtoll even though they share the same easterly derivation ($\theta = 272°$, $n = 49$). They are mostly cobble conglomerates (facies Ct4) and trough cross-bedded red sandstones (facies Ct5). The conglomerate facies, shown in Figure 54, forms two mappable horizons which are informally called the An Uaile and Rienachait conglomerates. The former is 42 m thick at An Uaile and the latter about 24 m thick at Rienachait. Both are mainly composed of rounded acid gneiss cobbles up to about 20 cm across. The conglomerates have erosional bases and grade upwards into trough cross-bedded sandstone (Ct5). The Rienachait conglomerate on the north coast of Stoer cuts 0.7 m into the underlying laminated sandstone of facies Ct8 (Stewart 1991*c*, fig. 3.6). The top of a conglomerate is defined where less than 50% of the rock has an average grain size of over a centimetre. Rare, small black pebbles of fine-grained dolerite can be found in the conglomerates, derived from the chilled margins of Scourie dykes. There are also rare pebbles of fine-grained red sandstone, of uncertain origin. Piper & Poppleton (1991) record pebbles of acid porphyry from both conglomerates. No samples of these pebbles now exist and the writer has been unable to find them in outcrop.

The breccio-conglomerate (facies Ct1) that appears in the bottom of the palaeovalley [NC 05383115] is mixed with cobble conglomerate, facies Ct4, which here has an epidotic matrix.

The trough cross-bedded facies (Ct5) is like that forming the bulk of the Bay of Stoer Formation (facies BS1, described below) but differs in lacking contortions and quartzite pebbles. It forms at least three well-defined fining-upward units above the Rienachait conglomerate. Each is about 45 m thick, coarse at the base, grading into rarely exposed red shale (Ct3) at the top. Each of the three cycles is finer than one beneath it, maximum pebble sizes at their respective bases being 4 cm, 2 cm, 1 cm.

The facies occupying the palaeovalley at Clashnessie are evidently fluviatile, in marked contrast to those around Clachtoll that represent alluvial fans and swamps. These two different facies assemblages are repeated in the Stoer Group elsewhere, suggesting that tectonic warping of a pre-existing landscape led to the closure of some valleys to regional drainage while others remained open.

The Clachtoll Formation on the northern coast

The northern coast of Stoer peninsula provides excellent exposures of the laminated sandstone constituting facies Ct8 (the Port Cam facies of Stewart 1988*a*, fig. 8.7). The facies consists of fine pale red sandstone weathering to a pale greyish-red colour. It is cross-bedded with foresets 1–4 mm thick, sharply defined by changes in grain size. The foresets have $p \approx 10\,000$ and are asymptotic to the bases of the cross beds. Most beds are less than a metre thick though some reach 6 m. When the erosional flat tops to the cross-beds are brought to horizontal the foresets dip 10°–20° to directions between NE and SE. The facies, believed to be aeolian, has frequent intercalations of fine to medium-grained sandstone with desiccated red shale bands (Fig. 55). Some of these intercalations are pebbly, with obviously erosional bases, suggesting that the dune field was subject to periodic incursions of storm water. Blocks of the laminated sandstone up to 20 cm in size are sometimes incorporated in the massive, pebbly beds.

The laminated sandstone facies forms two mappable units, one 150 m thick in the Clachtoll Formation, the other within the Bay of Stoer Formation. The former has a non-erosional contact with the underlying tabular sandstones (Ct2) about 300 m east of Port Cam [NC 04883290]. The top of facies Ct8 on the east side of Port Cam [NC 04563292] is defined by an extensive flat erosion surface, overlain by muddy sandstone (Ct7). The laminated sandstone is also in contact with the Rienachait conglomerate (Ct4), that reaches the sea about 200 m east of Port Cam [NC 04743298]. The conglomerate is 8 m thick at this point, with a strongly erosional base on the laminated sandstone and a pebbly top overlain directly by laminated sandstone (Stewart 1988*a*, fig. 8.6).

The laminated sandstone facies reappears within the Bay of Stoer Formation on the west side of Port Cam [NC 04413286], but

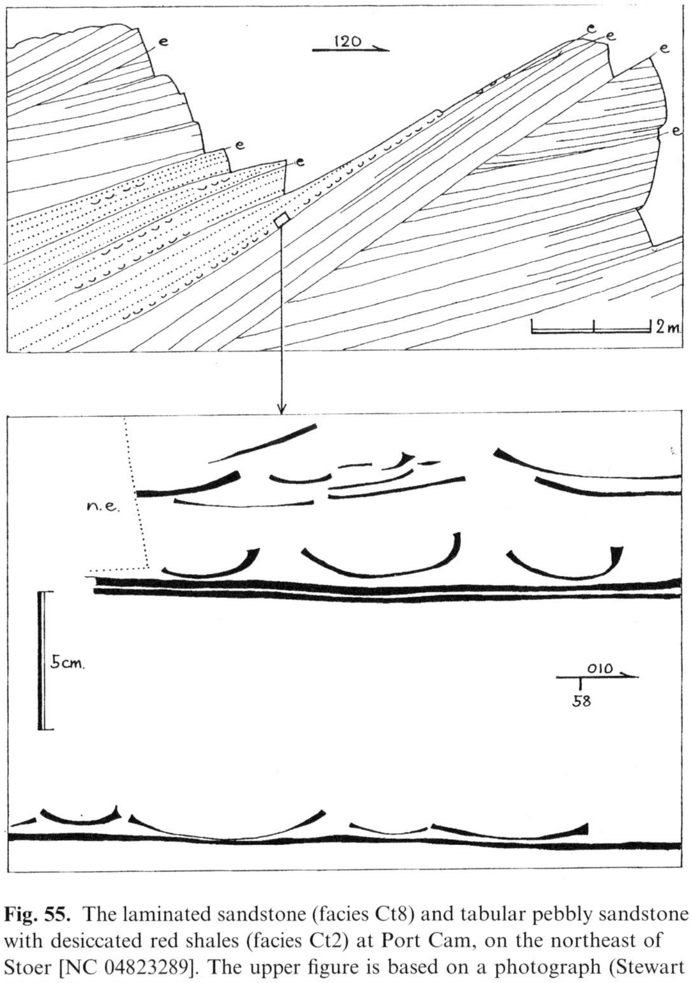

Fig. 55. The laminated sandstone (facies Ct8) and tabular pebbly sandstone with desiccated red shales (facies Ct2) at Port Cam, on the northeast of Stoer [NC 04823289]. The upper figure is based on a photograph (Stewart 1991*c*, fig. 3.7) and the lower one on a tracing. n.e. = no exposure; e = erosion surface. The laminated sandstone, shown by fine lines, is pale red, with a maximum grain size of 2 mm. The tabular sandstone (stippled) is medium grained and greyish red, with rare pebbles. The tabular sandstone contains red shale bands up to 0.5 cm thick, often desiccated as shown in the lower figure. The stratigraphically highest beds figured are erosively overlain by coarse pebbly sandstone (facies Ct2) containing blocks of reworked laminated sandstone (facies Ct8).

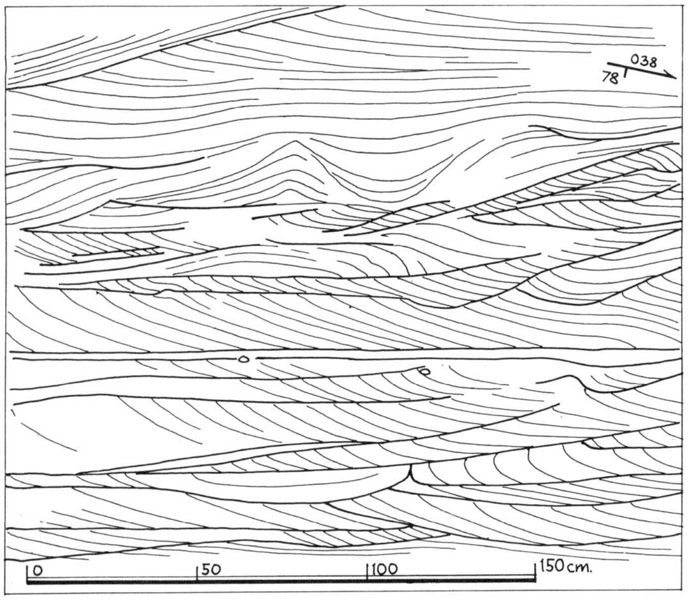

Fig. 56. Trough cross-bedding in pale red sandstone of the Bay of Stoer Formation on the type section. The base of the scaled drawing is 1.1 m above the stratigraphic base of the formation [NC 03712834]. Thick lines are truncation planes. The average grain size is about 0.5 mm; the two pebbles shown are 1.5 cm in diameter.

the base is cut out by a small strike fault. The top of the facies, 100 m to the west [NC 04353294] is marked by an erosion surface with relief of 0.5 m overlain by typical pebbly sandstones of the Bay of Stoer Formation.

Inland exposures are scattered so that interdigitation of the laminated sandstone facies with the Clachtoll and Bay of Stoer Formations (Fig. 5) can only be inferred from outcrop mapping.

The Bay of Stoer Formation

Pale red weathering sandstones form the dominant lithology in the formation and can be called the Bay of Stoer facies, BS1. They have the following characteristics;

- trough cross-bedding (Fig. 56);
- soft sediment contortions (Fig. 56 and Stewart 1988*a*, fig. 8.8);
- erosion surfaces with relief of up to a metre (Fig. 58);
- pebbles of gneiss *and* fine-grained metasedimentary quartzite;
- millimetre-thick concentrations of detrital hematite and ilmenite that commonly form drop structures (Selley 1964).

The stratotype for the base of the formation at Bay of Stoer is shown in Figure 51. Palaeocurrents in the trough cross-bedded sandstones above the basal erosion surface at Bay of Stoer flowed towards the east ($\theta = 83°$, $n = 61$), opposite to those in the underlying Clachtoll Formation. However, 1.5 km farther south, at Bay of Clachtoll, palaeocurrent and palaeoslope directions at the base of the formation are significantly different to those elsewhere. Slump structures in a shaly interval about 15 m above the base of the formation (Fig. 57) indicate a palaeoslope towards the NNW, whereas trough axes in the adjacent pebbly sandstones flowed northwards. About 10 m stratigraphically higher palaeocurrents flowed to the south and only at a level 30 m above the base of the formation do they swing round to the east as seen on the type section. The characteristic pebble suite of the Bay of Stoer Formation is also absent at A'Clach Thuill.

There are two possible explanations for these changes in stratigraphy across Bay of Clachtoll. The first is that Stoer Group stratigraphy may vary rapidly across the strike, in an east–west direction, and later strike-slip faulting has juxtaposed different successions. The second possible explanation is that the sediments on the block south of Bay of Clachtoll are relatively attenuated because it was moving upwards during sedimentation. The last alternative is best able to account for the peculiar palaeocurrent directions in the Bay of Stoer Formation on *both* sides of Bay of Clachtoll.

The sandstones of the Bay of Stoer Formation are punctuated by cycles of muddy sandstone (Ct7) followed by red shale (Ct3). Figure 58 shows five such cycles, together 17 m thick, that are exposed on the shore about 70 m stratigraphically above the base of the formation [NC 035285]. One of them contains small nodules of chalcocite and hematite (Fermor 1951). The same five cycles reappear little altered in Clachtoll, to the south (Fig. 58), and again on the northern coast of Stoer between Port Feadaig and Port Cam [NC 04293291 to NC 04323293]. Another group of cycles, totalling 8 m in thickness, appears immediately below the Stac Fada Member at its type locality (Fig. 59). A close examination of the muddy sandstone at the base of the cycles on the type section shows that:

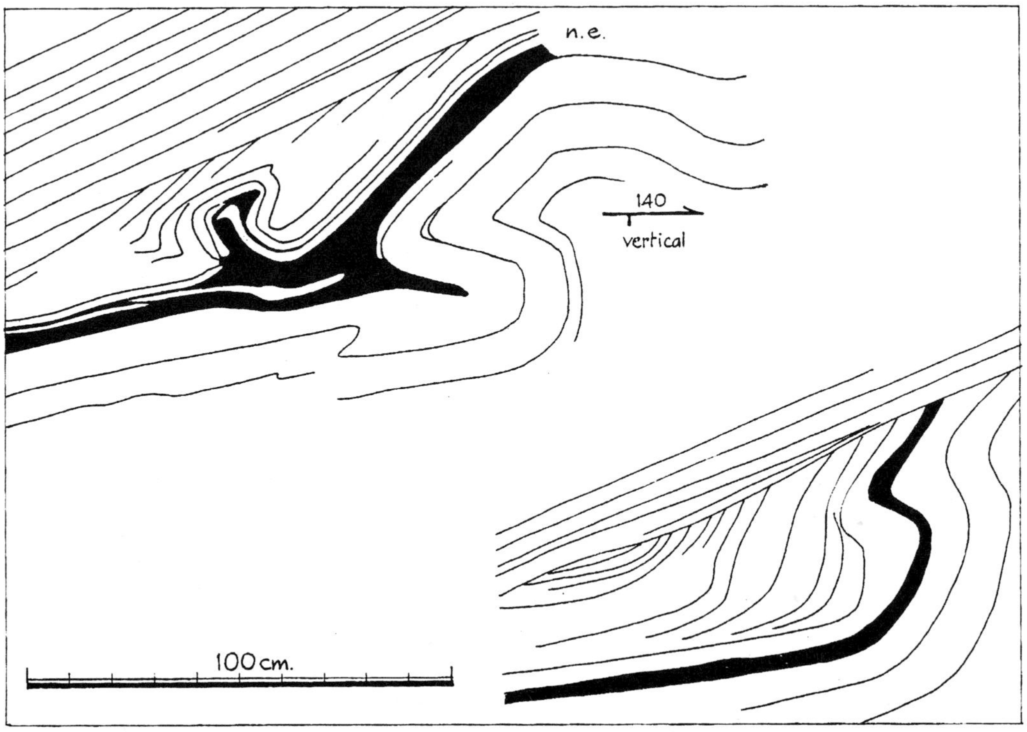

Fig. 57. Slump structures in the lower part of the Bay of Stoer Formation at Clachtoll [NC 03692722]. The deformed red siltstone horizon (black) is the same in both the scaled drawings, which are of joint faces about 9 m apart. The faces are almost exactly perpendicular to the slump axes, which indicate a palaeoslope towards 320°. n.e. = no exposure.

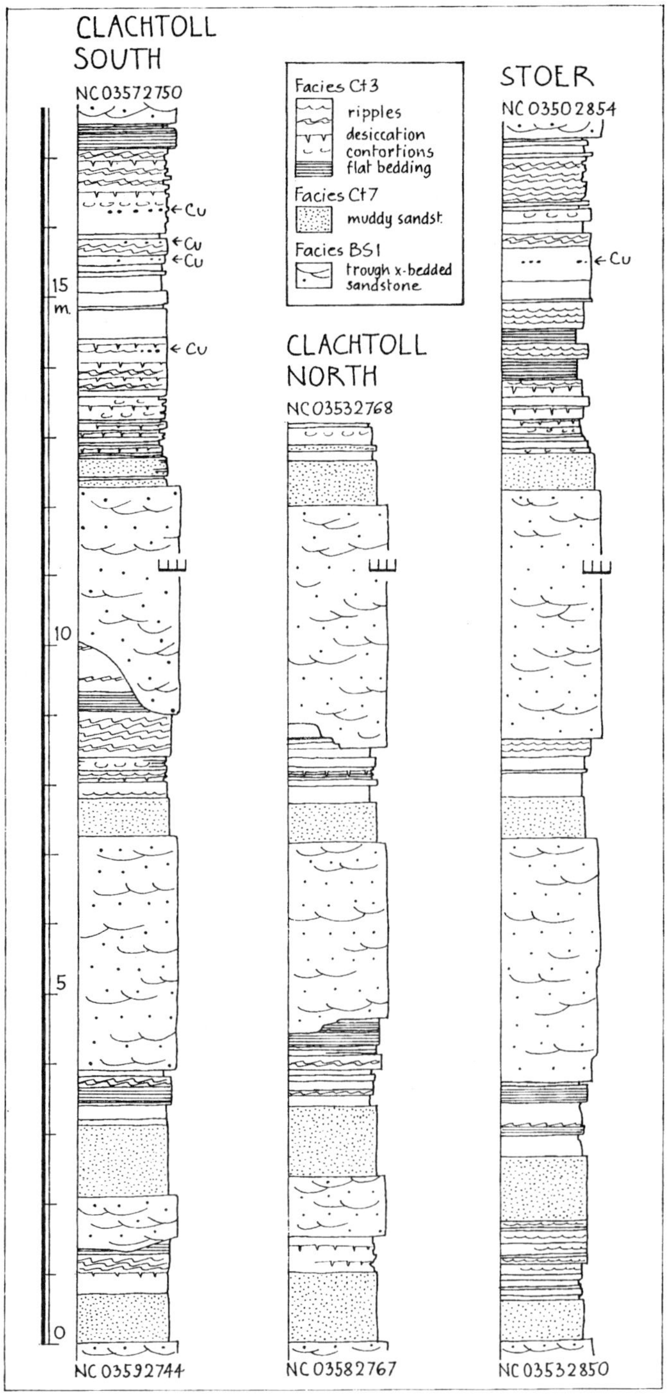

Fig. 58. Graphic logs of three correlative sections in the Bay of Stoer Formation, containing lacustrine intervals. The sections are about 70 m above the base of the formation. The grain size scale spans 4–1 ϕ units (0.06–0.5 mm). Horizons with cupiferous nodules are indicated (Cu). The Stoer (type) section is a kilometre from Clachtoll South, but sections Clachtoll North and South are only 200 m apart.

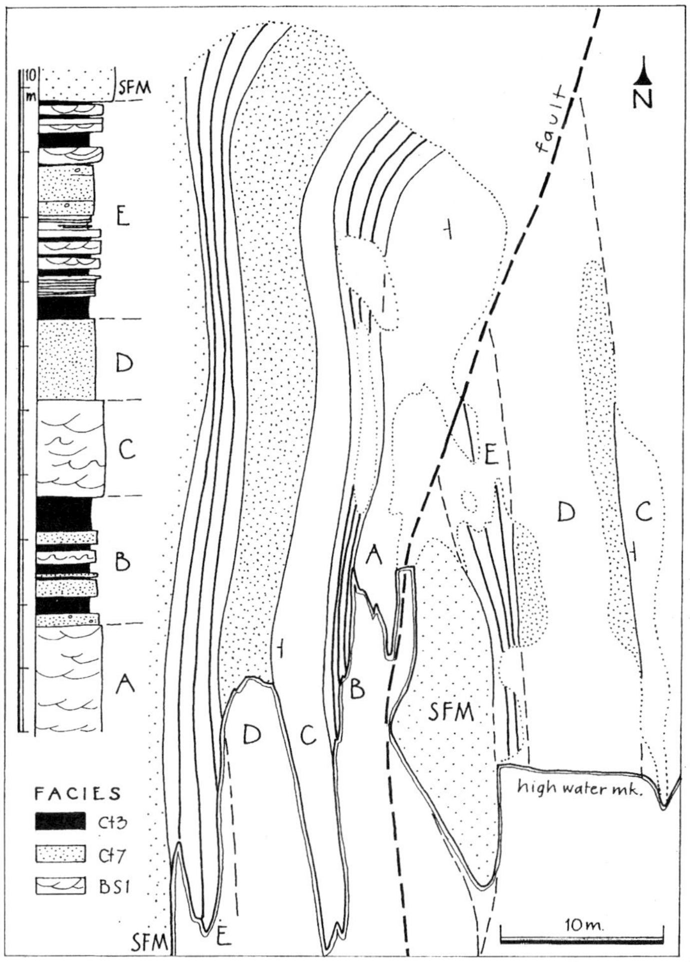

Fig. 59. Graphic log and map showing muddy sandstone–shale cycles immediately beneath the Stac Fada Member (SFM) on the beach at the type locality. See Plate 2 for location. The main outcrop of the member forms the western edge of the map but its base is repeated by a strike fault, as shown. The down-faulted part of the member, which extends southwards to the islet Stac Gruinn, is a repetition of the intrusive tongue shown in Fig. 60 at 35–50 m.

- the bases of the muddy sandstone beds are flat;
- the uppermost centimetre of the sandstone (BS1) underlying two of the muddy sandstone beds contains quartz pebbles up to a centimetre;
- matrix-supported pebbles 2 cm in diameter occur in some beds less than a metre thick (Fig. 59);
- matrix-supported quartz and gneiss grains become smaller and sparser upwards in some beds.

The Stac Fada Member (Bay of Stoer Formation)

The promontory of Stac Fada [NC 033284] marks the base of a giant cycle, consisting of muddy sandstone (Stac Fada Member) followed by shales (Poll a' Mhuilt Member). The muddy sandstone (Ct7) forming the stac is about 11 m thick at Stoer and can be mapped over a distance of 55 km, from Stoer south to Rubha Reidh. It was formally named the Stac Fada Member by the writer in 1969. It is unique in containing centimetre-sized lapilli of dark green, vesicular volcanic glass. Accretionary lapilli are present in the top few metres of the member at Stoer. The geology and geochemistry of the Stac Fada Member have been studied by Lawson (1972), Sanders & Johnston (1989, 1990), Stewart (1990, 1991*a*) and Young (2002) but its origin remains controversial. The visitor to the type section should note that the lower part of the Stac Fada Member and the underlying beds have been repeated by a strike fault (Fig. 59), a fact that has confused some (Sanders & Johnston 1989).

The Stac Fada Member east of the strike fault, outcropping above high water mark, is only about 5 m thick stratigraphically and thins rapidly to the south. At low tide the well-exposed upper contact can be seen to converge with the lower contact so that a few metres south of the map area the muddy sandstone is only 2 *centimetres* thick. Farther south still the member thickens and offshore forms the islet Stac Gruinn. The base of the member is exposed at intervals both above and below high water mark and, though slightly erosive, corresponds exactly in stratigraphic level to the base of the main outcrop of the Stac Fada Member west of the strike fault (Fig. 60). The member east of the strike fault probably represents no more than the lowest part of the fully developed Stac Fada Member to the west, for it lacks accretionary lapilli.

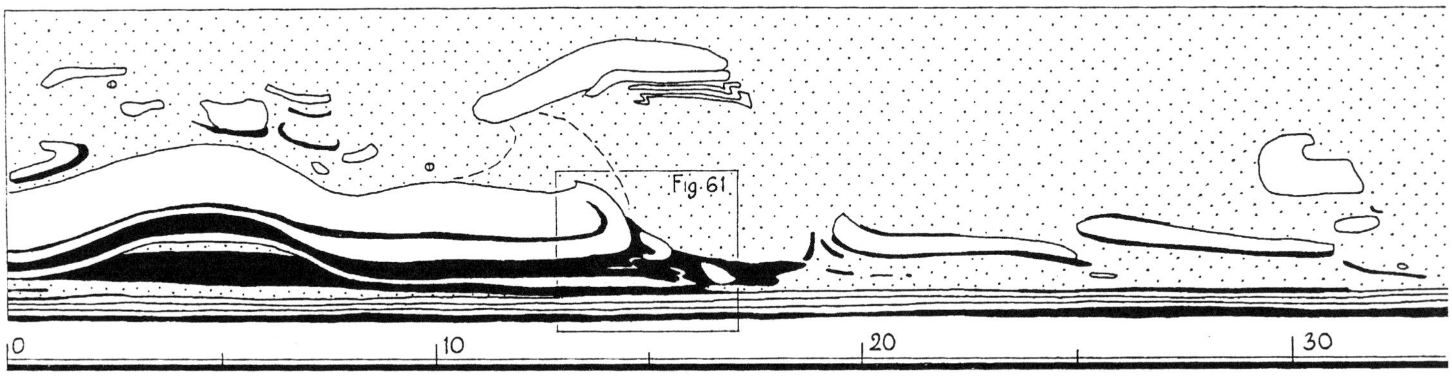

Fig 60. True-scale profile showing the lower half of the Stac Fada Member at Stac Fada [NC 03322849]. The observer looks down dip to the west. The Stac Fada Member is stippled, sandstone unornamented and shale black.

The Stac Fada Member in the main outcrop (Fig. 60) disrupts the underlying sandstones and shales by preferentially injecting the latter. The alluvial trough cross-bedded sandstones involved in the disruption were originally deposited on an essentially flat, nearly horizontal surface, so the way they are warped, folded and locally draped over lenses of volcaniclastic sandstone is clear evidence of post-depositional disturbance. So, also, is the presence of large rafts of sandstone 'floating' in the lower half of the Stac Fada Member. The sediments beneath the Stac Fada Member at all other localities lack siltstone, and disrupted bedding is consequently absent.

Critical features bearing on the origin of the member are listed below and located by co-ordinates (in metres) which refer to an origin at the bottom, left-hand corner (0, 0) of the profile, Figure 60:

(1) The disrupted strata were soft when deformed for not only are there overfolds (14, 3; 17, 5) but also boudinage structures (36, 3; 54, 4; 60, 1; 66, 0) and disharmonic folds (5, 3 to 10, 2).
(2) The main décollement is about a metre above the base of the measured section (0, 1), with about 2.5 m of strata above this level having been deformed.
(3) The mudflow material was intruded along shale units in the underlying beds. Consequently most of the sandstone rafts have veneers of shale along their top and bottom surfaces. One of these shales (20, 2 to 25, 1) contains a band of the Stac Fada Member 4 cm thick throughout its length.
(4) Wedges of Stac Fada material (0, 1; 40, 2) have sharp tops and bottoms, suggesting (but not proving) that they are intrusions from a single mudflow rather than separate mudflows. Mudflow material interfingers with shale at the top of one wedge (53, 0).
(5) The highest raft of sandstone (44, 6 to 59, 3) may be the upper part of the underlying sandstone unit broken off and inverted. The underlying sandstone can be seen from the NE or SW to have been turned up into the vertical (with strike 034°) and then snapped off, as if by northwestward movement of the mudflow. The presence of gneiss blocks pressed into the present top of the raft (46, 6) recalls the blocks in the underlying sandstone (38, 3 etc.).
(6) Isolated gneiss boulders occur in the sandstone tongue (38, 3, etc.). They are associated with a sheet of coarse sandstone only a few centimetres thick, sandwiched between dark-brown fine sandstone and siltstone. The coarse sandstone is poorly sorted like the Stac Fada Member but does not everywhere contain macroscopic clasts of volcanic glass. It seems that the boulders and the coarse sandstone were emplaced together, but how?

The original relationship between the boulders and the overlying bed has been obscured by compaction so that the possibility that the boulder bed was deposited *before* the sands overlying it is difficult to disprove. However, it is most unlikely that a flood capable of moving boulders should have carried so little sand or gravel. Consequently, I still favour the interpretation advanced previously (Barber *et al.* 1978, p. 32), that the boulders and the muddy sand were intruded into the sequence, probably as a hyperconcentrated, watery flood immediately preceding the main mudflow. The turbulence of the sheet flood lifted the metre-thick sandstone bed (and possibly others), allowing introduction of the sediment and boulders. The bed then closed, extruding most of the muddy sand but not the boulders. Hyperconcentrated floods have densities in the range 1300 to 1800 kg m^{-3} (Costa 1988), as compared with about 2000 kg m^{-3} for unconsolidated, damp sand. The sand would therefore have an effective density of only about 500 kg m^{-3}, easily lifted but still apt to fall back under gravity. By contrast a typical mudflow would have a density in the range 1800 to 2300 kg m−3 (Costa 1988), little different to that of the sand, making intrusion easy, but with little or no tendency for the strata once lifted to fall again.

The overfolded sandstone beds near the base of the Stac Fada Member mentioned in (1) above have axes oriented at about 150° from grid north, with monoclinic symmetry indicating movement to the SW or west (Fig. 61). The upturned sandstone bed in (5) above shows mass flow towards the NW. These directions are quite different to the easterly directed palaeocurrents shown by cross-bedding in the sandstones (Fig. 5). The deposition of the Stac Fada Member was therefore preceded by a tectonic upheaval of some kind.

The topmost 2.5 m of the Stac Fada Member at Stoer is fine-grained and contains abundant accretionary lapilli (Lawson 1972, Plate 6), best seen on the sand-scoured exposures occasionally exposed in the gully below the waterfall Steall a'Mhunain. They are flattened in the plane of the bedding, with an average intermediate diameter $Y = 4.5$ mm, and $X/Z = 1.7$. Many of the lapilli were hard enough to break on impact with the ground and yet survive resedimentation, either because they were cemented in the ash cloud by secondary minerals such as calcium sulphate (Gilbert & Lane 1994), or were frozen. There is no obvious erosion surface between the lapilli-bearing portion and the rest of the Stac Fada Member at Stoer, but the evidence from Enard Bay suggests a significant time break at this level.

The hypothesis that the Stac Fada Member is a single mass-flow deposit has been challenged by Young (2002). He suggests that it is made up of four separate mass-flow events with episodes of fluvial deposition between them. The main evidence is the supposed existence of an erosion surface within the Stac Fada Member (Young 2002, fig. 5d). This surface, on the profile Figure 60, descends from the sandstone snout at 36 m to reach the basal décollement at 34 m. However, it is only detectable by the presence of sandstone clasts in the muddy sandstone above it so that its existence is questionable. Young (2002) provides insufficient geochemical data to show that the compositions of the four mass-flow deposits are significantly different.

Poll a' Mhuilt Member (Bay of Stoer Formation)

The mainly shaly sediments overlying the Stac Fada Member are about 100 m thick and can also be mapped regionally. They were

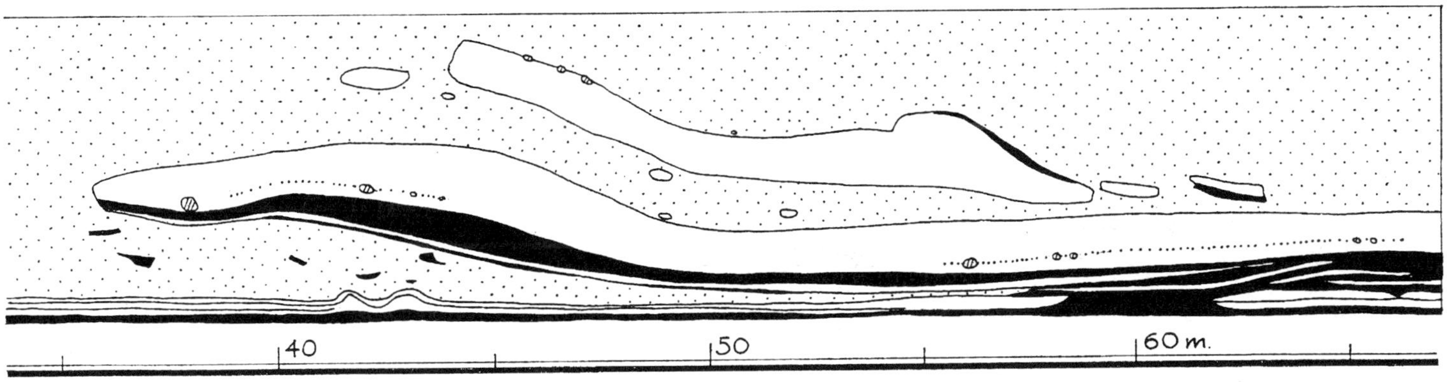

Fig. 61. Sand beds overfolded by mass movement of the Stac Fada Member at Stac Fada [NC 03322849]. The observer looks NNW, towards 330°. For location see Fig. 60.

called the Poll a' Mhuilt Member by Stewart (1991*a*), after the pronounced hollow they form a kilometre north of the coast section. Both the Stac Fada and Poll a' Mhuilt Members are contained in the Bay of Stoer Formation because of their lithological similarity to the muddy sandstone (Ct7) and shale (Ct3), and also because they are arranged one above the other as in the lacustrine cycles lower in the Bay of Stoer Formation.

The stratigraphy of the Stac Fada Member (SFM) and Poll a' Mhuilt Member (PMM) is summarized in Figure 62. The latter can be divided into seven informal units, **A**–**H**.

Unit **A** consists of some 10 m of greyish red fine sandstones with a few thin silt beds. Symmetrical ripples and thin carbonate layers are present in the uppermost 3 m. The carbonates have been described by Upfold (1984) as tufted stromatolitic mats and compared with *Conophyton*. However, material collected by Dr J. Bertrand-Sarfati from the same locality discloses several features at variance with this interpretation (Fig. 63). First, unlike *Conophyton* the tufted structures lack internal lamination. Second, they do not occur at particular stratigraphic levels, as they would if they formed part of a mat, but scattered through the sediment. The tufts, which might be described better as nodules, show clear signs of replacing and including the surrounding graded silt laminae. The laminae are 1–3 mm thick, with abundant pseudomorphs of small gypsum crystals. As an alternative interpretation Dr Sarfati suggests that the tufts were originally anhydrite nodules that grew prior to sediment compaction.

Unit **B** contains two limestone beds separated by red and green siltstone, altogether about 2 m thick. The limestones appear in the field to be intraclastic but under the microscope the 'clasts' are separated by stylolitic seams. Downie (1962) described 'spore-like bodies' from unit **B** at Stoer.

Unit **C** is a dark-grey (N3–N4) clastic laminite which has yielded cryptarchs, described and figured by Cloud & Germs (1971). They noted the presence of *Protosphaeridium* Timofeev 1966, a sphaeromorph with diameters from 16 μm to 43 μm, and described the new species *Favososphaeridium variabilis*. The species consists of single vesicles 10–30 μm in diameter. Unit **C** is similar to many described from the Devonian sediments of the Orcadian basin and other ancient lacustrine deposits. The laminae are formed by alternations of silty and organic-rich sediment, averaging 0.4 mm in thickness, with $p > 100\,000$. They have a TOC content in the range 0.24–0.37 wt% (R. L. Upfold, pers. comm.). Thin layers of calcite are common, so too are pyrite grains. The boron content of illite is significantly greater than in the unit beneath, a trend that continues up the section (Stewart & Parker 1979). Rare calcite pseudomorphs less than a millimetre long suggest the former presence of gypsum.

Unit **D** is made of dark greenish-grey calcareous laminated siltstone, well exposed on the walls of the eastern sea cave. The laminae are millimetres thick in the lower part of the unit ($p > 100\,000$), but graded beds about a centimetre thick appear near the top. Organic matter is absent but pre-compactional pseudomorphs after gypsum are relatively common and were collected by members of the Geological Survey in 1887 and again while searching for fossils in 1899 (Geikie 1900). The gypsum has been replaced by calcite, quartz and barytes – the Ba content of shales in this unit may exceed 2000 ppm. At some horizons the needle-shaped crystals have well-defined prism faces and reach several centimetres in length. They tend to lie parallel to bedding and are evidently pre-compactional.

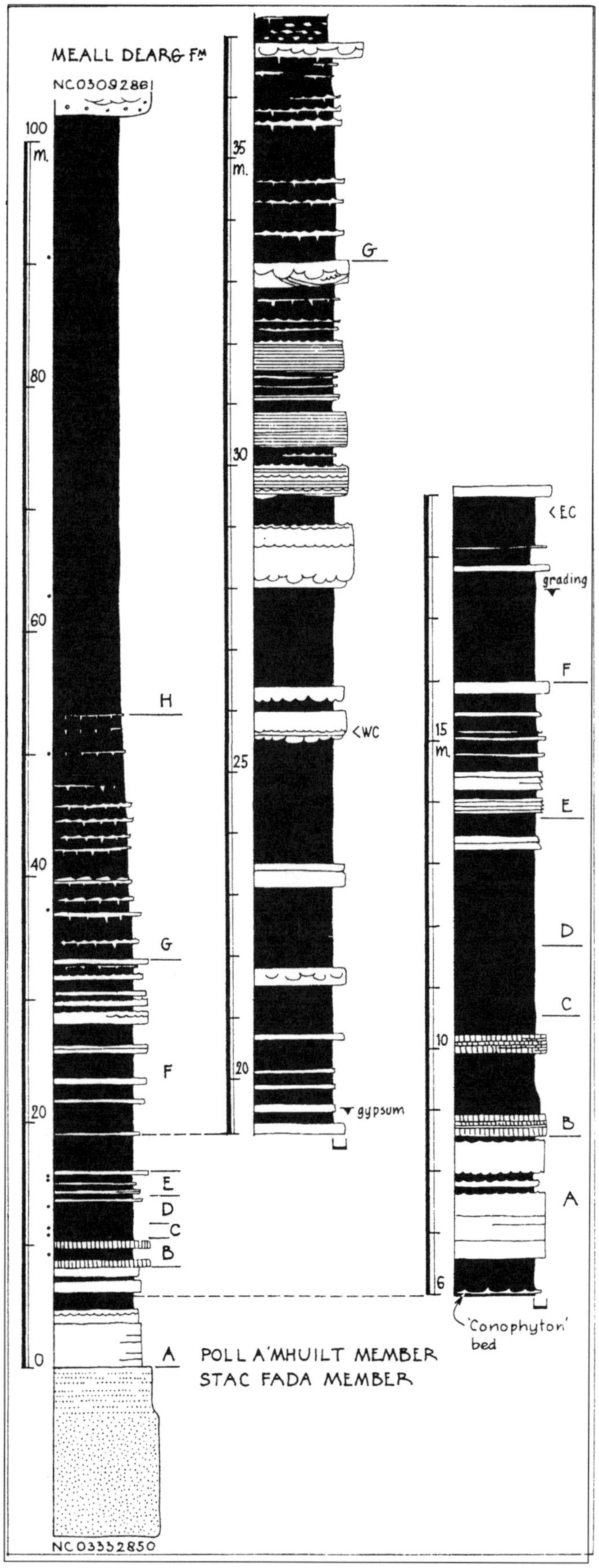

Fig. 62. Graphic log of the Stac Fada and Poll a'Mhuilt Members on the north side of Bay of Stoer. The left hand column shows the whole sequence, with boron sampling horizons shown by dots. The remaining columns show the Poll a'Mhuilt Member in more detail, from 6 m (the '*Conophyton*' bed) to 37 m above the base. The highest levels at which graded beds and gypsum pseudomorphs have been detected are indicated, together with the horizons forming the roof of the eastern sea cave (EC) and the western sea cave (WC). The grain size scale spans 4–3 ϕ units (0.06–0.1 mm).

Upfold (pers. comm.) has identified two distinct processes of gypsum replacement. In the first the gypsum was converted to anhydrite after burial and the anhydrite then replaced by lath-shaped, radiating crystals of calcite and quartz. The second process involved the replacement of gypsum by calcite, chert and chalcedony. Pseudomorphs in the same lamina may show either of the two replacement sequences.

Unit **E** consists of interbedded red shale and centimetre-thick fine sandstone beds. Most of the sandstones are graded, though this is only apparent on sand-polished surfaces in the eastern cave. Their soles are covered by millimetre-sized gypsum pseudomorphs.

Unit **F** is composed of interbedded red shale and sandstone. The sandstones are somewhat coarser than in the unit beneath, up to half a metre thick and never graded. Flat bedding in some beds passes upwards into ripple-drift lamination. Oscillation ripples are also present. Palaeocurrent directions from current lineation and ripple-drift were towards the east. Graded beds and gypsum crystals are detectable only in the lowest 3 m of this unit.

Unit **G** is predominantly made of ripple laminated and desiccated red shale, much of it massive. Samples for isotopic dating have been taken from a levels 2.3 m and 3.65 m below the top of this unit by Moorbath (1969: site LMS 26), O'Nions *et al.* (1983: samples T1 & T3) and Turnbull *et al.* 1996 (site PM).

Unit **H** which is 49 m thick, appears completely massive in all outcrops, whether or not scoured by the sea. However, polished slabs show that it is made up of laminated siltstone so intensely desiccated and reworked by water (and possibly wind) that it has completely lost its stratification. It is a fine-grained analogue of the muddy sandstone facies Ct7.

Three kilometres north of the type section, near Cnoc Breac [NC 036317], progressive excavation of a large gravel pit over the past 30 years has brought to light the Stac Fada Member and units **A**, **C** and **D** of the Poll a' Mhuilt Member. Bright green and indigo blue copper carbonates coat fractures. Only half a kilometre farther north, the Stac Fada and Poll a' Mhuilt Members are cut out by the angular unconformity at the base of the Torridon Group.

The Poll a' Mhuilt Member has been interpreted as a lake sequence, partly on the basis of the boron content of illite (Stewart & Parker 1979). Unit **A** is envisaged as having been deposited on a sabkha surrounding the lake while unit **B** formed on the lake shore. The lake deepening to a maximum in units **C** and **D**, started to fill and get shallower in units **E** and **F** and was ephemeral in units **G** and **H**. The boron content nearly doubles between the limestone of unit **B** and the gypsiferous siltstones of unit **D**. According to Stewart & Parker (1979) the lake initially had an extensive shallow part in which soluble salts were concentrated by evaporation. This water was periodically discharged into the deeper lake basin, depositing the thin graded siltstones with highly saline pore water from which gypsum crystallized (unit **D**). The reason for the disappearance of the lake is unclear. Lack of rain is an obvious cause, but another is shrinkage of the drainage basin, perhaps due to faulting, or tectonic warping.

The Meall Dearg Formation

The massive siltstones of the Poll a' Mhuilt Member, unit **H** are erosively overlain by red sandstones on the cliff a kilometre west of Stoer church [NC 03112862]. These sandstones belong to the Meall Dearg Formation (MD), named after the hill that they form near the type section along the coast.

The dominant lithology of the formation is planar cross-bedded medium-grained sandstone in sets up to a metre thick (Fig. 64). The sandstone is greyish-red on fresh surfaces, but pale reddish-brown when weathered. It can be called the planar cross-bedded facies MD1. It contains intercalations of thinly banded medium-grained sandstone with SE-trending wave ripples and desiccated shale drapes. Ripple-marked surfaces thousands of square metres in extent are commonly exposed. These intercalations constitute the parallel laminated facies, MD2. In contrast to the underlying Bay

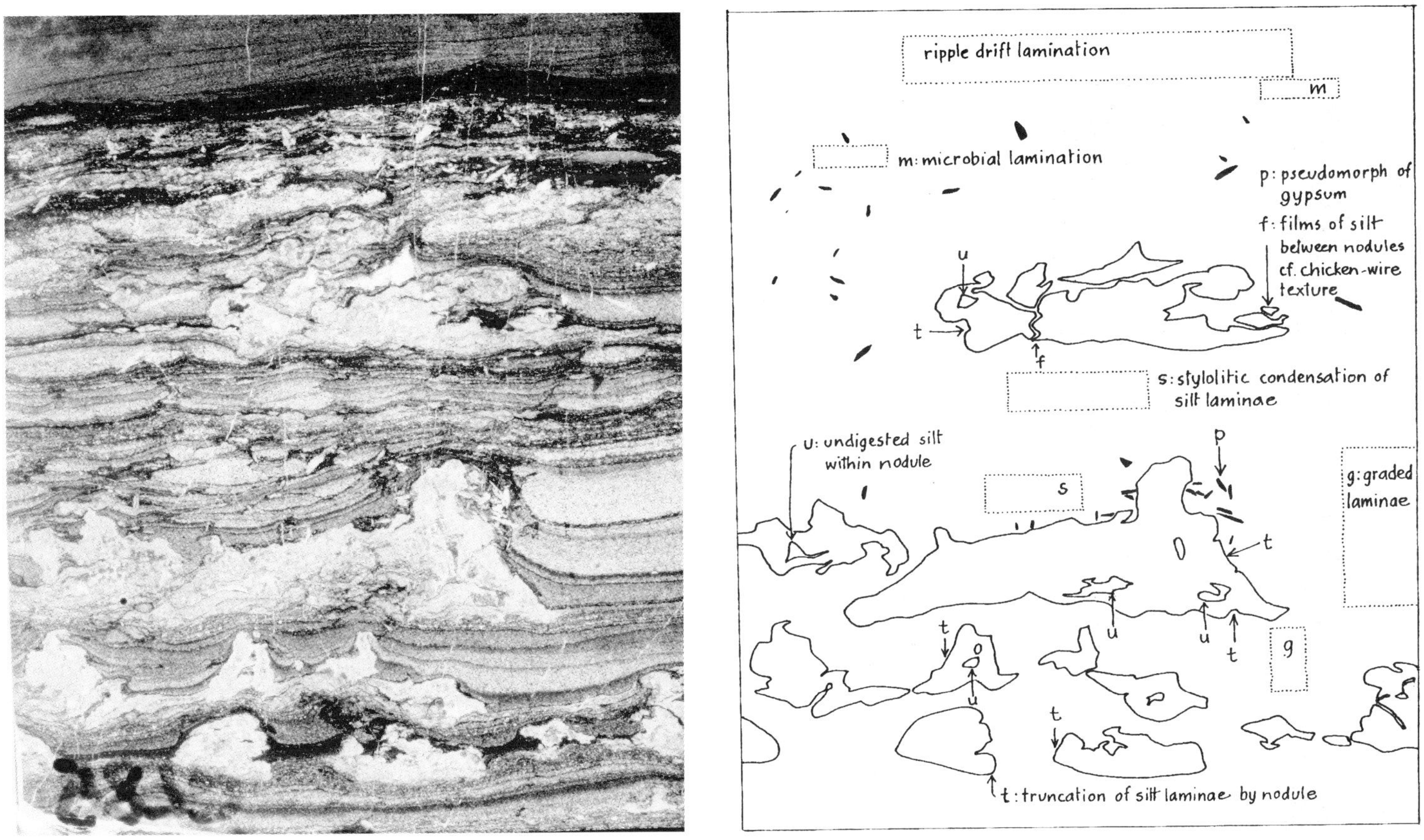

Fig. 63. Carbonate nodules in the Poll a'Mhuilt Member, unit **A**. This is the 'calcite bed' described and figured by Upfold (1984). The annotated diagram shows calcite nodules outlined and gypsum pseudomorphs in solid black. Most of the observations in the diagram are by Dr Janine Betrand-Sarfati (pers. comm.).

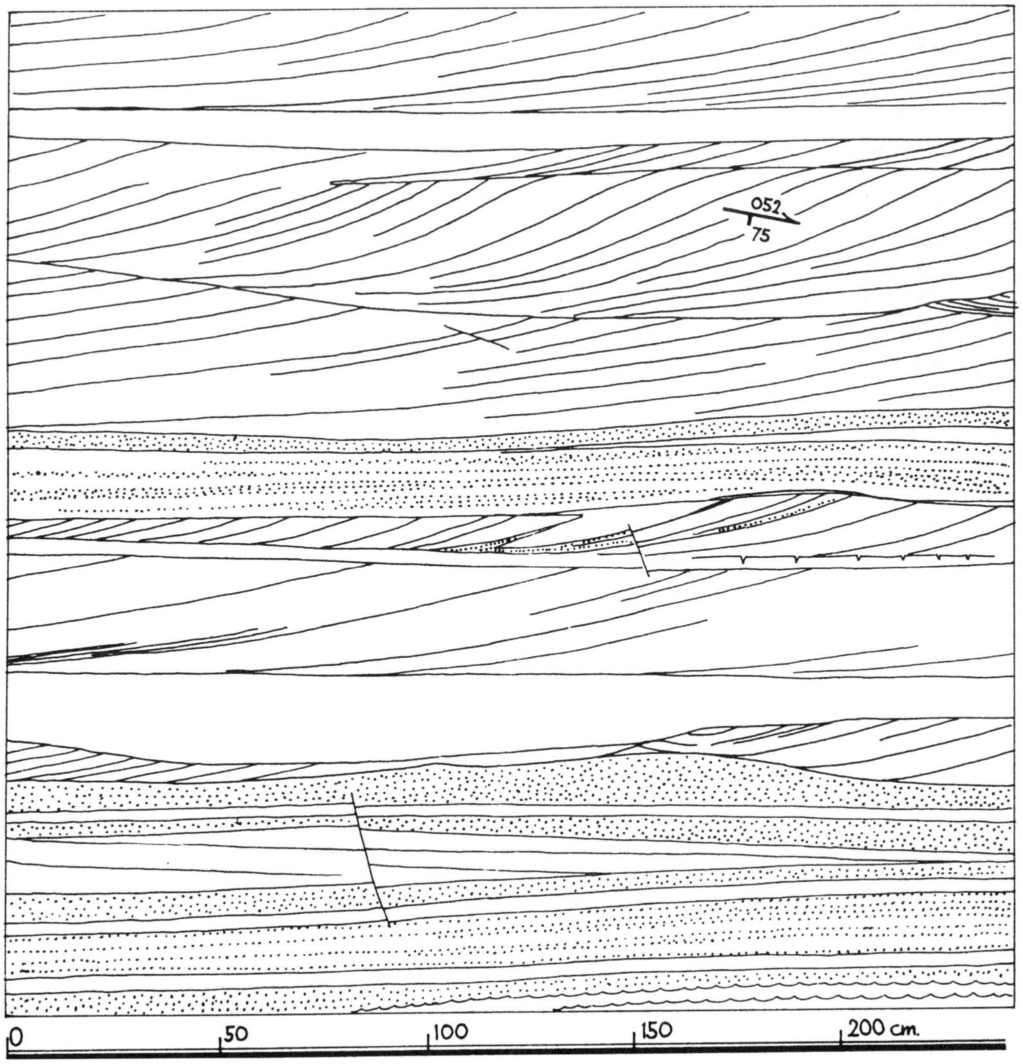

Fig. 64. Cross-bedding in the Meall Dearg Formation on the type section [NC 02982855]. The stippled bands are relatively clay rich and consequently dark red. The rest of the rock is greyish red when fresh, or pale to moderate reddish brown when weathered. Note the soft sediment faults, desiccation cracks and wave rippled surfaces. The beds in the upper part of the scaled drawing traced 7 m to the north are cut by a structureless slump sheet containing disoriented blocks of partially consolidated sand.

of Stoer Formation, contorted bedding, heavy mineral bands and pebbles are lacking from facies MD1 and MD2.

The lowest 7 m of the Meall Dearg Formation on the type section are anomalous in being trough cross-bedded and pebbly. The pebbles resemble those in the Bay of Stoer Formation. Material from the Stac Fada Member is not present. This pebbly interval (facies BS1) has been excluded from the Bay of Stoer Formation merely for ease of mapping. It can be traced northwards for 4 km until it is cut out by the unconformity at the base of the Torridon Group. The pebbly interval is sharply overlain by the planar cross-bedded lithology (facies MD1) more typical of the Meall Dearg Formation. Palaeocurrent directions in the pebbly interval and above are towards the west ($\theta = 295°$, $n = 53$), counterposed to those in the Poll a' Mhuilt Member and facies BS1 of the Bay of Stoer Formation (Fig. 5).

The composition of the pebble suite in the lowermost Meall Dearg shows that facies BS1 was being eroded to the east. The facies was evidently not completely blanketed by the Poll a' Mhuilt Member. The absence of detritus from the Stac Fada Member indicates that it, too, had only a limited extent to the east.

The Meall Dearg Formation is believed to have a fluviatile origin, like much of the Bay of Stoer Formation.

The Torridon Group

The Torridon Group unconformably overlies the Bay of Stoer Formation of the Stoer Group on the north coast of Stoer peninsula at Port Feadaig [NC 042329]. Both Groups have a westerly dip, the Torridon Group about 10° less than the Stoer. The lowest Torridon Group beds belong to the Diabaig Formation, erosively overlain by the Applecross. The Diabaig Formation and its contact with the Applecross are exposed only at low tide.

The Diabaig Formation occupies a palaeo-hollow in the unconformity at least 10 m deep and 100 m wide. It consists of desiccated and ripple-marked shales, mostly red, with occasional sandstone beds about 10 cm thick. Just above the unconformity the shale contains boulders of Stoer Group sandstone up to a metre across.

On the flank of the palaeo-hollow the Applecross Formation directly overlies the Stoer Group [NC 04243301]. This contact can be seen at any state of the tide. The lowest 2 m of the Applecross Formation here contain porphyry pebbles typical of the formation, along with rounded decimetre-sized blocks of sandstone from the underlying Bay of Stoer Formation. The next 18 m of strata consist of coarse to very coarse red sandstones with beds of durable pebbles up to 4 m thick, like facies FA2 at Cape Wrath. The pebble beds often have erosive bases. Maximum pebble size is about 3 cm at the base of the interval, but pebble size and pebble bed thickness decline upwards through the interval. Clasts of red shale or siltstone are common, often concentrated into bands up to a metre thick.

The top of the pebbly Applecross sandstones is seen on the south side of Bay of Culkein, about 300 m SE of the jetty [NC 03843300]. The pebbly sandstones are followed by about 50 m of coarse, cross-bedded red sandstones with rare pavements of durable pebbles, and red siltstone bands up to 20 cm thick. These sandstones resemble

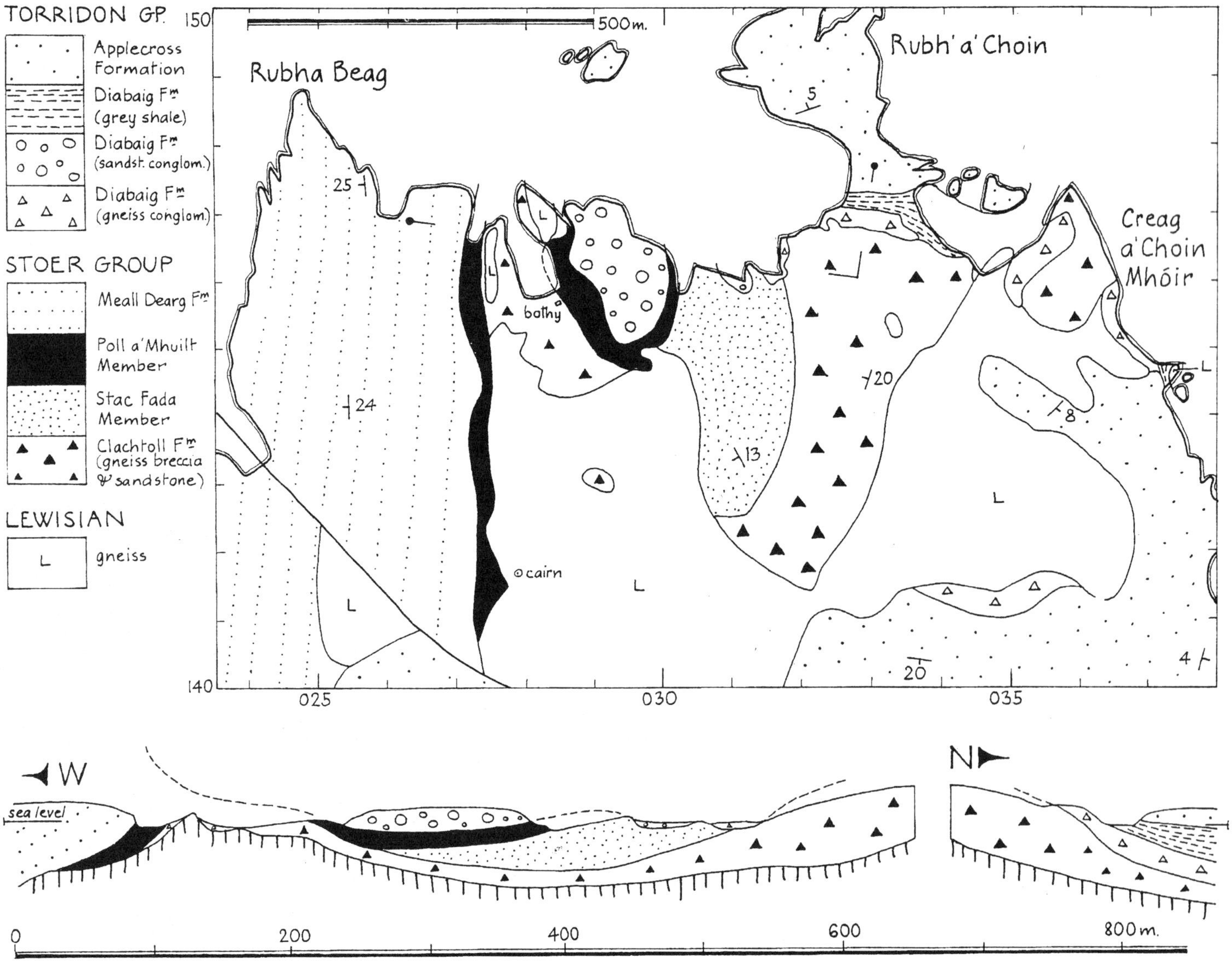

Fig. 65. Lithofacies map and true-scale section of an area on the south side of Enard Bay, based on 1:2500 mapping by the author. See Plate 2 for location.

facies FA4 at Cape Wrath. The Applecross sequence is truncated to the west by the Coigach fault. Palaeocurrents measured by Williams (1966*a*, p. 299) in the Applecross Formation on the south side of Bay of Culkein and in fault slices immediately east of the Coigach fault, flowed southwards ($\theta_r = 162°$, $n = 12$), which he interpreted to mean that these sediments belong to the Cape Wrath Member.

Enard Bay

Undifferentiated 'Torridonian' rocks on the south side of Enard Bay were mapped by Gunn for the Geological Survey in 1888. Their subdivision is due to A. J. Gracie who, when searching for a suitable doctoral research area near Achiltibuie in 1960, alighted by chance upon Enard Bay. The area was mapped during the years 1961-63 and the results published by Gracie & Stewart in 1967. A revised version of their map and section is shown in Figure 65. A geological map of the area by the author, scale 1:2500, has been deposited with the Geological Survey in Edinburgh. This map has been integrated into the 1:50 000 geological map of the Summer Isles, sheet 101W, published by the British Geological Survey in 1998. Most of the Enard Bay sub-area is a Geological Conservation Review site (Mendum *et al.* 2003).

The importance of the Enard Bay sub-area, which covers little more than a square kilometre, comes from the easily demonstrable superposition of typical Torridon Group sediments on the Stoer Group, the latter identified by the Stac Fada Member. The geology is extraordinarily complex due to the fact that both groups rest on landscape unconformities, and that the older unconformity is cut by the younger one within the map area. Each unconformity is associated with breccias, sandstones and shales. Some care is needed to distinguish the basal sediments of one group from those of the other.

Palaeomagnetic investigations by Stewart & Irving (1974) showed quite different palaeomagnetic directions in the sediments immediately above and below the unconformity separating the Stoer and Torridon Groups, similar to those found at Achiltibuie and elsewhere. Subsequent work on the Stac Fada Member and Rudha Beag sandstone at Enard Bay by Smith *et al.* (1983 supplementary publication, table 1) confirmed this conclusion.

The Stoer Group exposed in the map area is about 350 m thick, of which the lowest 150 m belong to the Clachtoll Formation. Small Lewisian inliers (hilltops) appear just south of the map area [NC 022134] near the top of the section, showing that basement relief in Stoer Group times was similar to the stratigraphic thickness, i.e. several hundred metres.

Hay *et al.* (1988, p. 827) record pumpellyite alteration and veining of the underlying gneiss at three localities: 500 m SE of Rubh' a' Choin [NC 03631466], about 150 m south of the ruined bothy [NC 028144] and again about 500 m SE of the bothy [NC 03291428]. At the first locality veins of red siltstone, pumpellyite and calcite penetrate down 1.5 m from the unconformity surface. The gneiss between the veins is reddened and shows signs of incipient weathering and corestone formation for as much as 1.5 m below the unconformity. Only 150 m distant [NC 03591472] gneiss blocks in facies Ct1 contain pumpellyite. Gracie (1964, p. 97–98) reported epidote (probably pumpellyite) covering joint planes, or disseminated in the matrix, at several localities in this facies [NC 03371454; NC 03431465; NC 03181467]. Hay *et al.* (1988) noted pumpellyite veins cutting the basal conglomerate of the Stoer Group at Rubh' a' Choin, and also 'detrital' pumpellyite with overgrowths in conglomerate clasts. However, it is not certain that any of the pumpellyite in the clasts predates deposition.

The massive breccia facies (Ct1) is well exposed about 450 m SE of Rubh' a' Choin (Fig. 66), on the headland that forms the northeastern extremity of the map area [NC 035147]. The facies is sharply overlain by bedded breccia with thin red sandstone interbeds (Ct2). The contact is well exposed on the headland (Fig. 65), a few metres below a prominent outlier of the Applecross Formation [NC 03591470]. At one point it is clearly erosional. Facies Ct2 contains 'epidotized' patches at NC 03431465 and NC 03181467, to the west. Facies Ct1 and Ct2 were called the massive breccia (I) and the Loch na Seana-chreig (IIa) facies, respectively, by Gracie & Stewart (1967).

About 100 m NW of the ruined bothy, on the western edge of the bay [NC 02781463], gently dipping brownish-grey interlaminated gravels and sandstones (Ct2) contain decimetre and metre-sized blocks of gneiss with drop-stone structure (Fig. 67). At first glance the sequence might be thought to be of glacial origin. However, the sandstones have symmetrical ripple marks and desiccation polygons, the latter best exposed in the SW corner of the bay. The lateral persistency of the sandstones is 30–100. There is no graded lamination or other evidence for a permanent lake. The gneiss blocks are lithologically like the gneisses seen underlying the sediments and also forming the ridge against which the sediments are unconformably truncated to the west. The blocks are only present in the sediment up to 10 m horizontally from this gneiss

Fig. 66. Massive breccio-conglomerate (facies Ct1) of the Stoer Group, unconformably overlain by the Applecross Formation on the headland about 400 m SE of Rubha a' Choin [NC 03591470]. The hammer head is on the unconformity; the hammer handle is 0.5 m long. Well-stratified sandstones (facies Ct2) in the top right of the photograph, 7.5 m to the right of the hammer, rest on an erosion surface in massive breccia (facies Ct1). The rocks in the foreground are breccias of the Diabaig Formation. The observer looks east.

Fig. 67. Gneiss blocks in gravelly red sandstone (facies Ct 2) at Enard Bay, showing drop-stone structure [NC 02781463]. The hammer handle is 0.5 m long. The observer looks westward.

ridge. The deformation of the bedding under and over the blocks is evidently due to differential compaction like that seen at the base of the Torridon Group at Rubha Dunan (Achiltibuie). Here, however, there is no steep cliff from which the blocks could have fallen, merely a ridge that could hardly have been more than 6 metres higher than the breccia at the time it was deposited.

Thinly bedded sandstones of facies Ct2, like those described above, probably underlie the unexposed low ground to the south, occupying a north–south trending palaeovalley which has been exhumed near the coast to form the bay with the bothy (cf. Gracie & Stewart 1967, fig. 3). There is a large outcrop of Ct2 about 250 m south of the bothy [NC 02901430].

The sandstones of facies Ct2 are separated from the gneiss by a fringe of massive breccia that can be traced round the headland to the west, into the next bay, where it underlies red siltstones of the Poll a' Mhuilt Member (see below). The breccia is tentatively regarded as belonging to facies Ct1 but is unusual in having a silty or fine sandy red matrix, rather than a coarse sandy one. A similar breccia immediately to the east [NC 028146] drapes the gneiss hill which has been interpreted as a Precambrian *roche moutonnée* by Davison & Hambrey (1996). However, the unconformity surface here, as elsewhere, is jagged rather than smooth (Stewart 1997, fig. 2).

The bedded breccia facies (Ct2) grades upwards into the Stac Fada Member, called the structureless sandstone facies (IIb) by Gracie & Stewart. The Member is 32 m thick, three times more than elsewhere. The lowest 20 m are massive and relatively rich in gneiss clasts, suggesting that the Member reworks facies Ct2. The uppermost 12 m by contrast, contain conspicuous accretionary lapilli. The base of the lapilli-bearing section is defined locally by a bed of red shale and fine sandstone, up to 0.8 m thick, showing that the Member was deposited during two distinct episodes with a substantial time lapse between. The topmost 4–5 m of the Member has been strongly reworked, with the production of thin graded beds.

The Stac Fada Member passes up into massive red siltstone with thin limestone bands that appear to be discontinuous owing to decalcification along joint planes [NC 03011458]. This limestone is probably correlative with unit **B** of the Poll a' Mhuilt Member at Stoer. Farther west, about 120 m north of the bothy [NC 02351468], the same limestone encrusts gneiss breccia and contains minute green vitreous tephra like those in the Stac Fada Member. The encrusting limestone resembles, at least superficially, the Keweenawan stromatolites described by Elmore (1983). Farther west still, in the floor of a small bay 175 m NW of the bothy [NC 02721467], about 25 m of the siltstone are exposed. The siltstone rests on massive gneiss breccia (Ct1) partly cemented by limestone, and is overlain by Rudha Beag sandstone forming a scarp. Note that the headland which gives its name to the sandstone is now called Rubha Beag.

The Rudha Beag sandstone, facies III of Gracie & Stewart (1967), might be expected to correlate lithostratigraphically with the Meall Dearg Formation at Stoer, but there are certain differences. The Rudha Beag sandstone consists of about 150 m of fine to medium grained reddish-grey (5R5/2) sandstone with tabular planar cross-bedding. Set thickness ranges from decimetres up to 5 m with $p \approx 100$. Foresets are a few millimetres thick with $p \approx 5000$. They are asymptotic to the basal erosion surface, which is generally planar but locally has decimetre relief. Convex upward and downward reactivation surfaces are common. Forsets dip 16°–20° from truncation surfaces and never more than 30°. They had a mean dip direction of 244° ($n = 17$) before tectonic tilting. There are no grains over a millimetre in size, nor are there contortions, ripples, silty interbeds or drapes in the Rudha Beag sandstone. However, there are interbeds of horizontally laminated sandstone up to 2 m thick.

These features are compatible with the sediments of transverse dunes of either aeolian or fluvial channel environments. However, convex upward aeolian reactivation surfaces comparable in size with those figured by McKee (1966, fig. 7) are not present.

In the Meall Dearg Formation at Stoer the tabular planar cross-bedded sets average only 30–40 cm in thickness. The beds were originally much thicker, but erosion has removed the upper parts. Reactivation surfaces, marked by changes of foreset inclination, are common. There are frequent intercalations of wave-rippled sandstone between the cross-beds. The general dip direction of the foresets at Stoer is westerly ($\theta = 295°$, $n = 53$), rather different to that in the Rudha Beag sandstone. The Meall Dearg Formation is clearly fluviatile in origin and it seems likely that the differences between the two rock units simply arise from a higher sedimentation rate at Enard Bay. This might explain the preservation of thicker sets, with consequently longer and steeper foresets than at Stoer. In short, the two rock units probably belong to the same formation, but for security the informal name Rudha Beag sandstone is retained.

The coarse red sandstones exposed between Dubh Lochan and the sea [NC 02221314], which have scour hollows with overhanging sides (Gracie & Stewart 1967), small convex upward reactivation surfaces, and rounded pebbles of gneiss and pink quartzite, are definitely fluviatile but apparently overlie the Meall Dearg

Formation and may belong to the Bay of Stoer. This is locality 1 of Hambrey *et al.* (1991, p. 104).

The 'Enard Bay facies' of Gracie & Stewart (1967) closely resembles the Diabaig Formation in its type area, and is likewise overlain by pebbly sandstones typical of the Applecross Formation. That part of the basal breccia (Db1a) which is composed exclusively of Rudha Beag sandstone blocks is unusual, however, in presenting the typical features of a rotational slide, or slump. The eastern margin of the deposit can be fixed accurately about 300 m from the present Rudha Beag sandstone scarp. Farther east the basal breccia is derived from underlying rock types, for example the Stac Fada Member. The sandstone breccia must originally have extended westwards to the foot of the scarp, but has been eroded from the intervening area due to differential compaction over a basement ridge. The original slip, using the data of Anderson & Dunham (1966, p. 186–194), was about 100 m high, the upper 75 m consisting of Rudha Beag sandstone. The sandstone moved on the underlying, water saturated siltstone which is about 25 m thick near the existing Rudha Beag sandstone scarp. By comparison with recent landslips it may be deduced that when movement occurred the Rudha Beag sandstone and the partly compacted siltstone beneath it were porous, rain-soaked and consequently weak. The collapse of the sandstone scarp may also account for the shearing and disruption in the uppermost metre of the underlying siltstone, for example 120 m north of the bothy [NC 02861463]. A landslide of almost identical size and nature to that envisaged above is figured by Heim (1919, vol. 1, p. 421) from the Pleistocene of Switzerland.

The clasts in the basal breccia of the Torridon Group (Db1a) accurately reflect the lithologies once exposed at the unconformity. The boundaries of the Stac Fada Member, for example, can be mapped to within 10 m from the clasts in the overlying breccia. The maximum distance moved by clasts at Enard Bay is about 100 m. On the coast, 600 m SE of Rubh' a' Choin [NC 03651460] the gneiss and sandstone clasts in facies Db1a, together with the matrix, all contain a green mineral, probably pumpellyite.

A few metres of mainly grey shales and tabular sandstones belonging to the Diabaig Formation (Db2) are developed south and SE of Rubh' a' Choin [NC 034146]. The topmost metre is red and contains a thin bed packed with centimetre-sized clasts of gneiss, fine red sandstone (probably from the Stoer Group) and orthoquartzite. Shaly beds beneath the Applecross Formation on the east coast of the map area [NC 037145] are all red. Re-examination of the channels reported from these beds by Gracie & Stewart [NC 03690451] suggests that they were formed by penecontemporaneous extension and attenuation of the beds rather than by erosion. The 'channels' reach 80 cm depth and are filled by fine-grained, laminated sandstones and shales essentially concordant with the 'channel' sides and floor. They recall the smaller monoclinal structures (type I) in facies Db2 at Achiltibuie (*q.v.*).

The Rubha Dubh Ard Member of the Applecross Formation is not definitely identifiable at Rubh' a' Choin. The lowest 25 m of the Applecross Formation though relatively fine grained do not grade up into the fine brown and grey tabular beds typically found at the top of the member. The overlying pebbly Applecross Formation, however, is quite similar to that normally found above the Rubha Dubh Ard Member. Detrital zircons from these beds have been dated by Rainbird *et al.* (2001).

East of the Garvie River, the Applecross Formation exposed in low cliffs is different to that in the Enard Bay sub-area. Stratigraphic sections are limited by faulting to about 10 m thickness but all of them tend to fine upwards. The coarsest beds, exposed on Rubha Lag na Saille [NC 049136], are thickly bedded, contorted red sandstones with pebble beds up to 20 cm thick containing durable pebbles up to 1.5 cm. Higher beds are trough cross-bedded in decimetre-thick cosets. The highest beds, well exposed around Garvie Point, are tabular beds with ripple marks, occasional desiccation polygons and frequent grey shaly partings. All these sediments could plausibly belong to the Rubha Dubh Ard Member. Differential compaction over the palaeotopographic high at Rubh' a' Choin during early Applecross time may have attenuated the Rubha Dubh Aird Member there with respect to the area east of the Garvie River. Upward movement of the west side of the Garvie River fault in early Applecross time could have had a similar effect.

Loch Veyatie to Canisp and Suilven

The area is shown on the one inch to the mile geological map of Assynt published by the British Geological Survey in 1923 but the stratigraphy is out of date.

Much of the Torridon Group has been removed by erosion in this sub-area to reveal the palaeotopography of the underlying basement. Three palaeovalleys are evident, all trending NW parallel to foliation in the gneiss, and today occupied by Loch Veyatie, Cam Loch and Lochan Fada. The last of these palaeovalleys is about 90 m deep, allowing for post-Cambrian fault movement (Peach *et al.* 1907, p. 306). The palaeovalleys were completely filled by the Diabaig Formation. Only at Canisp and Suilven is the Applecross Formation in direct contact with the basement (Stewart 1972, figs 7 & 8).

The Diabaig Formation is well exposed between Lochs Veyatie and Cam. About 450 m north of Creagan Mòr [NC 191144] the gneiss is overlain by 14 m of conglomerate composed of well-rounded gneiss cobbles up to 15 cm diameter. The conglomerate is interstratified with sandstone. It is overlain, apparently conformably, by tabular red sandstones (facies Db1b). About 15 m of these tabular sandstones are exposed a short distance to the SE [NC 189142]. About half of them are very coarse, with angular gneiss pebbles up to a centimetre, and the remainder trough cross-bedded in sets less than a decimetre thick. A further 20 m of these sandstones are exposed to the SE, overlain near the summit of Creagan Mòr [NC 192140] by 16 m of grey shales and grey, pink weathering sandstones. These grey beds are at the top of the Diabaig Formation, for typical Applecross sandstones follow above. Gneiss cobble conglomerate also outcrops on the north shore of the Cam Loch [NC 203145] about 90 m below the Applecross Formation at the top of Creagan Mòr. Grey sandstones and micaceous siltstones appear in the topmost 30 m of the Diabaig Formation about a kilometre north of Cam Loch [NC 203152; NC 197154; NC 191158], and again about 1.5 km north of the Loch [NC 199160; NC 204160]. No grey beds are known to occur in the formation farther north.

Gneiss conglomerate with cobbles up to 10 cm in diameter occurs at the bottom of the palaeovalley centred on Lochan Fada, near the south end of the loch [NC 207162]. The conglomerate also contains pebbles of banded quartzite and fine-grained, dark red sandstone. Similar fine-grained red and grey sandstone pebbles have been noted in conglomerate 0.5 km south of Lochan Fada [NC 196165]. They may have come from the Stoer Group 22 km to the west.

One kilometre NE of Lochan Fada [NC 216170] conglomerate is exposed a few metres below the Applecross Formation, a level that is topographically 130 m above the conglomerate locality near the south end of the loch. The conglomerate facies is evidently present at numerous stratigraphic levels within the Diabaig Formation.

Palaeocurrents in sandstones interbedded with conglomerate at Lochan Fada [NC 207162] came from the NW. To the west, at Suilven [NC 147178], palaeocurrents in tabular, pebbly Diabaig sandstones interbedded with gneiss cobble conglomerates have a similar direction. The Diabaig skirting Suilven is well exposed but cobble conglomerate is uncommon and grey shale absent. The overlying Applecross Formation, of which about 380 m are exposed, fines upwards from coarse sandstone with thick pebble beds at the base, to coarse sandstone with few pebbles at the top.

The gneiss cobble conglomerate is very similar to that in the Stoer Group (facies Ct4) which also contains rare fine-grained red sandstone pebbles (though no quartzites). If the similarity is real then all the rocks here described as Diabaig, with the exception of the grey beds immediately beneath the Applecross Formation, belong to the Stoer Group. The hypothesis is thought unlikely because of the apparently conformable succession from conglomerate up into the Applecross Formation, and the fact that the palaeocurrent direction is counterposed to that in the conglomeratic sequence at Stoer. However, only palaeomagnetism can provide a decisive test.

Inverpolly Forest

The area considered extends from Loch Bad a' Ghaill in the west to Loch Veyatie in the east. The western half of the area is included in 1:50 000 sheet 101W, published by the British Geological Survey in 1998. The eastern part is covered by the one-inch to the mile Assynt special sheet published by the Geological Survey in 1923. The central part is shown in Figure 68. The area embraces the lower part of the Torridon Group and its unconformable contact with the Lewisian basement. The Diabaig Formation is formed of breccias, conglomerates and red sandstones. Insignificant thicknesses of grey shale are present near Loch Veyatie. The lowest part of the Applecross Formation is formed by the Rubha Dubh Ard Member, here significantly thicker than in the type section near Achiltibuie.

The top of a Proterozoic gneiss hill, covered by a selvage of Diabaig breccia and sandstone, is well exposed immediately SW of Loch An Doire Dhuibh [NC 132102]. The unconformity surface has been dissected by joint-controlled differential erosion, leaving high areas 10–20 cm wide separated by sediment filled channels about 5 cm deep. Residual weathered gneiss has not been seen.

Massive gneiss breccia forms mappable outliers between Loch Sionascaig and Loch Veyatie. Its stratigraphic thickness is probably small. The breccia has resisted erosion more effectively than the overlying sandstone, perhaps due to unusual cementation (cf. Peach *et al.* 1907, p. 307). At one point, about a kilometre NE of Lochan Dearg [NC 169139], the gneiss blocks (up to 0.5 m in size) are draped by interbedded red sandstone, as in the basal Diabaig Formation at Achiltibuie and the basal Stoer Group at Enard Bay. The lack of a clearly exposed upward passage from massive gneiss breccia into the Applecross Formation in this area means that the massive breccias can only be provisionally assigned to facies Db1a of the Diabaig Formation.

Conglomerates made up of well-rounded gneiss cobbles, with a few red sandstone interbeds, occupy palaeotopographic lows in the gneiss at Loch Gainmheich [NC 135114] and on the east side of Loch Sionascaig [NC 126130; NC 127130; NC 130131]. The overlying beds are not exposed but the dip of the conglomerates is like that of the Diabaig Formation. They are therefore tentatively assigned to facies Db1a. This unusual facies is also extensively developed between Loch Veyatie and Canisp.

The ice-polished outcrops east of Loch Sionascaig [NC 12801307] show about 5 m of conglomerate overlying fresh acid gneiss. About 70% of the clasts, however, are made of fine-grained dolerite, perhaps derived from a Scourie dyke in the basement a short distance to the north. The basic clasts are generally 10–20 cm in diameter, often with altered rims. The largest clasts, of quartz and acid gneiss, reach 30 cm diameter. Pebbles of pink sandstone, like that forming the Stoer Group, are quite common. But white and grey quartzite pebbles, that might have been expected from erosion of the Bay of Stoer Formation, are rare. A block of acid gneiss measuring over a metre in size, seen near the base of the conglomerate, is particularly interesting in showing evidence of weathering, possibly Precambrian. Intersecting, joint controlled zones of reddening and alteration traverse the block, leaving fresh oval corestones in between. The corestones are about the same size as the pebbles in the surrounding conglomerate. Pebble imbrication

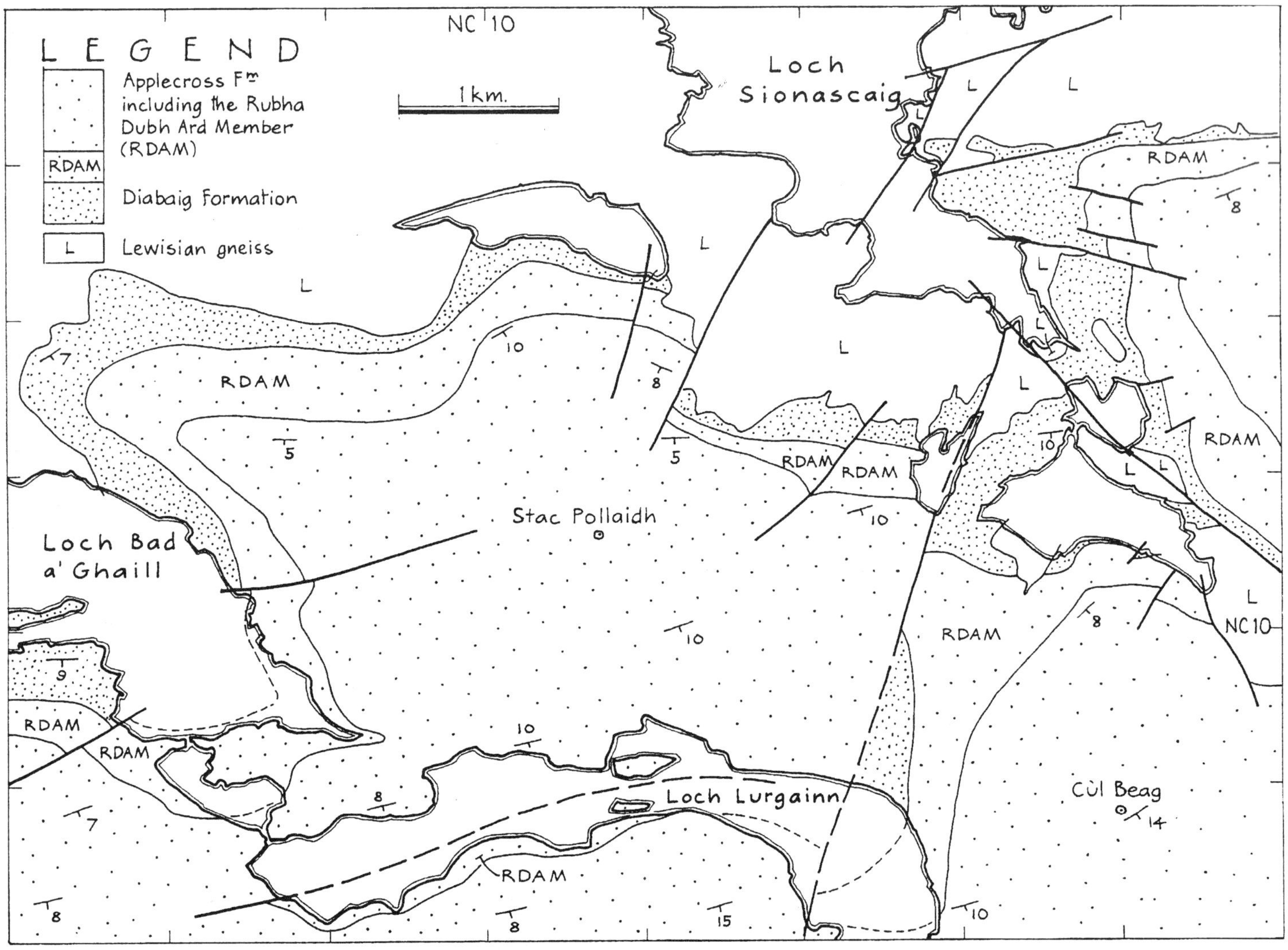

Fig. 68. Geological map of part of Inverpolly Forest, based on 1:25 000 mapping by the author. See Plate 2 for location.

at this locality suggests palaeocurrents flowed from the south, unlike those in the conglomerates north of Loch Veyatie which were counterposed.

The Diabaig Formation is about 160 m thick immediately west of Loch Bad a' Ghaill. The lowest 30 m are trough cross-bedded sandstones in sets up to 10 cm thick [NC 054103; NC 056104; NC 061102]. In addition to gneiss clasts these sandstones also contain well-rounded pebbles of red sandstone and red quartzite up to about 10 cm in diameter. The sandstone and quartzite pebbles may have come from the Stoer Group about 4 km to the west. Similar trough cross-bedded sandstones are found in the Diabaig on the eastern shore of the loch, immediately beneath the Applecross Formation [NC 081105; NC 084102] and also west of Loch An Doire Dhuibh [NC 131109]. Trough axes show that palaeocurrents flowed towards the NE and east.

About 16 m of grey sandstones and siltstones with ripple lamination are exposed near the southern margin of the wood flanking Loch Veyatie [NC 198124]. Similar exposures are seen in the woods 2 km to the NW [NC 181130]. Both sequences are overlain by the Rubha Dubh Ard Member.

Easily accessible exposures of the Rubha Dubh Ard Member (Applecross Formation) can be found immediately north of Loch Bad a' Ghaill. The member here is about 100 m thick, much more than at Achiltibuie. The base is seen by the roadside about 900 m SE of the islands [NC 08541025]. The coarsest sandstones at this locality contain grains up to a centimetre, including porphyry and fine-grained quartzose sandstone. A 4 m exposure gap separates the member from the underlying Diabaig Formation a short distance to the NW [NC 08511029]. Trough cross-bedding predominates in the lower part of the member; however, contorted bedding is uncommon. Towards the top of the member beds of ripple-drift laminated red sandstone become commoner, and some of the highest include grey (N3) siltstone. Exposures of the member in bluffs to the east and north consist of sandstones that weather light brown (5YR5/6) or reddish orange (10R6/5), noticeably paler than the greyish-red weathering sandstones in the overlying Achduart Member.

There are good exposures of the Rubha Dubh Ard member in the Doire Dhuibh, about 2 km NE of Linneraineach. The base of the member rests on sandstones of the Diabaig Formation near the shore of Loch An Doire Dhuibh, about 100 m east of the small island [NC 13801052]. The top of the member is exposed at the foot of bluffs that rise above the woods [NC 14101027]. The member is about 100 m thick in the Doire Dhuibh. The topmost 8 m contains grey rippled silty beds [NC 139103]. Similarly thick sections of grey sandstone and shale are also seen south of Loch Doire na h-Airbhe [NC 100120] and on the south shore of Loch Lurgainn [NC 087082]. Grey shales, probably at the same horizon, were once exposed by the roadside near Linneraineach (NC 127088; Peach *et al.* 1907, p. 308).

From Loch An Doire Dhuibh the Rubha Dubh Ard Member can be easily traced round the northern side of Cùl Mòr to a point on the southern side of Loch Veyatie about 2 km west of Elphin [NC 198124]. This is about 18 km from the type locality.

The beds overlying the Rubha Dubh Ard Member at Loch Bad a' Ghaill and Doire Dhuibh are coarse (up to 0.5 cm) but become pebbly only about 10 m above the contact. Pebbles reach 2 cm. They are scattered throughout the rock and also concentrated in seams up to 10 cm thick. These coarse sandstones presumably belong to the Achduart Member, the top of which has not been mapped.

The palaeocurrent vector mean in the Rubha Dubh Ard Member at Loch Bad a' Ghaill is directed southeastwards ($\theta = 132^\circ$, $n = 11$) whereas in the overlying Applecross Formation at Loch Lurgainn it is easterly ($\theta = 085^\circ$, $n = 26$). The latter vector mean almost certainly averages data from the Achduart Member and higher beds. These palaeocurrent data are from Williams (1969a, fig. 12).

The Rubha Dubh Ard Member has been tentatively correlated with the Cape Wrath Member by Williams (1969*a*, fig. 7) but the geochemical evidence suggests otherwise. The Applecross sandstones of Inverpolly Forest (part of Coigach) have Rb partitioned between K-feldspar and plagioclase in a completely different way to the Cape Wrath sequence (Stewart & Donnellan 1992, fig. 6), and the same is true for Y in apatite and zircon (see also p. 55).

Isle Ristol to Badentarbat and the Summer Isles

The coast section from Isle Ristol to Badentarbat jetty exposes the top of the Applecross Formation and 1200 m of the overlying Aultbea Formation (Fig. 69). A geological map of the area, revised by the author, is included in 1: 50 000 sheet 101W (Summer Isles) published by the British Geological Survey in 1998. Contorted bedding in the Applecross Formation at Reiff, about 3 km north of Isle Ristol [NB 96281497] has been described in detail by Owen (1996*a*).

The Applecross Formation exposed in the this area is only about 250 m thick due to a roll-over anticline related to the Coigach fault (Stewart 1993*a*, fig. 4). The axis of the anticline passes just west of Eilean Mullagrach. The contact between the Applecross and Aultbea Formations immediately east of Isle Ristol is gradational over about 40 m as it is at Aultbea. The Applecross lithology is defined as sandstone with a maximum grain size over 0.5 mm and scattered pebbles up to a centimetre in size. The Aultbea lithology is defined as having a maximum grain size less than 0.5 mm and generally no pebbles. The contact, defined at the level at which the two lithologies are of equal importance, is exposed on the coast about 750 m west of Dornie [NB 98091037]. The reduction in the grain size of the sandstone in the boundary zone is accompanied by a rapid increase in the total feldspar content. Most of the increase is represented by Rb-bearing plagioclase, indicating a change in source area (Stewart & Donnellan 1992). Some of the highest Aultbea beds on the section contain ferruginous spots, with white rims, up to 1.5 cm in diameter.

A horizon with ripple-bedded fine sandstone and red fissile siltstone about 150 m above the Applecross–Aultbea boundary is exposed on the coast SW of Dornie (Fig. 70). The top of the underlying sandstone contains fragments of desiccated grey shale some of which is phosphatic. Chemical analyses of the shales and sandstones at this locality have been published by Stewart (1995*b*, table 4).

A horizon with grey fine sandstones and siltstones also crops out on the east side of the island Tanera Beg [NB 970072] where it has yielded organic walled microfossils (Zhang Zhongying *et al.*

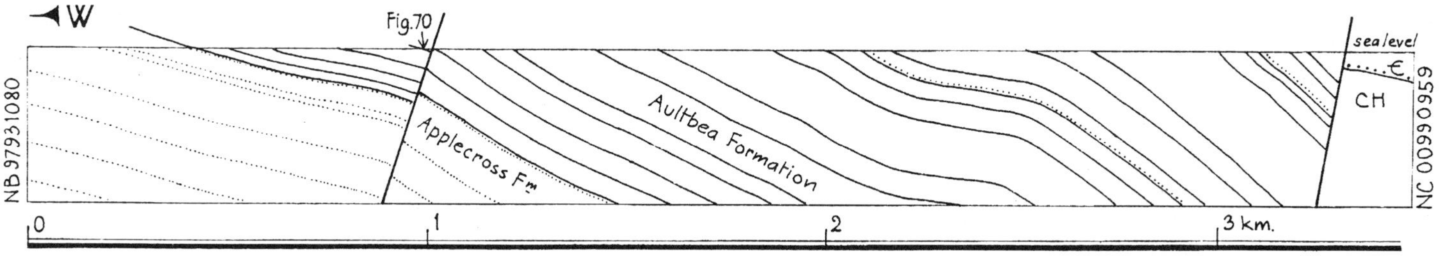

Fig. 69. True-scale cross section of the Aultbea Formation on the coast between Isle Ristol in the west and Badentarbat jetty in the east. The Applecross Formation, and two pebbly horizons in the Aultbea, are stippled. The Cambrian (C) probably overlies the Cailleach Head Formation (CH) at the eastern edge of the section.

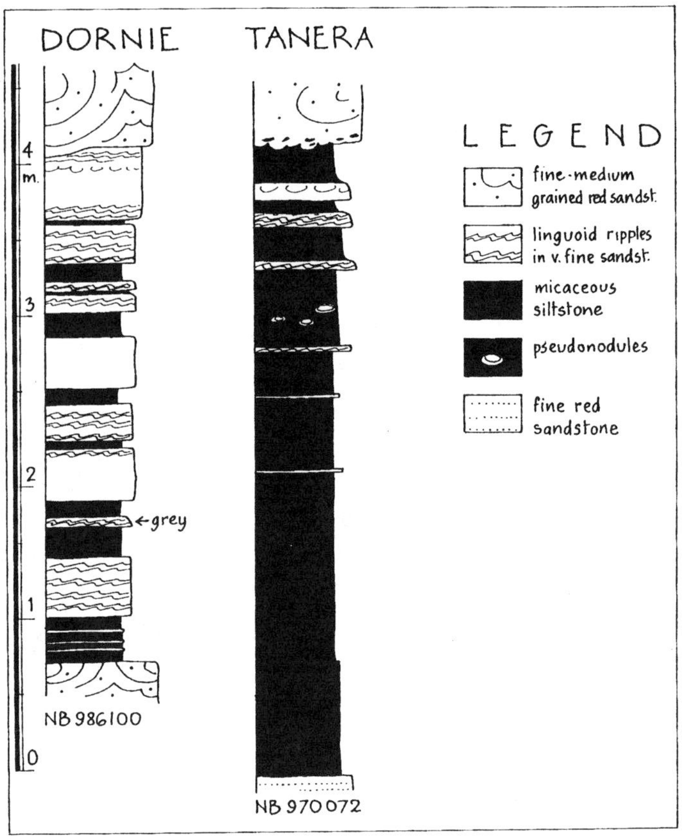

Fig. 70. Stratigraphic logs of fine-grained units in the Aultbea Formation near Dornie and on the island of Tanera Beg. The strata at Dornie are all red except for the graded bed indicated. The micaceous siltstones and very fine sandstones on Tanera are all grey.

1981; Zhang Zhongying 1982). A stratigraphic log is shown in Figure 70. It is uncertain if this is the same as the fine-grained horizon cropping out near Dornie, for in Tanera the sandstones overlying the siltstone unit are sparsely pebbly whereas at Dornie they are not.

Grey shales were also recorded by the Geological Survey (Peach *et al.* 1907, p. 320–321) at the NE end of Eilean Dubh [NB 976039] and the west side of Glas-leac Beag [NB 925051].

Achiltibuie

The area was mapped as 'Torridonian' by the Geological Survey in 1888. A 1: 10 000 geological map of this area by the author has been deposited with the Geological Survey in Edinburgh. The essential details appear on the 1: 50 000 sheet 101W (Summer Isles) published by the British Geological Survey in 1998. The coast sections at Rubha Dunan and near Achduart are Geological Conservation Review sites (Mendum *et al.* 2003).

The sandstone boulder conglomerate at the base of the Torridon Group, where it unconformably overlies the Stoer Group at Rubha Dunan, was originally recorded as Triassic (Peach *et al.* 1907, p. 182), but Lawson (1965) pointed out that several such 'Triassic' conglomerates are, in fact, Torridonian. Palaeomagnetic measurements by Irving (Irving & Runcorn 1957) showed a shift in direction of magnetization between the red sandstones (Stoer Group) forming the headland and the sandstone boulder conglomerate about 200 m to the SE. This led me to suggest that the intra-Torridonian conglomerates detected by Lawson overlay an angular unconformity corresponding at Rubha Dunan to the magnetic break (Stewart 1966*b*). Later palaeomagnetic and stratigraphic studies have confirmed this hypothesis (see below).

The Stoer Group at Achiltibuie is identified by the presence of the Stac Fada Member, and by the distinctive muddy sandstone facies (Ct7). The Applecross Formation of the Torridon Group contains its usual suite of durable pebbles, including porphyry. The results of a detailed examination of the lowest Applecross sediments have been published by Nicholson (1993).

The lowest beds of the Stoer Group are massive gneiss conglomerates (Ct1) fringing the gneiss palaeohill south of Achlochan [NC 024069]. The next 70 m are tabular red sandstones (Ct2) exposed along the southern shore of the peninsula Rubha Dunan. These sandstones contain centimetre-sized angular gneiss fragments, and occasional channels up to about 40 cm deep filled by lateral accretion deposits. In the highest beds, west of the fault [NC 02250674], grain size declines and bedding surfaces frequently show symmetrical ripples and desiccation cracks. Only 250 m from the extremity of Rubha Dunan [NC 02100677] the sequence contains a red shale band 0.9 m thick, above which the finely laminated (aeolian) facies (Ct8) makes its appearance. The finely laminated sandstones have an average grain size of about 0.2 mm and a maximum of 0.5 mm. Millimetre lamination ($p = 2000$–5000) forms low angle cross beds. The laminated beds are frequently interrupted by others of similar grain size, virtually indistinguishable in thin section, but poorly laminated or completely massive. The latter have erosional bases and contain gneiss pebbles up to 3 cm in size. Rarely, small blocks of the laminated facies can be found incorporated in the massive beds. A contact between the two sub-facies is particularly well seen on the low cliffs 200 m from the headland [NC 01990678] where the pebbly sandstone is trough cross-bedded.

The summit of a gneiss hill forms a small inlier at the headland Rubha Dunan [NC 018068] where the stratigraphic level is roughly 50 m beneath the Stac Fada Member, and a similar distance above the Horse Island strata described in the next paragraph. It is also 100 m above the local base of the sequence seen farther east, which gives a measure of the palaeorelief in Stoer Group times.

About 150 m of Stoer Group sediments are exposed on Horse Island. They overlie basement gneiss on the east side of the island and strike northwards into a basement hill. The lowest sediments (Ct2) are erosively overlain by a conglomerate unit formed of well-rounded gneiss cobbles of up to 30 cm diameter (Ct4). The conglomerate is about 10 m thick and fines upwards into sandstones with scattered centimetre-sized pebbles (Ct2). The western half of the island is occupied by the muddy sandstone facies (Ct7), which to the north passes laterally, over some 2 m, into facies Ct2 near the basement hill [NC 02170498].

The Stac Fada Member outcrops on the coast 1.4 km north of the point Rubha Dunan [NC 02220823]. It rests directly on an irregular gneiss surface. The member contains vitreous lapilli (but no accretionary lapilli) up to 8 cm long, as well as small angular gneiss fragments. It is sharply overlain by reddish-grey, very hard siltstone. The Stac Fada Member is only 2 m thick and evidently wedges out against a basement hill as it does at Enard Bay.

The lowest beds of the Torridon Group are massive breccias (Db1a), derived both from the Lewisian gneiss and from the Stoer Group. Breccia of this kind fringes the gneiss hill near Achiltibuie village. The presence of sandstone clasts in this breccia evidently puzzled the first surveyors who refer in their memoir to 'brecciated conglomerate' (Peach *et al.* 1907, p. 313). The most spectacular exposures of the basal breccias, however, are to be seen on the south side of Rubha Dunan [NC 02420678]. The low sea cliff (Fig. 71) exposes the side of a palaeohill formed out of Stoer Group sandstone (Ct2) that dips 15°–25° west. The hill is buried by tabular bedded red sandstones (Db1a), dipping 20°–25° south, enveloping red sandstone blocks up to 4 m across (Fig. 72). The blocks are petrographically indistinguishable from the Stoer Group close by. They evidently rolled down a steep hillside and plunged into shallow water, for the sands that covered them are rippled and desiccated. The boulder bed is about 6 m thick and appears to form a fringe, no more than 25 m wide, flanking the basement hill. Above the boulder bed the sandstones are interbedded with red shale. About 20 m stratigraphically above the base the red sandstones are succeeded by poorly exposed grey shales (Db2) with similar dip. The steeply dipping unconformity follows the present coastline westwards for about 150 m.

Fig. 71. Unconformity between red sandstones of the Stoer Group (dipping to the left) and the Torridon Group (dipping to the right) at Rubha Dunan, near Achiltibuie [NC 02420678]. The observer looks SE. The cliff in the centre of the photograph is 15 m high.

Fig. 72. Block of Stoer Group sandstone in the basal Torridon Group at Rubha Dunan, Achiltibuie [NC 02440678]. The hammer handle is 0.5 m long. Note how the beds are draped over another large boulder in the background, producing a

Palaeomagnetic studies by Stewart & Irving (1974) and Torsvik & Sturt (1987) show that the unconformity at Rubha Dunan corresponds to a large shift in magnetization vector, implying that the area drifted from a palaeolatitude about 10° or 20° north of the equator in Stoer Group times to a position 30° or 40° south of it when the Torridon Group was deposited. Note that these palaeolatitudes assume normal polarity.

Grey shales (Db2) about 40 m thick are well exposed on the coast south of Rubha Dunan. Blue-grey weathering phosphatic layers a few millimetres thick are common, for example 700 m south of Badenscallie [NC 037058] and 200–400 m NE of Rubha Dubh Ard [NC 042041]. The shales are millimetre-laminated micaceous siltstones with subordinate centimetre-thick pink sandstone interbeds, some of which are very coarse. Ripples are common, but desiccation cracks rare. There are occasional flat-bottomed channels in the shale up to 10 cm deep, filled by coarse sandstone, for example on the coast 400 m NE of Rubha Dubh Ard [NC 04480427]. The uppermost 5 m of the shales, beneath the Applecross Formation, contain grey greywacke beds up to 0.5 m thick that are well exposed near high water mark about 400 m NW of Acheninver Lodge [NC 03810562]. The tops of some greywacke beds have convolutions and linguoid ripples showing palaeocurrents flowing towards the SE. The bases of some beds are erosional.

Precompactional monoclinal folds are a peculiar feature of the shales at two localities. The first is on the coast NW of Acheninver Lodge [NC 03810562] and the other on the coast at Rubha Dubh Ard [NC 04500431 to NC 04210409]. The folds have amplitudes of 5–10 cm and axial directions 130°–140°. Representative examples are shown in Figure 73. There are two types of monocline. In type I shale is draped over a step cut in shale, or sand (Fig. 73a & b). In type II there is no step and the monoclinal flexure is truncated downwards by a near horizontal surface (Fig. 73c & d). Interpretation of the structures is complicated by penecontemporaneous erosion that has removed the tops of the folds, and by severe compactional deformation. Repeated deformation of the same layers may explain the relatively uncommon combination of type I and II structures shown in Figure 73c.

Type I structures are evidently due to filling of a shallow channel, partly by sediment deposited from suspension and partly by bottom traction. Type II structures suggest extension of a thin layer

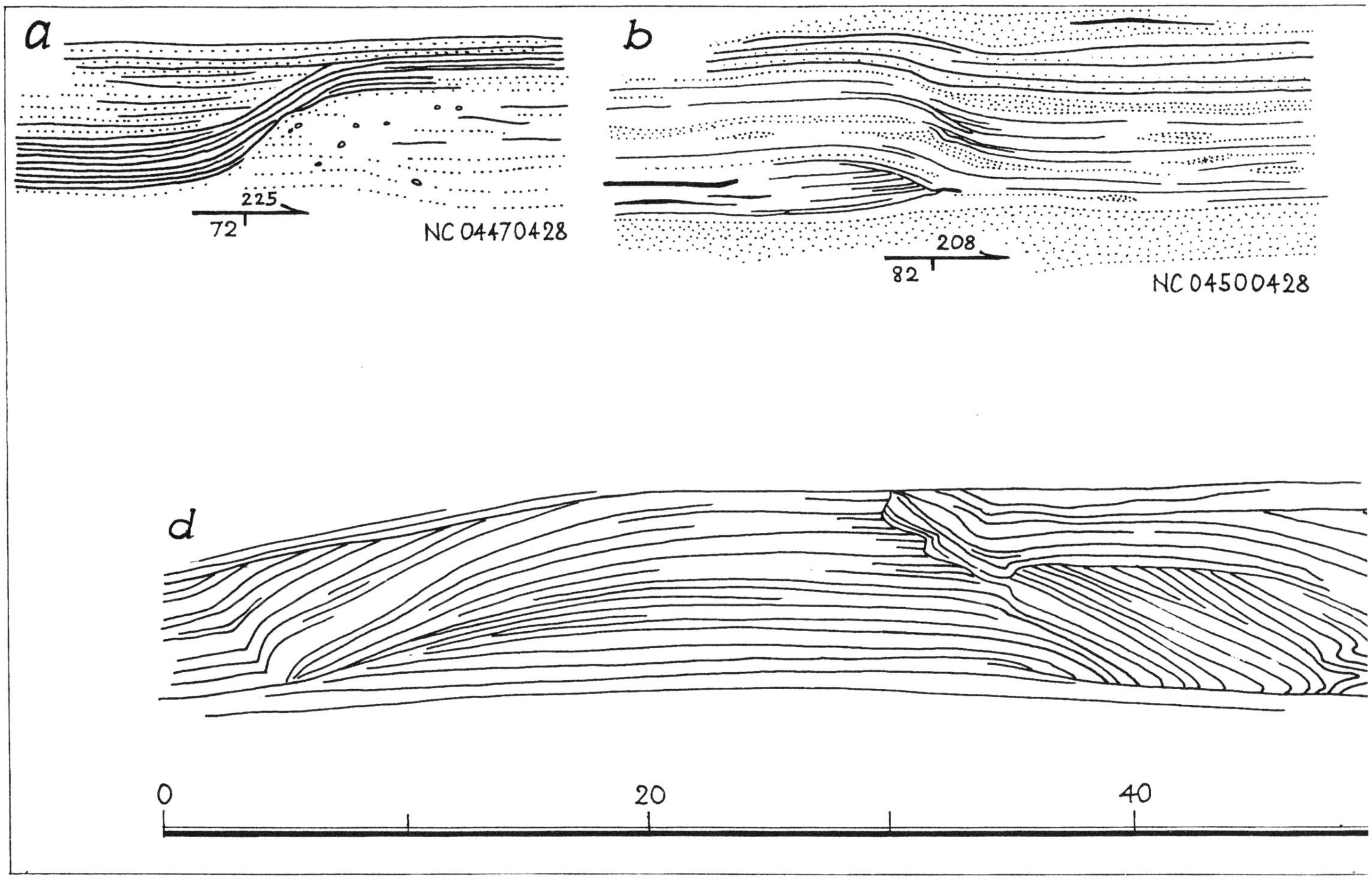

Fig. 73. Monoclinal structures in laminated grey siltstones (lined) and sandstones (stippled) of the Diabaig Formation (facies Db2) at Rubha Dubh Ard, near Achiltibuie. The strike and dip of the traced surfaces are shown.

of soft, wet sediment on small listric faults, followed by erosion and compaction. If this is correct then some type I structures may have formed over the hollow left after slip. However, the folds cannot have resulted from the down-slope creep of superficial sediment because they collectively lack the necessary asymmetry, and also because the hinge lines are too straight. Perhaps the superficial sediment was extended by the formation of conjugate shear planes, perpendicular to the movement direction, at the rear of gravitationally sliding sheets, the leading edges of which are not preserved.

The lower part of the Applecross Formation can be subdivided into two upward-fining units – the Rubha Dubh Ard Member (RDAM) and the Achduart Member (AM). The former is about 40 m thick and was considered as a possible lithostratigraphic unit at Loch Lurgainn by the Geological Survey (Peach *et al.* 1907, p. 316). Correlation with the Cape Wrath Member (Williams 1969*a*) is not now thought likely for the reasons given on p. 75. At the type locality the base is sharp and erosive on reddened Diabaig shale. The lowest 20 m of the Rubha Dubh Ard Member consists of trough cross-bedded, contorted coarse red sandstone with scattered durable pebbles, including red porphyry, up to 1.5 cm in size. The upper half of the member (Fig. 74) is brownish-grey tabular sandstone ($p \approx 5000$), generally medium to fine-grained. Small-scale trough cross-bedding, ripple cross-lamination, flat bedding with current lineation and shallow scours are common. Exotic clasts are still present but less than 0.5 cm in size. The Member is completely exposed at Rubha Dubh Ard [NC 04260374 to NC 04490355] which is here designated the stratotype. Palaeocurrents changed direction during deposition, flowing towards 160° ($n = 74$) in the lower part, towards 117° ($n = 72$) through most of the upper part and towards 070° ($n = 39$) at the very top (Nicholson 1993, fig. 6).

The base the Achduart Member which follows (Fig. 74) is formed of very coarse red sandstone with durable pebbles up to 2 cm in diameter. Grain-size diminishes upwards so that at the top sparsely pebbly medium-grained sandstone predominates. The member is about 100 m thick. My trough cross-bedding measurements show palaeocurrents flowing towards the SE ($\theta = 111°$, $n = 117$), significantly different to that in the member beneath, or in the undifferentiated Applecross Formation above ($\theta = 065°$, $n = 12$). Nicholson (1993, fig. 7) has shown that cross beds in the lowest 9 m of the member are arranged in cosets 1–2 m thick, dipping to the SSE. The base of the Achduart Member corresponds to the top of the underlying Rubha Dubh Ard Member; the top is exposed on the coast SE of Achduart [NC 05260347]. The coast section is the stratotype.

Stratigraphically higher beds of the Applecross Formation are shown in the section from Rubha Dubh Ard to Strath Kanaird (Fig. 75). They are all coarse grained with durable pebbles up to 3 cm diameter, frequently arranged in seams. The section shows that the maximum thickness of Applecross Formation present beneath the Cambrian is 1350 m.

Cailleach Head

The area considered here lies west of the Precambrian gneiss ridge forming the hills of Cnoc Sgoraig and Carn Dearg. Immediately to the west the Stoer Group overlies the gneiss, dipping westwards. Immediately to the east the Torridon Group overlies the gneiss, dipping eastwards. One kilometre west of the gneiss ridge the Stoer Group is truncated by the Coigach fault that brings down the highest beds of the Torridon Group to sea level. The throw on the fault is about 6 km (Stewart 1993*a*). A section of the area is provided in Figure 76. A geological map of the headland by the author, scale 1: 10 560, has been deposited with the Geological Survey in Edinburgh. The essential information also appears on the 1: 50 000 geological map of the Summer Isles, sheet 101W, published by the

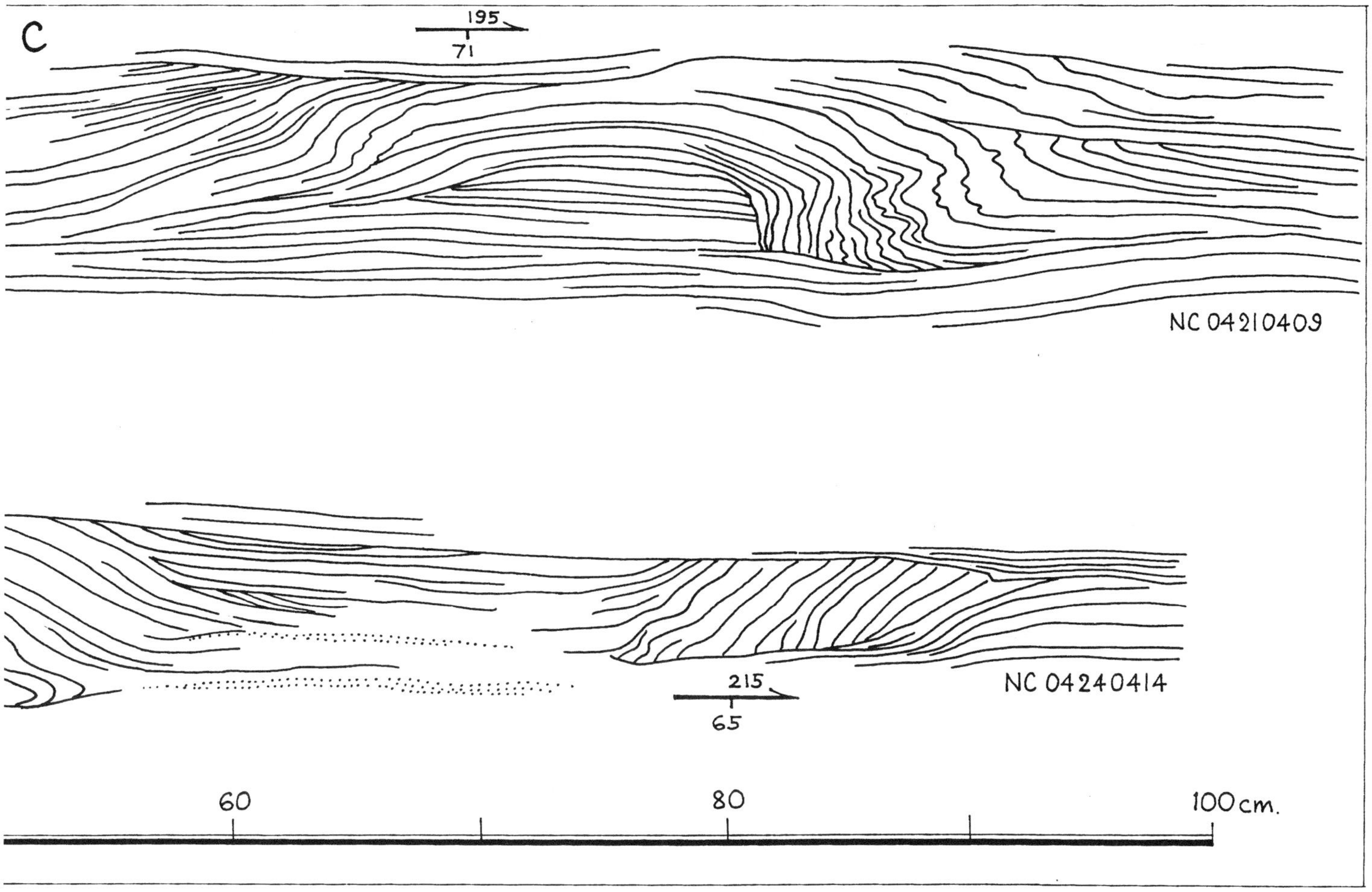

British Geological Survey in 1998. Cailleach Head is a Geological Conservation Review site (Mendum *et al.* 2003).

Along the SW coast of the peninsula about 150 m of Stoer Group are exposed, including the Stac Fada Member (Peach *et al.* 1907, p. 313). The stratigraphically lowest beds, and the unconformity with the gneiss, have been cut out by a fault in this section. Those remaining, which are poorly exposed, may belong to the Clachtoll Formation. They contain only gneiss pebbles and no quartzite. The palaeoslope, shown by overturned cross-bedding, was towards the NE. Some interbedded, well-laminated, trough cross-bedded sandstones in sets up to 2 m thick and 20 m in extent, are very like those seen below the Stac Fada Member near Stattic Point.

The Stac Fada Member is about 12 m thick, with a sharp, erosive base. It contains gneiss clasts up to 10 cm in size. There is the usual reworked top, but no accretionary lapilli. The beds overlying the Stac Fada, about 5 m thick, are very hard, red, fine to medium grained sandstones. These are faulted against the Applecross Formation that forms a prominent stack 6 m high on the beach.

The Applecross Formation in the stack contains porphyry pebbles up to 3 cm in size, and even larger clasts of dark red sandstone probably derived from the Stoer Group beneath. The sandstone forming the stac probably belongs to the Achduart Member and must have been faulted down at least 400 m. A few metres to the north, the Coigach fault, which appears to be vertical, brings white Cambrian quartzite down to sea level. The Cailleach Head and Applecross Formations near the Coigach fault are transected by quartzite veins up to 0.5 m wide, suggesting that the quartzite was only partially lithified when faulting occurred. A small exposure of quartzite *in situ* strikes 110° and dips 16° south, whereas the Cailleach Head Formation generally strikes 085° and dips 40° south. This means that the Cambrian is cutting down through the Torridon Group succession to the west, rather than to the east as it does near the Moine Thrust.

On the north side of the peninsula Stoer Group breccias fringe the bay Camas na Ruthaig, clearly an exhumed palaeovalley in the gneiss. The lowest breccias are massive, with angular gneiss clasts up to 30 cm, overlain by tabular bedded breccias with much smaller fragments of gneiss. These breccias are altogether about 200 m thick. An exposure gap, perhaps over a fault, separates the breccias from the overlying part of the Clachtoll Formation, which is about 140 m thick. This consists of finely laminated, medium grained red sandstone with large troughs. The troughs are up to a metre deep, with SW dipping foresets extending laterally for over 20 m. Occasional gneiss chips up to 0.5 cm suggest the troughs are fluviatile. A similar facies occurs below the Stac Fada Member on the south side of the peninsula (see above) and at the same stratigraphic level SE of Stattic Point. The sandstones are interrupted on the SW side of Camas na Ruthaig by a red siltstone unit 5 m thick [NH 001984].

About 200 m west of the red siltstone unit the Bay of Stoer Formation appears [NG 99909865]. About 100 m are exposed, grading down over a few metres into the sandstones beneath. The formation is easily identified by the presence of well-rounded pebbles of gneiss, dark-red quartzose sandstone and pink quartzite, some of which reach 6 cm in size. Cross-bedding shows that palaeocurrents flowed towards the NE.

The Stac Fada Member is not exposed on the north coast of the peninsula where it would have been expected, probably because it has been faulted out. If this supposition is correct the Bay of Stoer Formation is stratigraphically above the Stac Fada Member, as it is at Stattic Point.

West of the Coigach fault the headland is composed entirely of siltstone–sandstone cyclothems belonging to the Cailleach Head Formation. The lowest fifteen cyclothems are perfectly exposed in the cliffs SE of Cailleach Head lighthouse and are shown graphically in Figure 77. Although this is one of the finest cyclothemic sequences in Europe it has never received more than passing mention in the

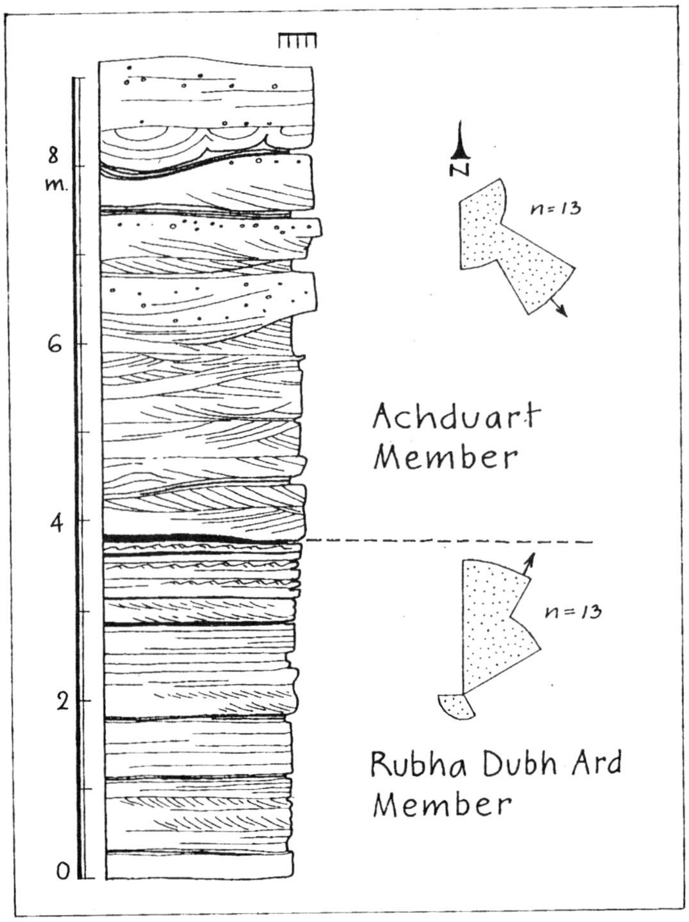

Fig. 74. Stratigraphic log of the boundary between the top of the Rubha Dubh Ard Member and the base of the Achduart Member on their type section about 500 m west of Achduart [NC 04490355]. The grain size scale spans +3 to −1 ϕ units (0.12–2 mm). Palaeocurrents (trough axes) in the Achduart Member are from above the 5 m level, below which they are oriented like those in the Rubha Dubh Ard Member. Durable pebbles over a centimetre in size are shown.

literature (Stewart 1988*b*), probably because it is more than 10 km from the nearest road. The sequence is also of historical interest in that the first Precambrian microfossils found in Britain came from this section. Dr John Horne, who mapped the area for the Geological Survey in 1885, collected the phosphatic concretions in which J. J. H. Teall found clusters of unicellular microfossils together with filaments (Geikie 1903, p. 56; Peach *et al.* 1907, p. 287–88 & Plate 52). Gregory (1917, p. 147) states that the sampled locality is on the shore 265 m NW of the Coigach fault, though phosphatic concretions occur sporadically throughout the formation. Downie (1962) has reported 'spore-like bodies, isolated and in clusters, cellular sheets of tissue, and filaments' from the shales containing the phosphatic nodules, but no further details have been published. He assigned them tentatively to the middle Riphean.

The type section of the Cailleach Head Formation starts at sea level 180 m NE of the lighthouse, where the lowest cyclothem exposed is cut by a small fault [NG 98649868]. The section continues south and SE along the coast to the Coigach fault [NG 99089733], encompassing about 630 m of strata. Continuous exposure ends about 500 m south of the lighthouse [NG 98619793] which is 270 m above the lowest cyclothem.

The base of the formation is concealed by the sea at Cailleach Head, and the top is cut out by the Cambrian Eriboll Quartzite. However, the base is exposed on the east side of Gruinard Island, where grey shales are well exposed (Peach *et al.* 1907, p. 321) which implies that there must be about another 500 m of the formation below that seen at Cailleach Head.

The cyclothems have an average thickness of about 20 m. The lower part of each consists of tabular bedded grey siltstones and sandstones (facies 1). The remaining upper part is trough cross-bedded, mainly red sandstone (facies 2). The alternation of these two facies is definitive of the formation. The facies can be subdivided as described below.

The base of each cyclothem is defined by a flat erosion surface overlain by dark grey siltstones, rarely mudstone or fine micaceous sandstone. The siltstone forms laminae averaging about 4 mm in thickness (subfacies 1a), generally interlaminated with pale brown or cream weathering ripple-laminated sandstone bands. Sedimentary veins frequently penetrate downwards from the sandstone into the siltstone. If the sandstones form more than 50% of a sequence it is called subfacies 1b. The lateral persistency (p) of siltstone laminae in subfacies 1a is over 10 000.

Subfacies 1c consists of tabular beds of light-red weathering, fine-grained sandstone. Typical sedimentary structures are flat bedding with current lineation, and repeated sets of tabular, planar cross-bedding. The bases of the sandstones often show drag marks and isolated shrinkage cracks, whereas the tops have symmetrical ripples, typically in interfering sets. In cyclothems X and XI large desiccation polygons cut the intervening grey siltstones of subfacies 1a or 1b. The lateral persistency (p) of beds in subfacies 1c is in the range 300–5000.

In the upper part of each cyclothem (facies 2), the sandstones are slightly coarser grained than those of facies 1 but only rarely reach medium grain size. There are two dominant subfacies, 2a and 2b. The first weathers to a moderate pink colour and locally contains intercalations of the grey subfacies 1b. Such intercalations are impersistent due to erosion at the base of the sandstone. The second subfacies (2b) consists of greyish red weathering sandstone in which the iron minerals are mainly concentrated in small spots. The sandstone typically shows rapid lateral passages into yellowish green and greyish red micaceous sandstone and siltstone.

Channels up to 40 cm deep and several metres across are quite common throughout facies 2. The largest is seen 15 m above the

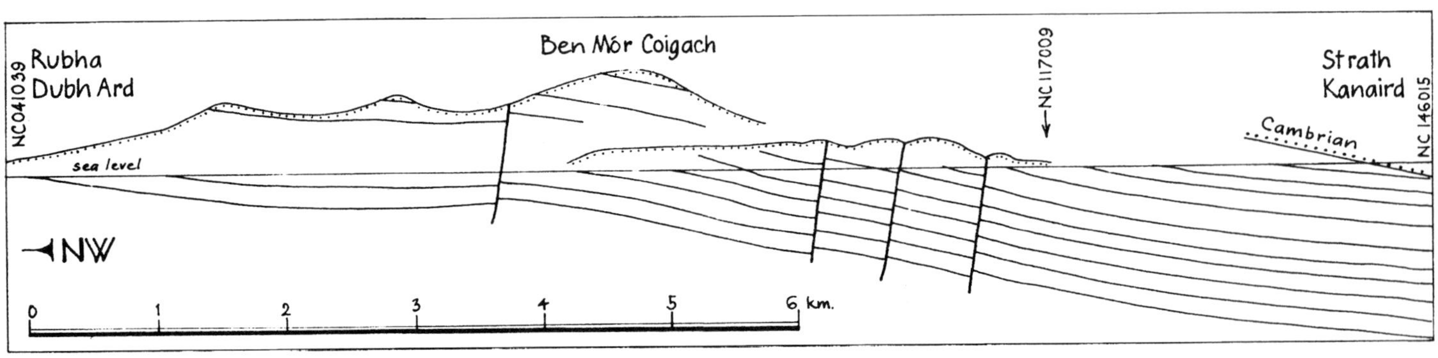

Fig. 75. True-scale section of the Applecross Formation from Rubha Dubh Ard to Strath Kanaird. Topography is shown by lines with dots. The base of the Applecross has been extended downwards to the SE from Rubha Dubh Ard using dip measurements along the coast. The dip of bedding in the mountains of Coigach, projected onto the section, is taken from Geological Survey maps.

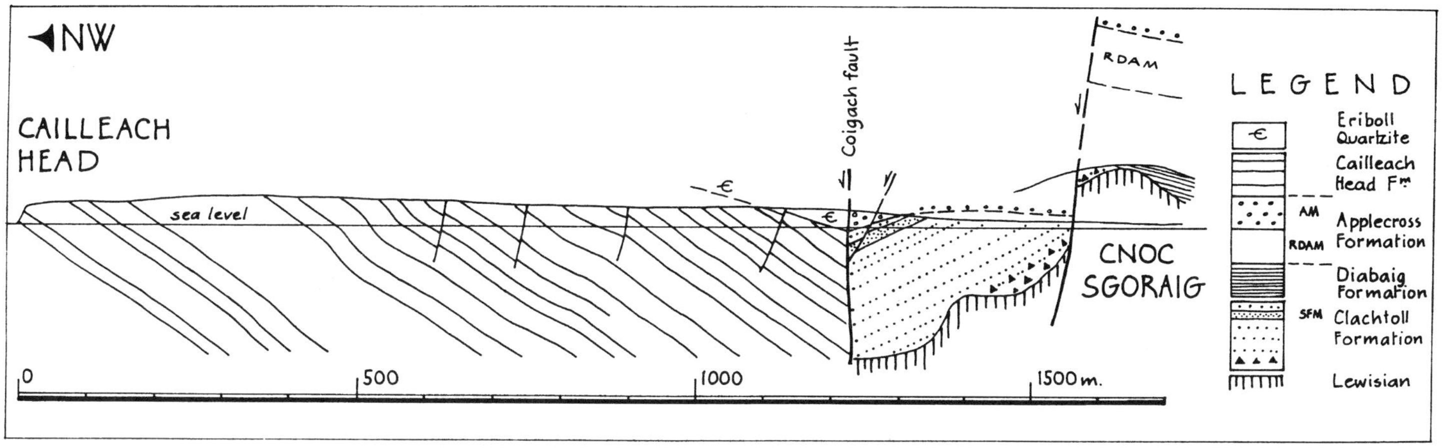

Fig. 76. True-scale section of the Cailleach Head Formation at Cailleach Head. The section line extends from Cailleach Head lighthouse [NG 98569853] in the NW, along the coast to the Lewisian basement 350 m north of Iola Creag Mhór Sgoraig [NG 99029686]. The southeastern extremity of the section, at Cnoc Sgoraig, is about a kilometre inland. The Stoer Group is identified by the Stac Fada Member (SFM). The Diabaig Formation and the Rubha Dubh Ard Member (RDAM) of the Applecross Formation NW of Cnoc Sgoraig probably thin over a basement hill eroded in the Stoer Group. Hence the pebbly base of the Achduart member (AM) rests on the Stoer Group close to the Coigach fault.

base of cyclothem IV, and is well exposed in the cliff beneath the lighthouse. A sandstone of subfacies 2a here fills a channel 5 m wide and about half a metre deep in the underlying subfacies 1b. Siltstone fragments up to 40 cm long are present sporadically throughout facies 2. The lateral persistency (p) of units is in the range 10–100.

The average palaeocurrent direction from cross-bedding, cross-lamination and current lineation in the cyclothems is towards the NE, with $\theta = 037°$ ($n = 52$).

The boundary between facies 1 and 2 is easily located on the sections by the appearance of trough cross-bedding, and by the abrupt change in the persistency factor. It is indicated on the graphic logs (Fig. 77) by a solid triangle. A few anomalies in the facies succession require comment.

The siltstone unit about 13 m above the base of cyclothem VII might be taken to indicate the base of a separate cycle. However, the sandstones immediately above and below have $p = 50–100$ which suggests that they belong to facies 2 rather than facies 1. This is also true for the sandstone at the top of the cyclothem. On the other hand the sandstones at the tops of cyclothems V and VIII have high lateral persistency and appear to belong to facies 1. The top of cyclothem V in fact, lacks an erosion surface and grades into the cyclothem above.

Phosphate occurs rarely as centimetre-sized nodules. It has been noted in facies 2b of cyclothem XIV and in facies 1c of cyclothem XV.

The lowest sediments in each cyclothem (subfacies 1a) were deposited from suspension below wave base, where bottom currents rarely arrived. Upwards through the subfacies, however, progressively more sandy sediment was introduced. The absence of carbonate, or macroscopic pyrite, suggests that the water body above the sediment had good circulation, was well oxygenated and dilute – probably a hydrologically open lake.

The flat-bedded tabular sandstones of subfacies 1c were deposited from fast-flowing, shallow flood water, crossing sandflats bordering the lake. When the lake level rose and temporarily covered the sandflats, the flood water was forced to decelerate and deposit sediment as straight-crested sand waves. Some of the sediment was carried beyond the sandflat and across the lake floor, thus accounting for the ripple-laminated sandstone bands in subfacies 1b. After the floods had ended waves reworked the sediment top into symmetrical ripples. No beach deposits have been identified.

Upstream from the sandflat the flood waters crossed a fluviatile braid plain. The currents were swifter than on the sandflat and formed the dunes responsible for the trough cross-bedding seen in facies 2.

Scoraig

A geological map of the Scoraig area by the author, scale 1:10 560, deposited with the Geological Survey in Edinburgh, has been integrated into 1:50 000 sheet 101W, published by the British Geological Survey in 1998.

Torridon Group sandstones about 2.5 km thick are exposed on the peninsula between Loch Broom and Little Loch Broom. The lowest beds are well exposed along the shore of Annat Bay. Basement gneiss fringed by massive breccia containing gneiss and red sandstone clasts (facies Db1a) is here overlain by a further 150 m of Diabaig Formation. The lowest 40 m of the Diabaig, in the bay called Féith an Fheòir [NH 014981], are formed of centimetre-thick beds of coarse-grained brown sandstone containing scattered red sandstone and quartzite pebbles that have come from the Stoer Group a kilometre (or more) to the west. The bedding planes of these brown sandstones, perhaps because they are silt-draped, weather out with unusual facility to give the beds an oddly platy appearance. Cross-bedded cosets, decimetres thick, with foresets of consisted strike for as much as 30 m interrupt the upper part of the section. A similar facies occurs in the Diabaig Formation near Balgy, on the south side of Upper Loch Torridon.

South of Féith an Fheòir the Diabaig Formation consists of the more usual tabular coarse red sandstone, in decimetre to metre-thick beds. Seams of pebbles derived from the Stoer Group (red sandstone and quartzite) persist. Trough cross-bedding indicates that the palaeocurrents flowed due east. Surfaces covered with wave ripples are common, especially near the top of the Diabaig where thin red shales and rare desiccation cracks also appear. These finer beds are seen on both coasts of the peninsula, 0.5 km NW of Achmore [NH 022971], and below Scoraig School House [NH 010958]. They grade over a metre into the overlying Applecross Formation which, at Annat Bay, forms a low cliff.

The lowest Applecross consists of very coarse, red tabular sandstones with small troughs, altogether about 60 m thick. The maximum grain size of these sandstones is 3–4 mm. Palaeocurrents flowed towards the NE. Porphyry and jasper pebbles are not visible in outcrop and neither is the contorted bedding so typical of the formation, but the general appearance of the beds is so like the upper part of the Rubh Dubh Ard Member at Achiltibuie that correlation is almost certain. Like the Rubh Dubh Ard Member these sandstones are overlain by a much coarser and rather thicker sandstone member. Its base is exposed on the shore of Annat Bay about 200 m north of Achmore [NH 024970] and again 500 m SE of Achmore [NH 030964]. The basal sandstones are very coarse, with decimetre-thick beds of pebbles, including exotic types up to 2 cm in

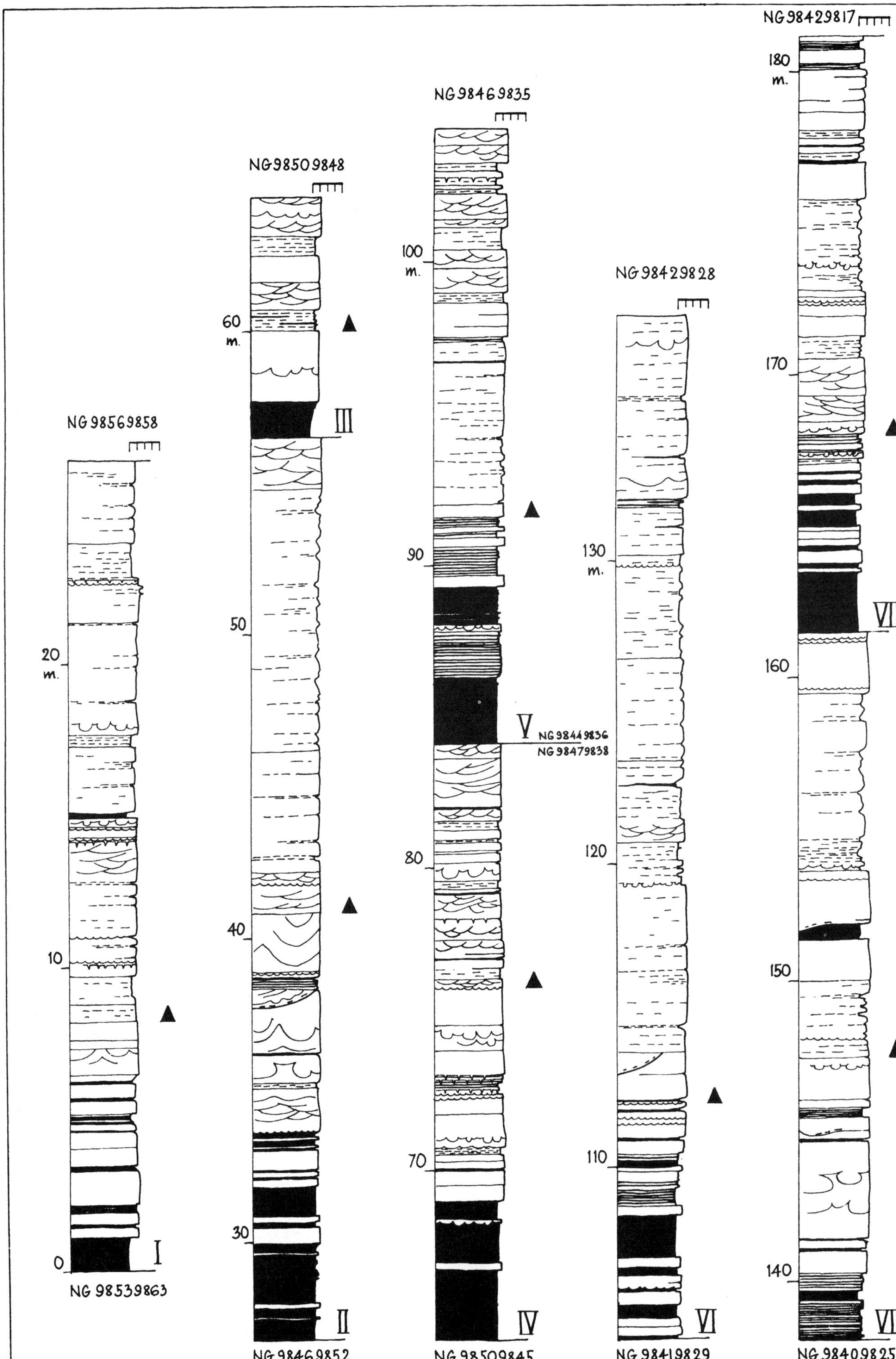

Fig. 77. Graphic logs of fifteen cyclothems (I–XV) in the Cailleach Head Formation exposed on the cliffs of Cailleach Head. The solid triangles indicate the base of facies 2 in each cyclothem. The grain size scale spans 5–1 ϕ units (0.03–0.5 mm).

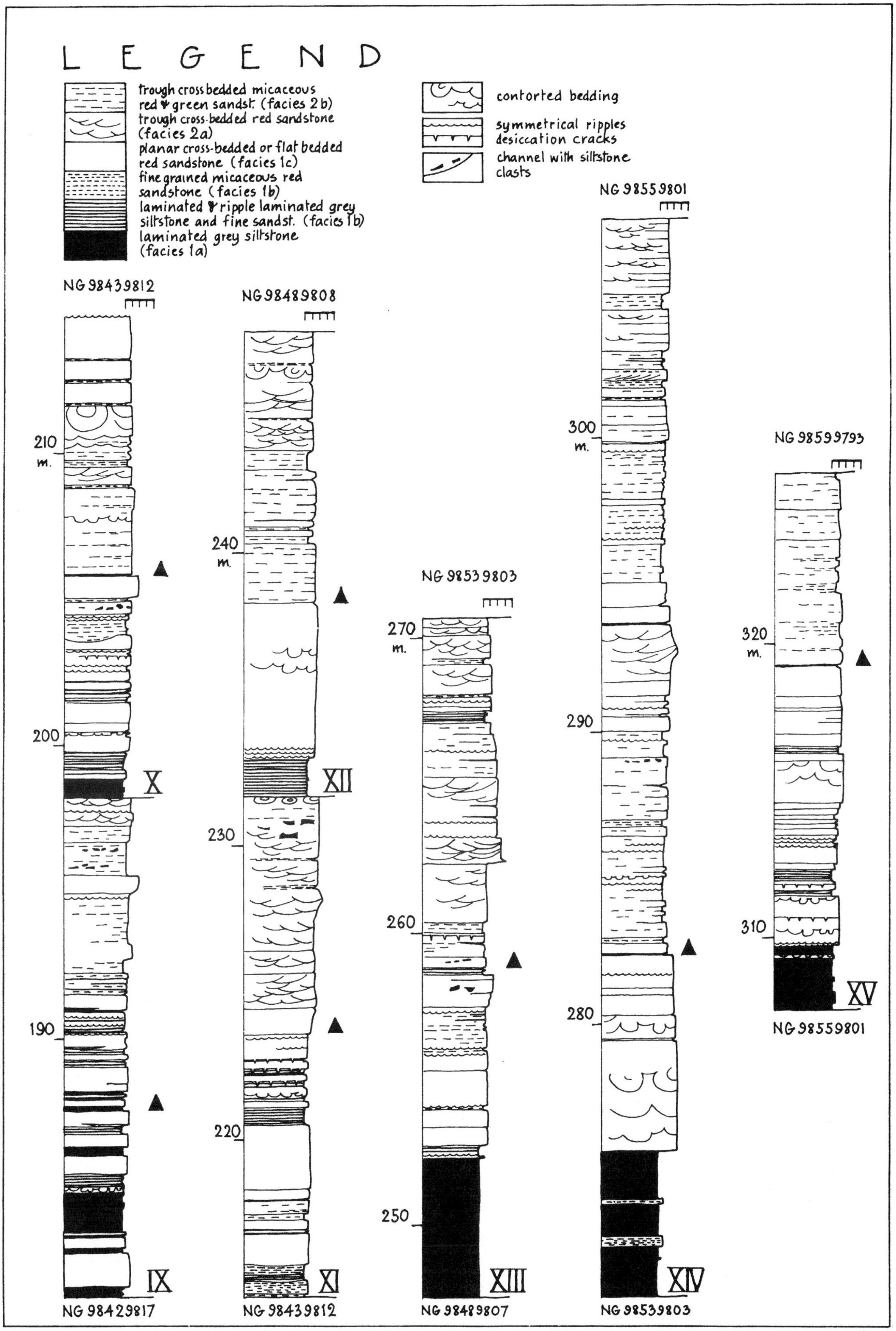
LEGEND
trough cross bedded micaceous red & green sandst. (facies 2b)
trough cross-bedded red sandstone (facies 2a)
planar cross-bedded or flat bedded red sandstone (facies 1c)
fine grained micaceous red sandstone (facies 1b)
laminated & ripple laminated grey siltstone and fine sandst. (facies 1b)
laminated grey siltstone (facies 1a)
contorted bedding
symmetrical ripples
desiccation cracks
channel with siltstone clasts
NG 98439812
210 m.
200
190
IX
NG 98429817
NG 98489808
240 m.
230
220
XI
XII
NG 98439812
NG 98539803
270 m.
260
250
XIII
NG 98489807
NG 98559801
300 m.
290
280
XIV
NG 98539803
NG 98599793
320 m.
310
XV
NG 98559801

size. Trough cross-bedding shows that the palaeocurrents flowed towards the east ($\theta = 090°$, $n = 30$). Stratigraphically upwards the pebbles get sparser and smaller. The member is about 120 m thick, similar to the Achduart Member at Achiltibuie with which it doubtless correlates.

South of Little Loch Broom the Rubh Dubh Ard Member cannot be traced, but the Achduart Member is about 150 m thick where it forms the top of the mountain Sàil Bheag. The member was described by the Geological Survey as a 'bright red sandstone series' and mapped southeastwards from Sàil Bheag for 6 km, as far as Loch na Sealga (see one-inch geological map sheet 92 Inverbroom). The mean palaeocurrent direction at Sàil Bheag is southeastwards ($\theta = 122°$, $n = 8$) according to Williams (1969*a*, fig. 12).

Very coarse pebbly sandstone that overlies the relatively fine-grained top of the Achduart Member can be seen by the side of the path from Badrallach to Scoraig, at Creag a' Chadha [NH 035932]. The pebbles reach 3 cm in size and form beds 20 cm thick. Quartz-muscovite pebbles from this locality, thought by Gregory (1915) to derive from the Moine Supergroup, have given K–Ar dates of about 1700 Ma (Moorbath *et al.* 1967). Trough cross-bedding here forms cosets up to 3 m thick, often contorted. The palaeocurrents flowed due east ($\theta = 090°$, $n = 47$), apparently no different in direction to those in the Achduart Member beneath. The contact between the Achduart Member and the overlying beds is gradational over some 10 m and corresponds to a line of crags that can be followed easily from Creag a' Chadha to the shore of Annat Bay.

No other fining-upward members like those described above have been found in the Applecross farther east, which maintains its pebbly nature until covered unconformably by the Cambrian about 3 km north of Dundonnell House [NH 112900]. A cross section of the rocks between Creag a' Chadha and Dundonnell shows about 2 km of beds belonging to the Applecross Formation (Stewart 1993*a*, fig. 5).

Stattic Point

The Stoer and Torridon Groups are both present in this area, the former identified by the Stac Fada Member. The area was mapped by H. M. Cadell for the Geological Survey in 1886. The Stac Fada Member was called a porphyrite (i.e. pyroxene andesite) on the field slip, though in the memoir it is described as a 'breccia made up of igneous fragments' derived from the Lewisian basement (Peach *et al.* 1907, p. 313). Sandstone boulder conglomerate at the base of the Torridon Group was thought by the Survey to be Triassic and mapped as such. It was not until D. E. Lawson and the writer visited the area in 1965 that it was recognized as part of the Torridon Group.

A geological map of the area by the author, on a scale of 1:10 560, has been deposited with the Geological Survey in Edinburgh and is integrated into 1:50 000 sheet 101W (Summer Isles) published by the British Geological Survey in 1998. The section in Figure 78 shows the main features of the geology.

The lowest Stoer Group sediments on the section are tabular red sandstones containing angular gneiss clasts up to 3 cm in size. They pass upwards into trough cross-bedded sandstones deposited by palaeocurrents that flowed towards the NE ($\theta = 043°$, $n = 11$). The troughs are 0.5–1 m deep and 2–5 m wide. All these beds are thought to belong to the Clachtoll Formation because they contain locally derived detritus. About 400 m SE of Stattic Point, roughly half way up through the Stoer Group, the cross-bedding becomes more planar, like that in the Meall Dearg Formation at Stoer, and the palaeocurrents swing round towards the SW ($\theta = 224°$, $n = 19$). This palaeoslope direction is corroborated by metre-sized slump folds overturned towards the south. A good example, in which the over-steepened foresets are clearly truncated by the overlying bed, can be seen on the shore about 160 m south of Stattic Point [NG 97389607]. Inland from the section line the Stoer Group buries palaeorelief of at least 200 m in Lewisian gneiss.

The Stac Fada Member is well exposed at intervals along the west coast. It is about 13 m thick, with a sharp base and top, but lacks the accretionary lapilli found at Stoer and Enard Bay. The overlying beds are shown in Figure 79. The red and grey siltstones and sandstones that form the first 10 m recall the Poll a' Mhuilt Member at Stoer. In particular, the laminated grey siltstone is like unit C. The overlying fluviatile red sandstones, which are at least 100 m thick, have an erosional base strewn with pebbles of gneiss and quartzite. Their contact with the Coigach fault to the west is concealed by the sea. Note that only the lowest beds are shown in Figure 79. These sandstones are believed to belong to the Bay of Stoer Formation (facies BS1) because of their contorted trough

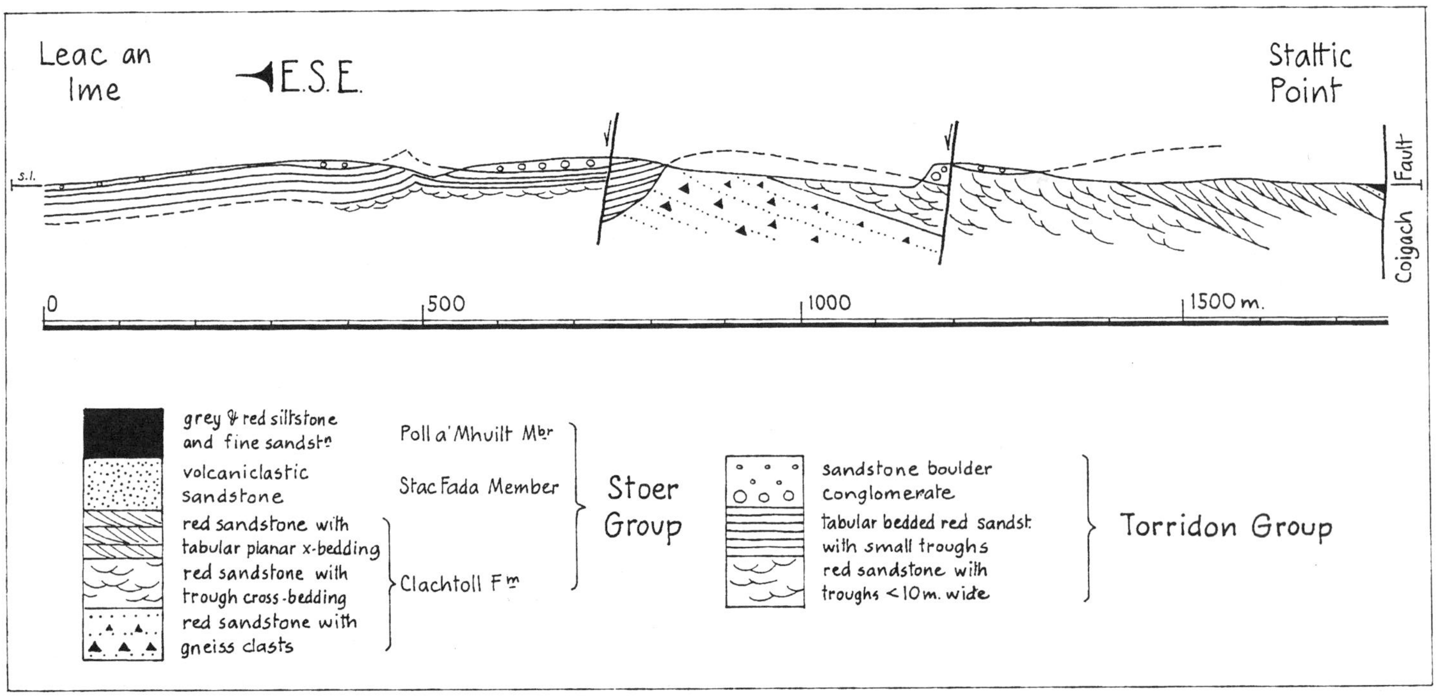

Fig. 78. True-scale section of the Stoer and Torridon Groups exposed on the cliffs between Leac an Ime [NG 98809562] and Stattic Point [NG 97349622]. Stratigraphically higher beds than those on the section are shown in Fig. 79.

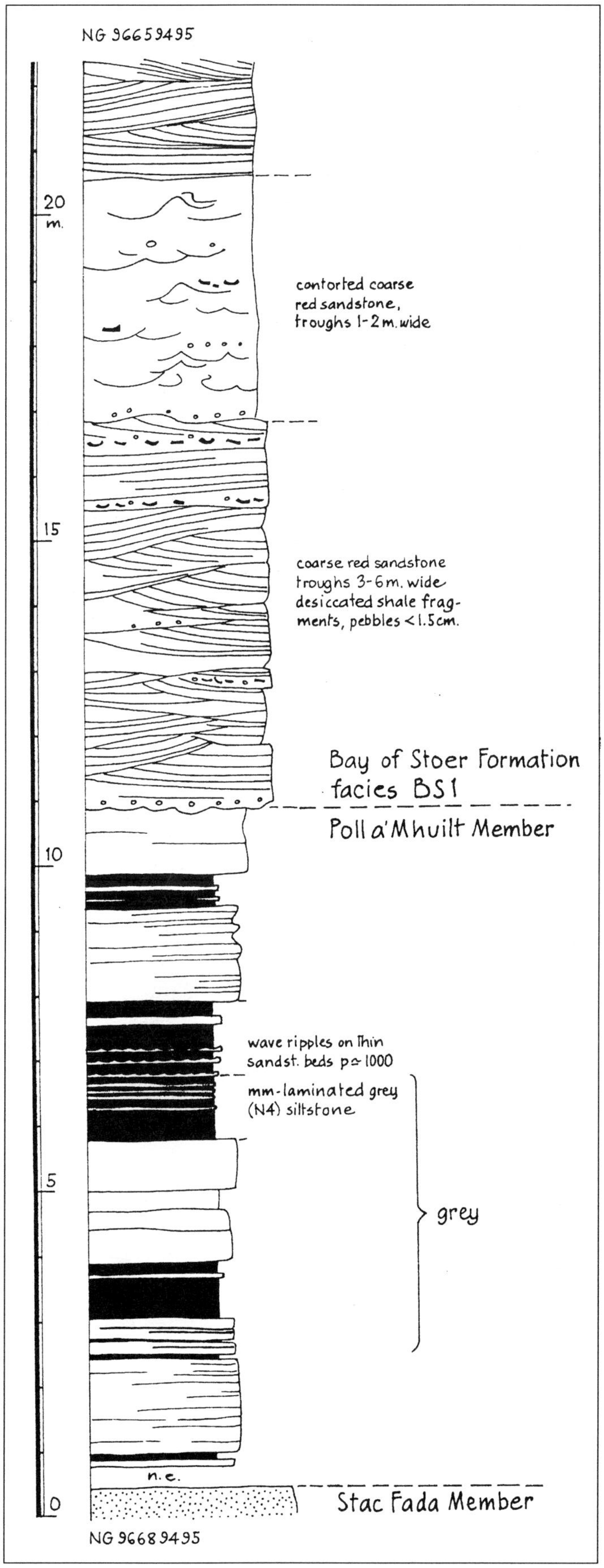

Fig. 79. Stratigraphic log of the strata overlying the Stac Fada Member near Stattic Point. The bed are all red except whether otherwise indicated. n.e. = no exposure.

cross-bedding, and the presence throughout of small quartzite and gneiss pebbles together with concentrations of ore minerals. However, the palaeocurrents flowed northwards ($\theta = 010°$, $n = 12$), quite different in direction to those in the Bay of Stoer Formation at either Stoer or Poolewe.

The substantial thickness of beds above the Stac Fada Member at Stattic Point as compared with their virtual absence at Cailleach Head only 2 km away, may be due to a component of vertical movement on the Little Loch Broom fault, which cuts the Coigach fault between Cailleach Head and Stattic Point. Downward movement of the south side of the Little Loch Broom fault in Stoer times or shortly after would have preserved a wider outcrop of Stoer Group sediments.

The unconformity between the Stoer and Torridon Groups is exposed at four points, the most remarkable of which is certainly the cliff at Uamh an Oir, about 500 m SE of Stattic Point (Lawson 1976, plates 1 & 2). Here massive sandstone boulder conglomerate (Diabaig Formation) rests on a stepped erosion surface cutting cross-bedded sandstone belonging to the Stoer Group. The sandstone blocks in the conglomerate are well rounded and reach 3 m in size. At least half the blocks are made of coarser sandstone than any in the underlying Stoer Group. A few of them contain decimetre-sized quartzite pebbles. They probably come from a lithology like the Bay of Stoer Formation though, strangely, there are none of the gneiss pebbles that might have been expected from such a source. The fabric of the conglomerate is clast supported with a sandy matrix. The clasts tend to lie on their sides, roughly parallel to the unconformity.

The second point where the unconformity can be seen is on the sea cliff about 850 m SE of Stattic Point [NG 98029570]. It is only accessible at low water spring tide. The unconformity surface dips steeply to the east and is overlain by red sandstone (Torridon Group) containing well-rounded quartzite and red sandstone pebbles up to 20 cm in size. The Torridon sandstones contain small troughs up to 20 cm deep and a metre across, formed by palaeocurrents that flowed towards the SE ($\theta = 141°$, $n = 12$). The difference in dip between the eastward dipping Torridon Group and the westward dipping Stoer Group beneath the unconformity is 29°. The eastward dip of the Torridon Group means that the sandstone boulder conglomerate that conformably overlies the Torridon Group sandstones in this area must be a stratigraphically higher unit than the very similar conglomerate at Uamh an Oir.

Inland the contact between the Stoer and Torridon Groups is well exposed on the hills Carn Dearg Ailean and Carn Dearg na h-Uamha. At first sight the contact appears to be an unconformity. However, outcrops of the Torridon Group on the western slope of the first named hill show red sandstone evidently brecciated while still unconsolidated – like the autoclastic breccia described by Spalleta & Vai (1984). The sandstone blocks, enveloped by liquefied sand, did not rotate sufficiently to destroy the bedding that now dips unusually steeply at about 40° eastward. True breccia, distinguished by the presence of pebbles of quartzite and fine red sandstone, bounds the autoclastic breccia above and below. Most of Carn Dearg na h-Uamha also consists of autoclastic breccia, with traces of original bedding dipping steeply northeastwards. True breccia, however, intervenes, just above the unconformity.

Clear evidence of tectonic instability during and immediately after deposition of the Torridon Group in this area is provided not only by the autoclastic breccia but also by the presence of large slump folds and a 180° palaeocurrent reversal, now to be described.

The anticline in the cliff 400 m west of Leac an Ime [NG 98369570], that forms such a prominent feature of the section viewed from the sea, was generated when the sediment (Torridon Group) was still soft. The orientation of its hinge (015°), together with small soft-sediment thrusts on the limbs of the fold, indicate compression in a direction between 105° and 125° on a slope inclined towards the NW. This compression was perhaps contemporaneous with the autoclastic brecciation – which can also be detected at intervals along the cliffs westwards from Leac an Ime. It may also be responsible for the steep dips at Carn Dearg Ailean and Carn Dearg na h-Uamha, noted above.

The pebbly red sandstone forming the slump fold shows trough cross-bedding in sets up to a metre deep and 10 m in width. The palaeocurrents flowed northwestwards ($\theta = 332°$, $n = 17$). Metre-scale intraformational slump folds at this locality confirm that the palaeoslope was towards the NW. The small-scale troughs described earlier from a stratigraphically higher level were formed by palaeocurrents that flowed northeastwards. Hence, there must have been a 180° reversal in palaeoslope and palaeocurrent direction during deposition of the Torridon Group near Stattic Point. Nevertheless, the distribution of rounded quartzite pebbles throughout the sequence shows that the southeasterly palaeoslope predominated, for such pebbles can only have come from the Stoer Group stratigraphically above the Stac Fada Member, somewhere west of Stattic Point.

Synsedimentary tectonism is also shown by dilatational sandstone dykes striking NNE, with vertical dip, cutting the Torridon Group. They can be seen at Leac an Ime [NG 98829563], Carn Dearg Ailean [NG 974954] on the shore near Badluchrach [NG 99169504] and about 50 m east of Stac Cas a' Bhruic [NG 98289567].

The tectonic instability described above may be due to episodic movement on a NNE trending fault a short distance west of Stattic Point, where the Coigach fault now is. The sandstone dykes are almost exactly parallel to the Coigach fault. The steep dips in the autobreccia could then be interpreted as the result of rotational slip on a westerly inclined palaeoslope.

The sediments described above as belonging to the Torridon Group cannot readily be assigned to any of its component formations. They contain only locally derived material, a typical feature of the Diabaig Formation, but there are none of the grey siltstones usually found in that formation. Other differences from the type Diabaig are the large-scale cross-bedding and frequent soft sediment deformation. Furthermore, none of the sediments at Stattic Point contain pebbles of porphyry or chert that would indicate the presence of the Applecross Formation. Indeed, the absence of exposures immediately east of the map area makes it impossible to prove conclusively that the beds are actually a conformable part of the Torridon Group. The correlation is based on their unconformable relationship to the Stoer Group, the palaeocurrent directions, lithification and colour. The sandstone boulder conglomerate is also remarkably similar to conglomerates at Rubh Réidh and Enard Bay that are demonstrably part of the Diabaig Formation.

Gruinard Bay

Several small outliers of the Stoer Group were mapped along the southeastern margin of Gruinard Bay by H. M. Cadell for the Geological Survey in 1886. They are shown, wrongly, as Diabaig Formation (Torridon Group) on the one-inch geological map (Inverbroom sheet, 1913). The outcrops have been described briefly by the writer (*in* Hambrey *et al.* 1991) and a 1:10 560 geological map deposited with the Geological Survey in Edinburgh. The most important outcrops are shown in Figure 80.

Immediately east of the Coigach fault the Stoer Group buries an unweathered gneiss surface with at least 200 m of relief. Many of the Lewisian hills visible today near the coast have, in fact, between exhumed from beneath the Stoer Group. The well-known viewpoint by the roadside at Little Gruinard (Hambrey *et al.* 1991, p. 103) is near the top of a palaeohill and the view to the north shows many more. Sediment-filled veins occur in the gneiss below the Stoer Group around Innis nan Gobhar [e.g. NG 94149026 and NG 93768990] and sporadically elsewhere.

The westernmost outlier shows a complete sequence from the basal breccia up to, and including, the Stac Fada Member. The stratigraphic log (Fig. 81) closely resembles that at Stoer (Fig. 43). Breccias and coarse red sandstones of the Clachtoll Formation form the lowest 80 m. The gneiss blocks are up to a metre in diameter near the unconformity, clast supported and mostly subangular or subrounded. Well-rounded, perfectly elliptical blocks also occur, though rarely. The breccia fines upwards into desiccated and ripple-laminated red shales with $p = 50–500$. Both straight crested and linguoid ripples are present in the shales, the latter indicating palaeocurrents flowing towards the west and NW.

The Bay of Stoer Formation, which forms the upper half of the sequence, rests erosively on the shales. The erosion surface is covered by well-rounded quartzite and gneiss pebbles up to about 10 cm in size, their long axes oriented at 125°. A metre or two above there is a well-rounded gneiss boulder 0.5 m in diameter, but at

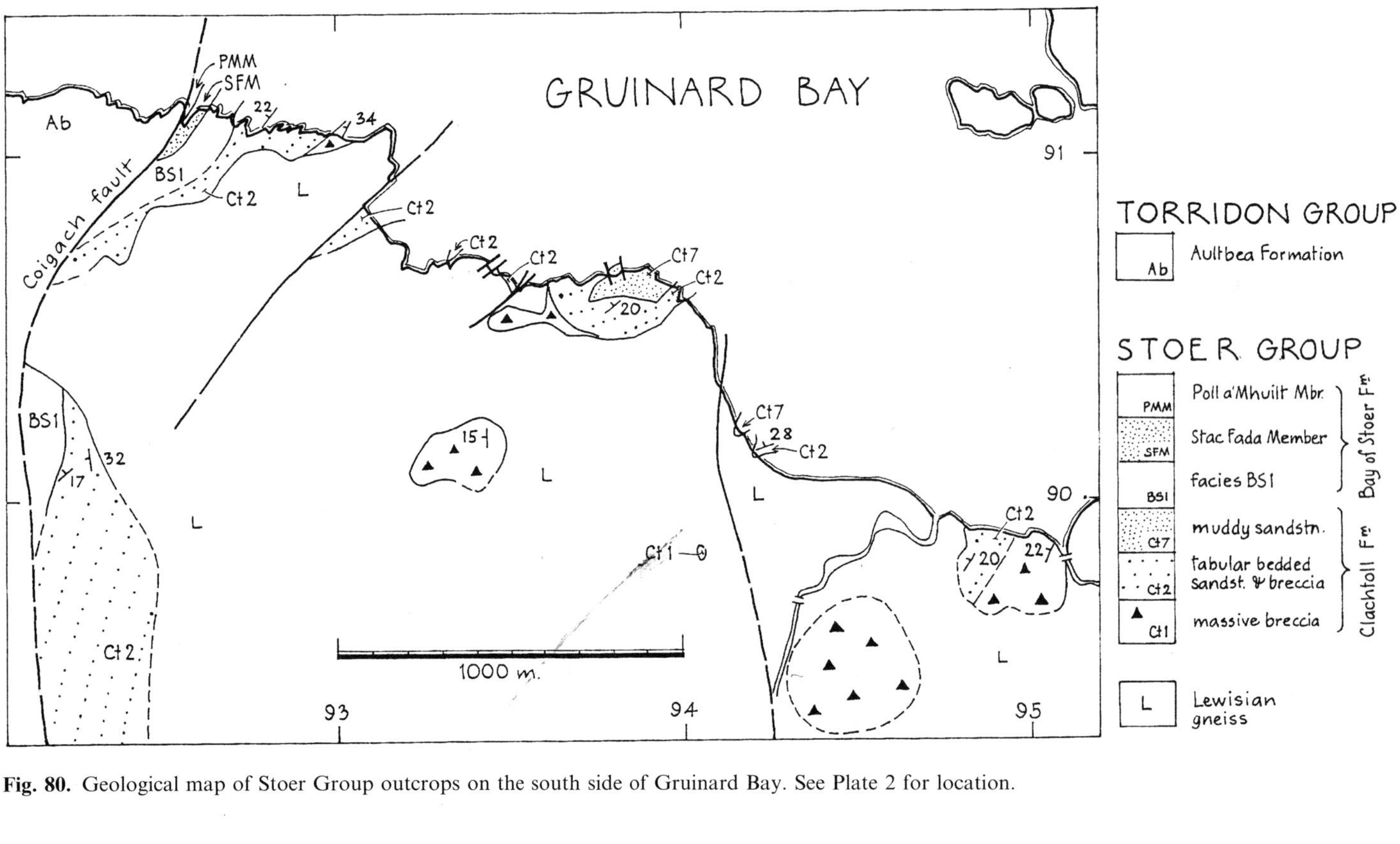

Fig. 80. Geological map of Stoer Group outcrops on the south side of Gruinard Bay. See Plate 2 for location.

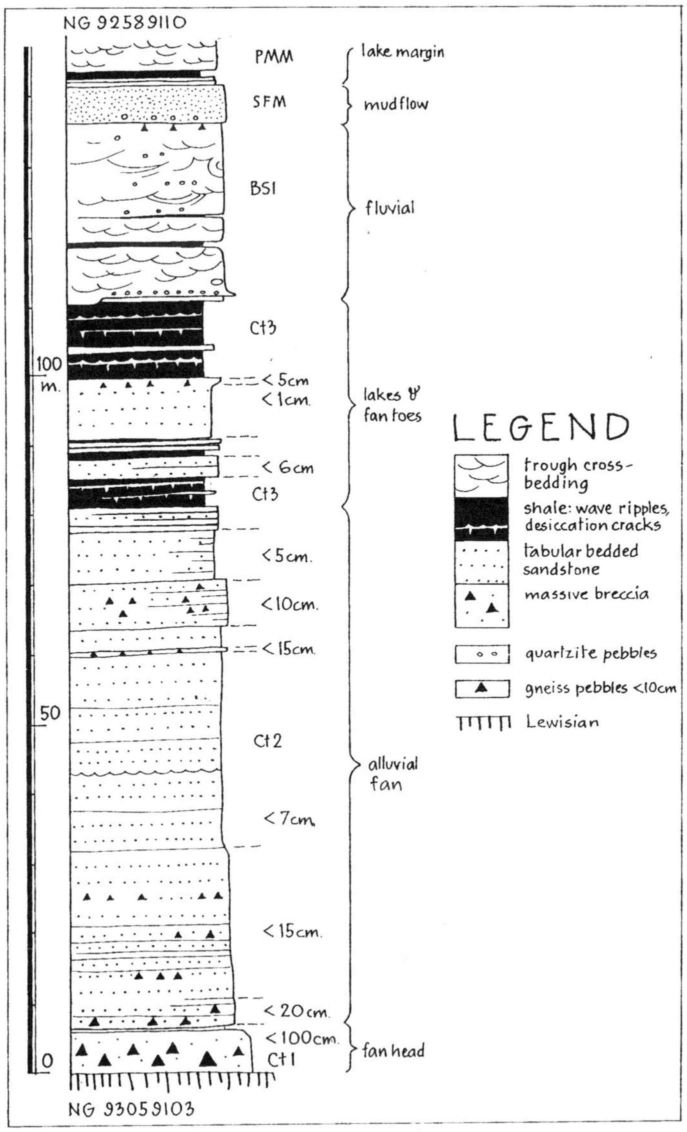

Fig. 81. Graphic log of the Stoer Group exposed on the coast at Gruinard Bay. The maximum size of gneiss pebbles in facies Ct1 and Ct2 is indicated. PMM = Poll a'Mhuilt Member; SFM = Stac Fada Member.

higher levels the pebbles are small and comparatively sparse. Small troughs in the sandstones show that the palaeocurrents flowed west or northwestwards, as in the Clachtoll Formation beneath. Heavy mineral bands with drop structures are common in the lowest 8 m of the Bay of Stoer Formation.

The Stac Fada Member has a very sharp base. The underlying sandstone maintains its normal grain size right up to the contact, except that angular blocks of gneiss up to half a metre in size are present *along* the contact. Such large blocks are rarely seen in the Bay of Stoer Formation below, and are absent from the Stac Fada Member above. The lowest metre of the member contains rounded red sandstone pebbles up to 20 cm in size, probably derived from the Bay of Stoer sandstones beneath. In addition, small quartzite and gneiss pebbles occur higher in the member, albeit rarely. The concentration of blocks near the base of the member suggests deposition from a hyperconcentrated flow rather than a mudflow. As at Stoer the member is abundantly charged with greenish volcanic fragments, generally under 2 cm in size. The topmost metre of the member has clearly been reworked, for the volcanic debris is segregated into bands. There are no accretionary lapilli. The reworked zone grades up into red siltstone. The topmost part of the section is formed by 7 m of medium-grained, very hard brown sandstone, trough cross-bedded in sets about 10 cm thick, but lacking pebbles. These sandstones probably correlate with the Poll a' Mhuilt Member (unit **A**) at Stoer. Higher beds are cut out by the Coigach fault.

Inland, exposures of the Stoer Group are poor, but the Bay of Stoer Formation with its characteristic quartzite pebbles is definitely present south of Cnoc Badan na h'Earbarge [NG 92209006]. Parallel laminated brown sandstone and gneiss breccia (Clachtoll Formation) underlie it to the east.

The Stoer Group outlier at Innis nan Gobhar shows an excellently exposed lateral passage from the Clachtoll breccia facies (Ct2), outcropping near the Lewisian, into massive muddy sandstone (Ct7) some 20–30 m distant from the Lewisian. The bedded sandstones of facies Ct2 also pass upwards into muddy sandstone over a stratigraphic interval of about 6 m near the eastern edge of the outlier [NG 93969060]. The bedded sandstones consist of alternating laminae of fine and coarse grain size, the latter containing gneiss fragments of millimetre, and rarely centimetre size. The lamination dies out upwards into the muddy sandstone while the matrix content seems to increase, so that the larger grains (up to 4 mm) are clearly matrix supported. Faint, pale, vermiform structures, probably deformed desiccation cracks, can be seen on the shingle-scoured surfaces of the muddy sandstone. Nests of millimetre-sized blade-shaped cavities are present throughout the muddy sandstone. The nests are a centimetre or so in size and elongated parallel to the bedding direction. They may record the former presence of gypsum.

There are several outliers of the Clachtoll breccia facies south and east of Innis nan Gobhar, only one of which shows the typical fining upwards from the Lewisian basement. This is at Inverianvie River. An eastward facing bluff by the river, about 30 m upstream from the road bridge [NG 95088977] shows the base of the deposit. About 10 m of massive conglomerate (facies Ct1) overlies the unconformity. It is remarkable for the excellent rounding of some of the acid gneiss blocks, two of which, both perfectly elliptical, measure 2.0 × 1.1 m and 1.8 × 1.5 m respectively. Most clasts, however, are only 15–20 cm across and quite comparable in shape with those in the equivalent facies at Stoer. An additional 40 m of stratigraphically higher beds are seen in the road cutting and on the shore nearby. The section as a whole fines progressively upwards and at the very top is interstratified with coarse sandstone (facies Ct2).

Aultbea and Rubha Mór

The area was mapped by Gunn in 1887 and subsequently chosen as the type area for the Aultbea Formation (Peach *et al.* 1907, p. 319). Both the Applecross Formation and the overlying Aultbea Formation are extensively developed across the peninsula of Rubha Mór. The former consists of very coarse, pebbly red sandstone, whereas the latter is mainly pebble-free, medium-grained red sandstone. Several strike faults have been detected in the area, but the uniformity of the lithologies makes their importance difficult to establish. A 1:50 000 geological map of this area, revised by the author, was published as part of the Gairloch sheet by the British Geological Survey in 1999. Two sections within the Aultbea sub-area have been designated Geological Conservation Review sites (Mendum *et al.* 2003).

The Applecross Formation that forms the northwestern part of the Rubha Mór is about 500 m thick. Pebbles on the coast NW of Slaggan reach 5 cm in size, but generally they do not exceed 4 cm. They are either scattered through the sandstone or concentrated into seams one pebble thick. On the east side of the sandy bay at Mellon Charles [NG 845908] a medium grained sandstone unit 14 m thick is overlain by a grey siltstone bed. The siltstone is grey (N4–N5), millimetre-laminated fine sandstone to coarse siltstone, with traces of ripple lamination, altogether about 1.5 m thick. About 15 cm below it there is a disrupted grey micaceous siltstone bed a few centimetres thick showing partial phosphatization. The siltstone is overlain by coarse sandstones of the Applecross Formation with pebbles of red porphyry, white quartz and quartzite, green chert and pink quartz–feldspar rock. The pebbles reach 3 cm in size just above the siltstone but their size and abundance diminish stratigraphically upwards. The colour of the fresh rock is moderate red, dominated by the feldspar.

The same siltstone bed reappears along the strike inland at Creag an Fhithich Mór [NG 85009194], overlain by pebbly Applecross Formation like that at Mellon Charles. The contact with the Aultbea Formation, apparently 165 m stratigraphically above the grey siltstone, is exposed 160 m SW of the south end of Loch Beinn Dearg [NG 85429200]. It is defined by the centre of the first 10 m interval, descending the section, in which sandstone with a maximum grain size surpassing 0.5 mm is predominant. A definition like this is required by the metre-scale interbedding of medium and coarse sandstone near the contact.

This section, from Creag an Fhithich Mór to the outlet of Loch Beinn Dearg, is designated the stratotype for the Applecross–Aultbea boundary. Palaeocurrents from trough cross-bedding and linguoid ripples are directed consistently towards the east or NE ($\theta = 076°$, $n = 19$). No change of direction is observed at the contact between the Applecross and Aultbea Formations.

By the same grain-size definition the Applecross–Aultbea boundary occurs on the coast a kilometre SE of Mellon Charles [NG 852906] at a point about 280 m stratigraphically above the grey siltstone bed. The apparently different stratigraphic intervals between the siltstone and the top of the Applecross Formation in the coastal and inland sections may be attributed to faulting. Normal faults parallel to the strike are definitely present in the inland section.

A possible correlative of the grey siltstone described above crops out in the topmost Applecross about 6 km to the north, on the coast north of Opinan [NG 87969760]. It is formed of millimetre to centimetre banded siltstone with linguoid ripples. The base of the bed is grey (N4) but otherwise reddish-grey.

The NE coast of Rubha Mór exposes some 2000 m of the Aultbea Formation with the top concealed beneath Triassic sandstone and the base cut out by several strike faults north of Opinan. The most continuous coastal section, 800 m thick, through the formation has its base at the end of the track leading north from Mellon Udrigle [NG 885973] and the top very near Mellon Udrigle [NG 892960]. This is the type section. The rocks are mainly medium-grained sandstones. The fresh rock is pale red or light brown in colour, but weathers to reddish brown, reddish orange or reddish pink. Contorted bedding and heavy mineral bands (mainly hematite and ilmenite) are typical and are abundant on the coast near Creag an Eilean [NG 889975]. An example is shown in Figure 82. The diagnostic features of the Aultbea Formation, based on the type section, are pale red sandstone with a maximum grain size of 0.5 mm, 95% contorted, with flat bedded and pebbly sandstones each forming less than 1% of any 100 m thick section.

The coast section north of Mellon Udrigle locally has coarser sandstones, some of which are sparsely pebbly, and a few fine-grained red intercalations usually showing an association of flat bedding and linguoid ripples. Three of these intercalations are shown in Figure 83. The flat bedded sandstones in the central graphic log (Fig. 83) were described as 'tiles' by the Geological Survey (Peach *et al.* 1907, p. 321). Close by is the small cave in Aultbea sandstone, roofed by Triassic breccia [NG 90009299], sketched by Nicol (1857*b*).

The palaeomagnetism of samples estimated to be 950 m to 1380 m above the base of the Aultbea Formation near Mellon

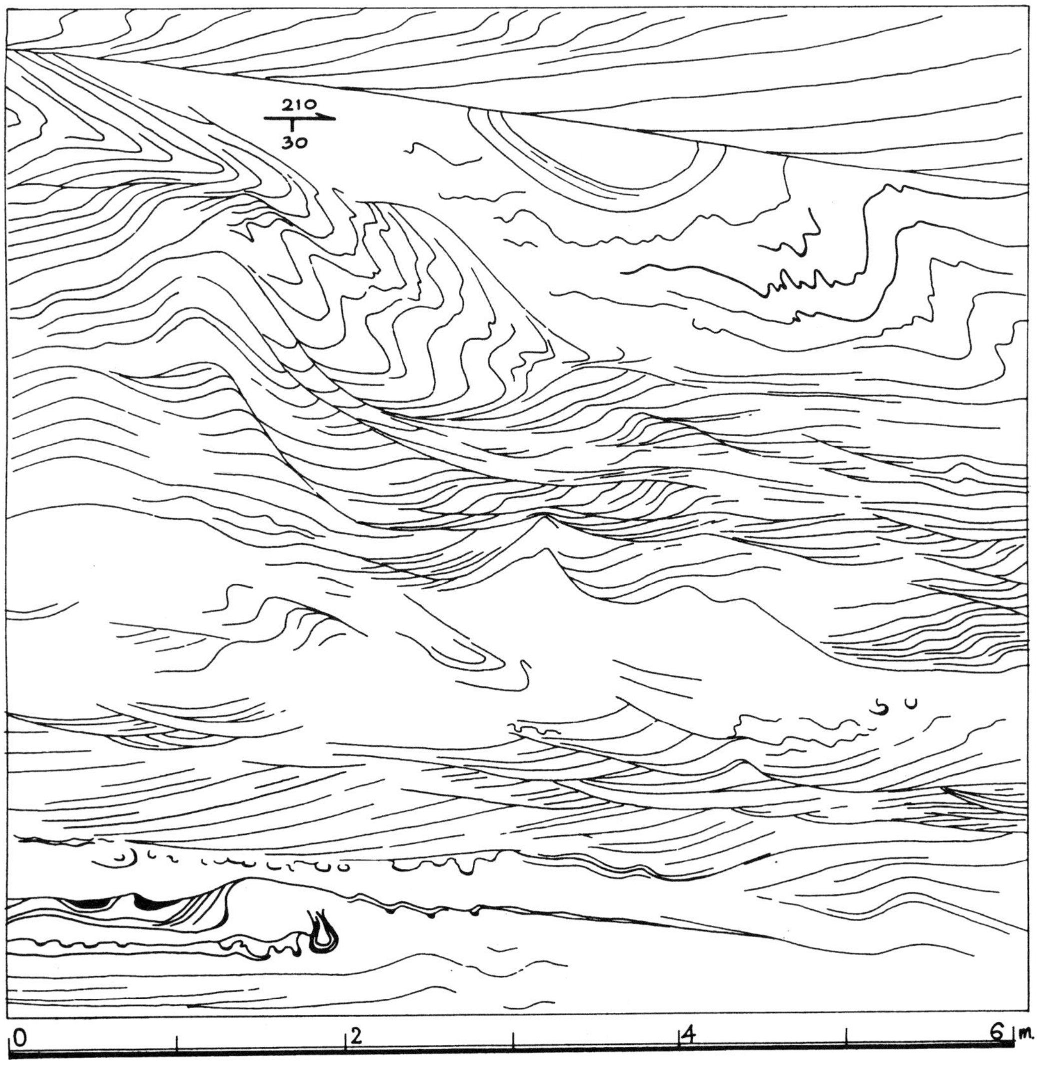

Fig. 82. Contorted bedding in the Aultbea Formation on the coast 2 km north of Mellon Udrigle [NG 88919758]. Fine lines represent grain-size changes, blank areas are structureless. Thick lines in the top right and lower left part of the scaled drawing are bands rich in opaque minerals. The observer looks down dip to the SE at a joint face which cuts bedding at 60°.

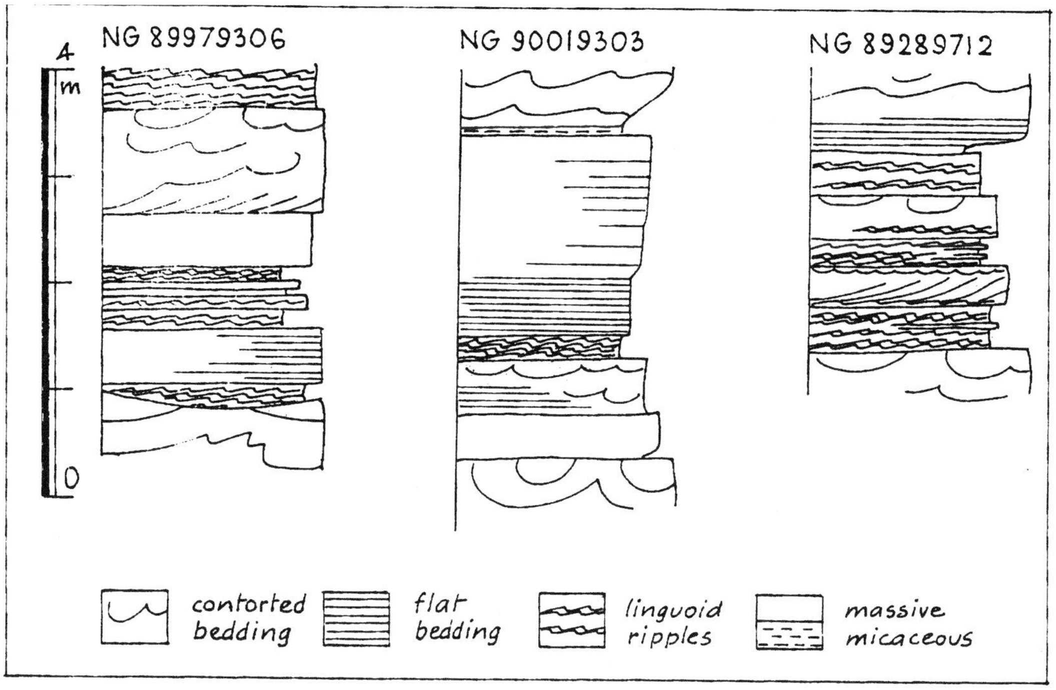

Fig. 83. Graphic logs of fine-grained intervals in the Aultbea Formation near Aultbea.

Udrigle has been measured by Smith *et al.* (1983). They report a direction of magnetization significantly different to that in the Applecross Formation, a fact attributed by them to polar wandering during the time elapsed between deposition of the two formations.

Poolewe

The Poolewe area was mapped for the Geological survey by Gunn in 1889. In his description (Peach *et al.* 1907, p. 317–318) the strata were assigned erroneously to the Diabaig Formation of the Torridon Group. The presence of the Stac Fada Member near the top of the succession, however, shows that it belongs to the Stoer Group. A 1:10 000 geological map of the area by the writer has been deposited with the Geological Survey in Edinburgh, and incorporated into the 1:50 000 Gairloch sheet published by the Survey in 1999. The Clachtoll and Bay of Stoer Formations are both easily recognizable, though their overall stratigraphic thickness is about 1.4 km, as compared with a kilometre at Stoer. Palaeomagnetic investigations by Smith *et al.* (1983) confirm the Stoer Group correlation. A stratigraphic profile of the beds is provided in Figure 6 and a map of the area around Loch Ghiuragarstidh, where facies inter-relationships are particularly well exposed, in Figure 84. A brief description of the lithofacies follows.

The breccio-conglomerate facies (Ct1) is formed of sub-angular gneiss blocks generally not more than 0.5 m across. It is always in contact with basement gneiss. Very coarse breccia exposed about 400 m east of Loch Ghiuragarstidh [NG 89408097] was noted by Gunn (in Peach *et al.* 1907, p. 317) and is shown in Figure 85. The largest block here is 4 × 1.4 m and is estimated to weigh over 30 tonnes. It lies on its side parallel to the palaeoslope like all the others. Sub-rounding of this block is probably due mainly to weathering rather than transport, though weathered gneiss is not seen at Poolewe. The breccia in contact with the gneiss is often massive. Upwards, clast size diminishes and the breccia is interleaved with coarse sandstone. Predominance of sandstone marks the passage into either the tabular sandstone facies (Ct2) or the trough cross-bedded sandstone facies (Ct5), both described below.

The tabular sandstone facies (Ct2) consists of coarse sandstones with scattered gneiss pebbles up to 10 cm in size. Small scale trough cross-bedding though common is subordinate to low-angle cross-bedding of uncertain origin. The trough cross-bedded sandstone facies (Ct5) is similar in pebble content and grain size to Ct2 but trough cross-bedding is the main or only sedimentary structure. Red siltstones (facies Ct3) up to 2 m thick are locally present in both facies but are not mappable. Trough cross-bedding shows that most of the palaeocurrents in Ct2 and Ct5 flowed west ($\theta = 263°$, $n = 38$), with a few exactly counterposed.

The conglomerate facies (Ct4) is formed of metre-thick beds of well-rounded acid gneiss cobbles about 20 cm in size, rarely as much as 50 cm, with relatively thin sandstone interbeds. Stratigraphically upwards through the facies clast size diminishes and interbeds become thicker, as in the breccia facies (Ct1). The facies is about 115 m thick where it crops out by Loch Maree [NG 887773] in the bottom of a deep palaeovalley. The underlying breccia (Ct1) at this point may date from the erosional episode that cut the palaeovalley.

Facies Ct6, which is peculiar to Poolewe, consists of fine to medium grained sandstone with millimetre to centimetre lamination parallel to bedding. The maximum grain size is about 2 mm. In the field the sandstone appears to be poorly sorted, like the muddy sandstone facies (Ct7), but thin sections show it has only about 15% of matrix. West of Loch Losguinn [NG 882814] facies Ct6 passes laterally to the south over some hundreds of metres into the muddy sandstone facies (Ct7) and to the north into the pebbly sandstones of facies Ct5. Low angle cross-bedding occurs very rarely. Red shale bands up to 3 m thick occur sporadically in this facies. The facies generally overlies either the trough cross-bedded facies (Ct5) or the tabular sandstone facies (Ct2).

The muddy sandstone facies (Ct7) is a fine-grained dark red sandstone with greywacke texture. Scattered quartz grains up to about 1.5 mm lie in a hematitic matrix that under the microscope forms 50–60% of the rock. The sandstone forms apparently massive beds as much as 5 m thick though usually they are much less. Graded red siltstone layers cut by decimetre-sized desiccation polygons separate the massive beds. Ripple-drift lamination associated with the shale shows palaeocurrents flowing consistently towards the east, counter to most of those in the other facies.

The facies described above are bounded laterally by basement, and like the Clachtoll Formation at Stoer contain only locally derived gneiss debris. All but facies Ct4 were deposited by alluvial fans prograding into swamps. The massive breccio-conglomerates represent the fan heads; downslope, and stratigraphically upwards, one encounters pebbly sandstones of either the tabular bedded sandstone facies (Ct2), or the trough cross-bedded sandstone facies (Ct5). Both of these pebbly facies grade into facies Ct6, representing the fan toes. The muddy sandstone facies (Ct7) formed as the storm water flowing across the fan surfaces was ponded in temporary shallow

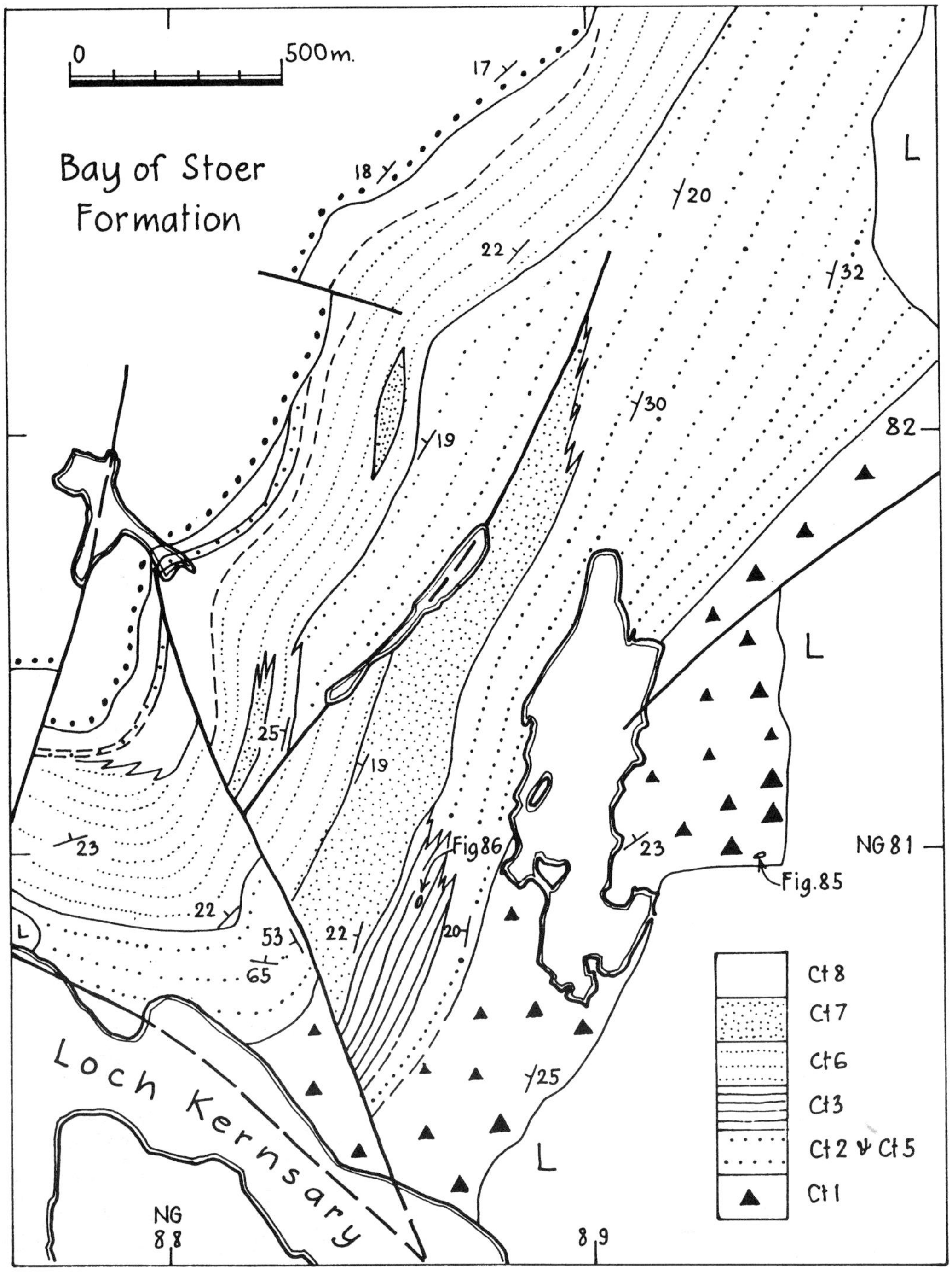

Fig. 84. Lithofacies map of the Clachtoll Formation north of Loch Kernsary. See Plate 2 for location. Facies Ct3 is generally red shale but here contains millimetre-thick graded sandstone laminae and is desiccated. The shale is interbedded with very coarse sandstone (facies Ct2 & Ct5) and sets of planar cross beds probably aeolian (facies Ct8), together forming a mappable complex. The Clachtoll Formation is underlain by the Lewisian gneiss complex (L) to the east, and overlain erosively by the Bay of Stoer Formation to the west.

lakes that dried out rapidly. The cobble conglomerate facies (Ct4) differs from those just described in containing clasts that have been brought a substantial distance by a trunk river.

The laminated sandstone facies (Ct8) is easily recognizable by its sharply defined, millimetre-thick cross beds with lateral persistency $p > 1000$. Alternate laminae typically have modal grain sizes of about 0.25 mm and 0.6 mm, with a maximum grain size of about 2 mm. The cross-lamination forms sets up to 4 m thick and in plan appears straight or gently curved (Fig. 86). At one locality [NG 882818] cross-lamination strike changes 50° in 18 m defining a trough with its axis directed towards the north. Foresets in the facies generally dip north to NW after correction for tectonic tilt. These directions are quite different to those in the same facies at Stoer, and also different to the palaeocurrent directions indicated by cross-bedding in adjacent facies. Erosion surfaces are sometimes covered with thin sheets of red shale or coarse sandstone with gneiss pebbles up to 2 cm in size. Desiccated shale films sometimes coat asymptotic foreset toes. The laminated facies (Ct8) is interpreted as aeolian, as at Stoer and Achiltibuie.

Facies Ct8 where it outcrops about 300 m west of Loch Ghiuragarstidh [NG 88618088] is interbedded with the trough cross-bedded sandstone facies Ct5, but also with units up to 3 m thick of graded and desiccated red siltstone laminae, deposited in a small, ephemeral lake. This complex forms a mappable unit, labelled as facies Ct3 on Figure 84.

The upper part of the Poolewe succession consists of medium-grained sandstones, typically trough cross-bedded, with well-rounded pebbles of acid gneiss and fine or medium grained quartzite. The sandstones and their pebble suite are identical to the Bay of Stoer Formation (BS1) at Stoer, except that the pebbles (see below) are larger and more feldspathic. Palaeocurrents at all levels were directed towards the NW ($\theta = 304°$, $n = 60$), significantly different to those in the Clachtoll Formation which flowed towards the west, and completely different to those in the Bay of Stoer Formation at Stoer that flowed towards the east.

The base of the Bay of Stoer Formation is well defined but not evidently erosional except at a point about 2 km due east of Inverewe [NG 882819] where it cuts through about 11 m of the

Fig. 85. The basal conglomerate of the Clachtoll Formation, facies Ctl, about 300 m east of Loch Ghiuragarstidh [NG 89408098]. The hammer handle is 0.5 m long. The gneiss block in contact with the hammer head is estimated to weigh over 30 tonnes.

(a)

(b)

Fig. 86. Facies Ct8 (aeolian) about 200 m west of Loch Ghiuragarstidh [NG 88588088]. In the upper photograph (**a**) the observer looks NE at a bedding plane which has been stripped to expose the foreset edges. The foresets originally dipped towards the NW. In the lower photograph (**b**) the observer looks NNW, along the strike of bedding, defined by a band of pebbly sandstone dipping about 20° to the west, partly concealed by the top left-hand corner of the map case. The map case is 34 cm long.

laminated facies Ct8, and about 4 m of very coarse sandstone belonging to facies Ct5.

The highest beds of the Bay of Stoer Formation contain the distinctive Stac Fada Member (SFM). This volcaniclastic sandstone unit is about 6 m thick where it crops out on the coast, just over a kilometre north of Inverewe. There is an isolated outcrop beneath Alexander Cameron's monument at Inverewe [NG 86208170], probably brought down by a major strike fault. Pebbly sandstones typical of the Bay of Stoer Formation that are exposed in road sections about 100 m NNE of the monument apparently overlie the Stac Fada Member normally. These sandstones are at least 70 m thick and suggest that the Stac Fada Member at Poolewe does not define the top of the Bay of Stoer Formation as it does at Stoer.

The pebbles of the Bay of Stoer Formation are mainly less than 5 cm in size, but may reach 10 cm. They are especially abundant in the topmost 260 m just beneath the Stac Fada Member. Pebbles of quartzite and gneiss as much as 20 cm in size can be seen in outcrops by the main road at Tournaig water tower [NG 87598393] and again about 800 m north of Inverewe House. Pebbles are also common in the lower 340 m of the Formation, SW of Loch Kernsary. A relatively pebble-free middle section about 100 m thick crops out east of Poolewe village and around Inverewe, giving a total thickness for the formation of about 700 m. If the strike fault passing Alexander Cameron's monument is supposed to repeat the upper part of the Bay of Stoer Formation then the formation thickness drops to about 650 m. This is still four times the thickness of the same formation at Stoer.

Stoer Group sediments are generally pale red to greyish red, but within a kilometre of the Loch Maree fault are grey, and locally veined by a green mineral like epidote. Gunn records epidote in a thin section of these sandstones (Peach *et al.* 1907, p. 317). Similar grey discoloration of the Stoer Group is seen on the other side of the Loch Maree fault, at Bac an Leth-choin. Stoer Group sandstones at Rudha Réidh, however, just over a kilometre from the fault, are red. It is significant that the Torridon Group sandstones that overlie the Stoer Group at Bac an Leth-choin are also red. The grey colour could have developed as a result of the partial reduction of iron oxide pigment by hot formation water moving up the Loch Maree fault under a cover of Stoer Group sediment several kilometres thick. The thermomagnetic decay curves, however, show that hematite is responsible for the magnetization measured in these rocks. Traces of magnetite were found at only two localities out of the fourteen sampled palaeomagnetically (Smith *et al.* 1983).

Torridon Group sandstones now unconformably overlie the Stoer Group at Bac an Leth-choin and must once have done so at Poolewe, at an erosional level only slightly above the present one.

Bac an Leth-choin

The Stoer Group, unconformably overlain by the Torridon Group, is exposed east and NE of the hill called Bac an Leth-choin. It was thought by the Geological Survey to be Diabaig Formation (Peach *et al.* 1907, p. 330). Outcrops are abundant along the NE facing Loch Maree fault scarp. A 1:50 000 geological map of the area incorporating the author's 1:10 000 mapping was published as part of the Gairloch sheet by the British Geological Survey in 1999. A graphic log of the succession is given in Figure 87.

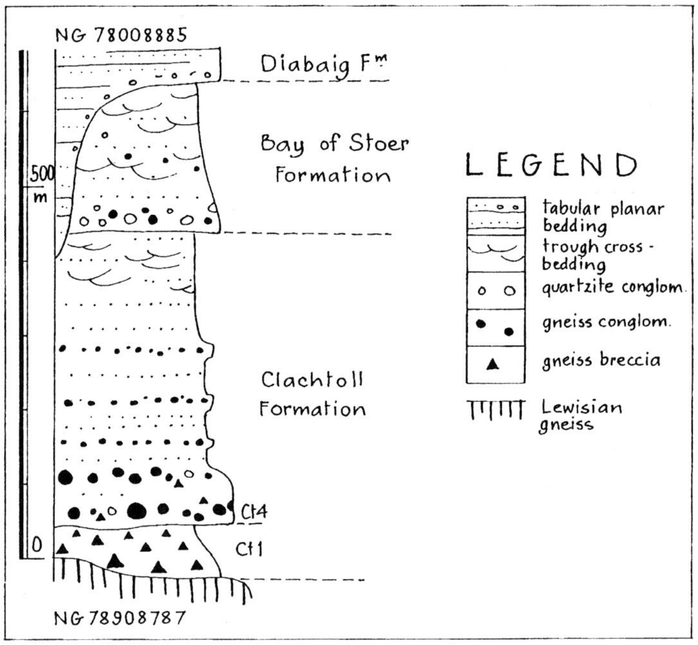

Fig. 87. Graphic log of the Steer Group near Bac an Leth-choin.

Angular gneiss breccia at the base of the sequence is seen in contact with the underlying gneiss in a small knoll about 730 m south of Loch na Féithe Dìrich [NG 78908789]. About 70 m stratigraphically higher, well-rounded gneiss cobbles up to a decimetre, or more, in size, become abundant. They form the base of an important upward-fining cycle about 400 m thick. The matrix of the conglomerate consists of coarse sand and small pebbles, some of which are quite angular. Although the vast majority of the clasts are made of coarse-grained acid gneiss, there are also white quartz, jasper and magnetite–quartz pebbles. Very rare fine-grained quartzite and marble clasts have also been found. They almost certainly come from the metasediments of the Loch Maree Group that now crop out in the basement about 10 km to the south. Stratigraphically upwards, the pebble size diminishes as matrix increases in importance so that the rock becomes a coarse grey sandstone with seams of centimetre-sized pebbles. Pebble imbrication suggests palaeocurrents flowing towards the NW. The grey colour, by analogy with the Stoer Group of the Poolewe area, is attributed to reduction of the iron oxide pigment by hot water coming from the Loch Maree fault zone.

Well-rounded cobbles again become abundant in the sandstone west of Loch na Féithe Dìrich [NG 782887], about 470 m stratigraphically above the local base of the Stoer Group. The lowest 4 m of this new unit is packed with decimetre-sized pebbles. The rock higher up is reddish-grey fine to medium-grained sandstone with pebble bands 1–2 m thick. Trough cross-bedding is typical and is locally contorted. The pebble bands become thinner and less frequent upwards. The stratigraphically highest exposure of these sandstones, by the footpath SE of Loch Ceann a Chàrnaich [NG 78008928], is pebble free.

The main difference between these two fining-upward cycles is that in the upper one, pink and white quartzite cobbles form about 20% of the pebble suite, whereas in the lower one quartzite is very rare. The abundance of quartzite pebbles in the upper cycle invites comparison with the Bay of Stoer Formation; the lower cycle is like Clachtoll Formation facies Ct4 exposed SE of Inveran, across Loch Maree.

The top of the Stac Fada Member is well exposed in a small stream (not shown on Ordnance Survey maps) that plunges down the south side of the deep valley called the Feadan Mòr [NG 773893]. The underlying beds are not exposed. Assuming no intervening faults, the Stac Fada Member is stratigraphically about 430 m above the base of the Bay of Stoer Formation to the SE, and 900 m above the local base of the Stoer Group. The Stac Fada Member contains vesicular glassy shards, now altered to coarse chlorite or to iron-rich sericite, set in a black ferruginous matrix. Quartz grains with deformation lamellae occur both in the matrix and in the altered glass.

The contact between the Stoer Group and the unconformably overlying Torridon Group (Db1) is exposed 700 m NE of the summit of Bac an Leth-choin [NG 77968881]. Roughly horizontal tabular red sandstone beds, containing red sandstone and quartzite pebbles derived from the Stoer Group, lie across the unconformity. These pebbly beds were originally assigned by the Geological Survey to the Triassic. The difference in dip across the unconformity is 27°.

Rubha Réidh

The sea cliffs east of Rubha Réidh lighthouse expose boulder breccia banked against a spectacular fossil 'cliff' of Stoer Group sandstone that is 30 m high. The breccia contains only sandstone clasts, so it is not surprising that it was thought to be Triassic by Gunn who mapped the area for the Geological Survey. The discovery that the breccia belongs to the Torridon Group is due to Lawson (1965, 1976). Irving & Runcorn (1957) had already shown that the magnetization of the sandstone beneath the unconformity at Rubha Réidh was similar in direction to that in beds at Stoer now called the Stoer Group, but the stratigraphic significance of this was not appreciated until later (Stewart 1966*b*). Irving & Runcorn did not sample the beds above the unconformity. However, later work by Stewart & Irving (1974) on the rocks both below *and* above the unconformity at Rubha Réidh showed that the directions of magnetization conform with those for the Stoer and Torridon Groups in their respective type areas. A geological map of the area, revised by the author, was published as part of the 1:50000 Gairloch sheet by the British Geological Survey in 1999.

The Stoer Group is about 100 m thick NE of Rubha Réidh lighthouse where it is least affected by faulting. It consists of fine to medium-grained red sandstone in two interbedded facies –one dominated by cross-bedding, the other parallel laminated and frequently rippled. These two facies are like MD1 and MD2 at Stoer.

The cross-bedded facies is formed of sets about a metre thick of essentially planar cross beds. The foresets are finely laminated ($p \approx 2000$). Foreset dips are generally low, inclined towards the SW ($\theta = 222°$, $n = 46$) after correction for tectonic tilting. Reactivation surfaces and metre-deep channels are common. Weak contorted bedding occurs only rarely.

The parallel laminated facies has laminae of high persistency ($p \approx 10000$), despite low angle discontinuities. Straight crested or slightly sinuous ripples with thin mud drapes frequently cover bedding surfaces of as much as 1000 m^2. Ripple crests trend consistently NE. These ripples appear to represent periods when shallow flood waters temporarily covered large areas. Desiccation polygons of metre size downwards are common.

The similarity of these two facies at Rubha Réidh to those forming the Meall Dearg Sandstone Formation at Stoer invites lithostratigraphic correlation, but the palaeocurrent direction at Stoer is to the NW, not SW as it is here. Furthermore, ripple trends at Stoer and Rubha Réidh are orthogonal.

The Torridon Group at Rubha Réidh is lithostratigraphically similar to that at Gairloch (*q.v.*). The only difference is that the lowest facies of the Diabaig Formation (Db1) at Rubha Réidh contains clasts of sandstone rather than gneiss. There is no evidence pointing to the marine coastal origin for this deposit proposed by Lawson (1976).

At Rubha Réidh the Diabaig Formation rests on the Stoer Group with an angular discordance of about 16°. The unconformity has relief of at least 50 m (Lawson 1976). Where it is transected by the present sea cliff 600 m NE of the lighthouse [NG 74519211] the relief is about 30 m. Massive breccia (Db1) containing tightly packed angular to sub-rounded sandstone blocks up to 3 m × 5 m is banked up against the exhumed unconformity at this locality, which is described in detail and figured by Lawson (1976). The sandstone forming the blocks appears to be identical to that in the underlying Stoer Group.

The unconformity reaches the sea again about 300 m SW of the lighthouse [NG 73849160] but is cut out by a small fault on the cliff. It is also exposed by the road 900 m south of the lighthouse [NG 74089095] due to fault repetition. To the west it descends the cliff and reaches the sea about 950 m south of the lighthouse [NG 73959087]. The basal breccia in this area is only a few metres thick, however.

The massive breccia passes upwards over a few metres into interbedded red sandstone and sandstone conglomerate. The sandstone clasts are centimetre to decimetre sized and relatively well rounded. They become smaller and less abundant upwards through the sequence. The sandstone interbeds have frequent black bands

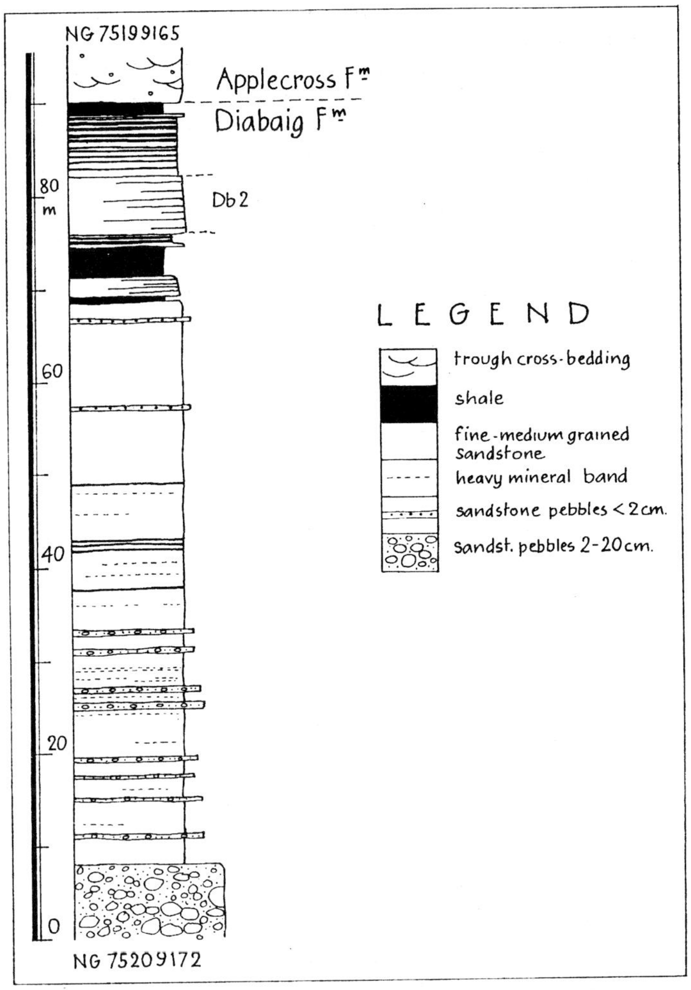

Fig. 88. Graphic log of the Diabaig and basal Applecross Formations on the cliff about 1.2 km east of Rubha Réidh lighthouse. The section is red apart from the part labelled Db2 which resembles the corresponding facies at Diabaig in being grey and desiccated.

(Lawson 1976, p. 75). A band on the cliff opposite Stac Dubh [NG 747920] is 0.5 m thick. Under the ore microscope about 90% of the black grains are seen to be martite, often with ilmenite lamellae arranged in a triangular pattern. The remainder are ilmenite. Grain size of the opaques is 100–150 μm; the rare detrital quartz grains are two or three times larger. Grain rounding is excellent. The matrix is clear, unstrained quartz, in optical continuity with the detrital grains. The martite and ilmenite could have come from the Stoer Group, which contains an almost identical suite.

Small-scale cross-bedding in the lowest part of the Diabaig Formation shows that the palaeocurrents flowed towards the SW, whereas from pebble imbrication the palaeoflow was towards the south (Lawson 1976).

The cliff about 500 m west of Camas Mòr [NG 752916] exposes about 60 m of interbedded sandstone and sandstone conglomerate (Db1), overlain by 20 m of red and grey siltstones (Db2). The sequence is shown graphically in Figure 88. The siltstones are followed erosively by pebble-free sediments like the Applecross Formation, perhaps the Allt na Béiste Member. Grey shale is exposed elsewhere near the cliff top west of Camas Mòr and also in a gravel pit by the road about 350 m SE of the lighthouse [NG 74079148].

Gairloch

The area stretches from near Rubha Réidh in the north, to Diabaig in the south. It was mapped by Gunn and Clough for the Geological Survey in 1889–90 and published as part of the one-inch to the mile sheets 91 & 100 in 1893 & 1888, respectively. A new edition of these maps in one sheet, scale 1:50000, with the Torridonian revised by the author, was published by the Survey in 1999. The Torridon Group is exposed extensively throughout the area, everywhere resting unconformably on unweathered gneiss with palaeorelief up to 300 m (Geikie 1880; Stewart 1972). A general description of the rocks is given in the NW Highlands memoir (Peach *et al.* 1907, p. 326–34). The sedimentology of the Stoer and Torridon Groups at Gairloch has been investigated by Lawson (1970).

The lowest part of the Torridon Group, the Diabaig Formation, was subdivided by the Geological Survey into three 'zones' (Peach *et al.* 1907, p. 324). My mapping shows that all three zones are valid lithostratigraphic units but the lower two are here considered as facies. The lowest unit (identical to facies Db1 at Diabaig) is formed of locally derived breccia and conglomerate in contact with the underlying gneiss, grading laterally into tabular red sandstones with angular gneiss fragments. It is overlain by grey shales and tabular grey sandstones forming the middle unit (like facies Db2 at Diabaig). The topmost unit consists of red sandstones with a few interbeds of red and grey siltstone. The sandstones are often trough cross-bedded and contain scattered rounded pebbles of quartz, jasper and fine-grained quartzite up to a centimetre in size. Towards the top of the unit contorted bedding becomes increasingly common. The sandstones of the topmost unit resemble those of the overlying Applecross Formation except for their finer grain size and slightly paler colours. They correlate with the Allt na Béiste Member which, as at Diabaig, is here considered part of the Applecross Formation. On the 1:50000 geological map of Gairloch (British Geological Survey 1999), however, they are still included in the Diabaig Formation, with the code TCD3.

Breccia clasts in the basal breccia (facies Db1a) are either sub-angular acid gneiss or metasediments, or well-rounded to sub-rounded metadolerite, reflecting the lithology of the underlying basement. An upward decrease in pebble size through the facies is apparent at most localities. Pebble size also generally diminishes laterally over tens or hundreds of metres into a more sandy, crudely stratified deposit ($p \approx 20$) showing frequent clast imbrication. Sedimentary structures in the sandstones include various kinds of low angle cross-bedding. Red, less commonly green or grey, silty partings with sand-filled cracks and ripple marks are seen in places.

Just east of Shieldaig Lodge, in the wooded crags above the road [NG 809723] the basal unconformity is exposed for over 100 m. The lowest sediments contain clast-supported sub-rounded to rounded Lewisian blocks of metadolerite (and rarely acid gneiss), ranging in size up to 1.5 × 0.7 m with the average around 0.4 m. The matrix is rich in chlorite and hornblende. Average clast size diminishes upwards over 25 m stratigraphically to less than 10 cm. Lawson (1970, p. 240) has reported a hematitic arkose pebble from this outcrop, probably derived from the Stoer Group. Metadolerite pebbles have selvages of hematite several millimetres thick in exposures about 450 m SE of Shieldaig Lodge Hotel [NG 809721]. The greatest stratigraphic thickness of breccias and sandstones belonging to member Db1 near Loch Shieldaig is about 50 m. The palaeorelief was at least 100 m.

Exposures of the basal unit (facies Db1) around Lochan nam Breac, noted by both Geikie (1880, p. 403 & fig. 3) and the Geological Survey (Peach *et al.* 1907, p. 330), show both massive and tabular breccias, almost flat-lying and about 30 m thick, apparently occupying the bottom of a gentle hollow in the gneiss. The hollow has a relief of about 180 m. The 350 m long bluff on the eastern side of the loch shows vertical transitions from massive breccia [NG 816782] into laminated red sandstone with only sporadic gneiss clasts, low angle cross-bedding, scours and heavy mineral bands [NG 815786]. The palaeocurrent direction from imbrication and cross-lamination was towards the NW, sub-parallel to the exposed face. Breccia exposed 1–2 km to the north gives a similar palaeocurrent direction.

Easily accessible exposures of bedded breccia (Db1b) can be seen along the shore near Gairloch Hotel (Fig. 89). Imbrication of clasts in the shore section agrees with ripple cross-lamination in showing northward flowing palaeocurrents.

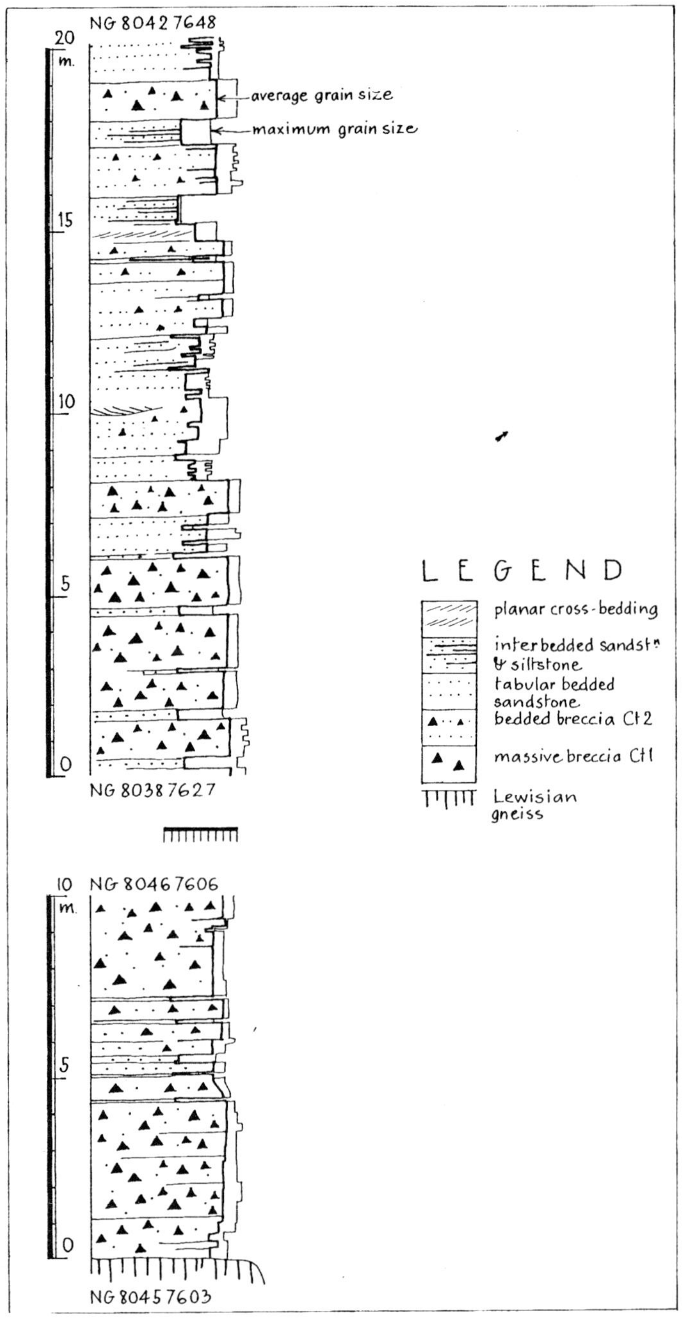

Fig. 89. Graphic log of breccias and sandstones exposed along the shore near Gairloch. The lower section is close by the Free Church and the upper section north from Gairloch Hotel. The two are separated by the beach Port an Daraich. The top of the upper section is about 8 m below the Applecross Formation exposed at Creagan nan Cudaigeam. The grain size scale spans +3 to −6 ϕ units (0.12–64 mm). The graphic log shows average grain size (thick line) and maximum grain size (fine line), measured every 50 cm. The maximum was that found within 0.5 m laterally from the traverse. Palaeocurrents from planar cross beds flowed northwards.

The grey shale facies (Db2) is seen at only a few points in the Gairloch area. Exposures by the roadside at Badachro [NG 784730] show several metres of shale with well-developed phosphatic laminae and lenses, all containing microfossils. In the river beneath the road, Abhainn Bad a' Chròtha, the shale is 50 m thick (Lawson 1970, p. 328) and passes laterally [NG 78357330] into grey sandstone and breccia of facies Db1.

Grey shales and sandstones in the Allt Mór, a tributary of the River Sand [NG 774805] are 40 m thick according to Lawson (1970, fig. 8-2). Phosphatic lenticles have yielded microfossils. The shale sequence (Db2) is overlain by red sandstone of the lowermost Applecross Formation, forming a waterfall. Fault repetition brings the same sandstone to view again about 300 m farther up the stream where there is another waterfall [NG 772808]. A third waterfall 200 m still farther up the stream [NG 77078093] is due to a similar sandstone which, however, is underlain by grey sandstone, not shale. This last sandstone is probably at a higher stratigraphic level in the Applecross than the other two.

A shale interval on the cliffs 1.2 km due south of Rubha Réidh lighthouse [NG 73979063], about 12 m thick, is anomalous in being mainly red. The grey shales present are finely laminated, with high lateral persistence, and contain hematite and calcite inclusions about a centimetre in size. Slight compaction of bedding around the former suggests they developed around early diagenetic pyrite. There is no compaction visible around the calcite inclusions, nor are they evidently pseudomorphs.

Grey shales in Abhainn Bràigh-Horrisdale [NG 817686] are about 60 m thick. Mappable grey shale also crops out in the River Talladale above Victoria Falls [NG 894709] and about 2 km NE of Baosbheinn [NG 875688]. The linear structures in grey shale in the river bed 300 m above Victoria Falls described by Jolly (1880) have not been examined but, from the description, resemble kink bands. The reference by the Geological Survey (Peach *et al.* 1907, p. 330) to a section of the Diabaig Formation 150 m (500 feet) thick in a stream 1.5 km north of the outlet of Loch Gaineamhach is misleading. The stream banks expose gently dipping beds of grey shale and sandstone with a thickness of only a few metres.

The lowermost part of the Applecross Formation, the Allt na Béiste Member, is very variable in thickness perhaps due to differential compaction of the underlying palaeovalley fill (cf. Stewart 1972, fig. 3). The thickest sections are seen on the coast west of Strath [NG 797772] and west of Big Sand fishing station [NG 745791]. The Strath section is about 190 m thick, though the upper half is not exposed. The Big Sand section is 160 m thick, all well exposed but with the base faulted out. Both sections contain shales and siltstones, mostly red, studied geochemically by Stewart (1995*b*) and shown in Figure 90. A similar thickness of strata belonging to the Allt na Béiste Member is seen on the cliffs about 1.2 km south of Rubha Réidh lighthouse. The 'fine red grits' (i.e. the Allt na Béiste Member) noted on the one-inch geological map north-west of Baosbheinn [NG 8268 to NG 8469] outcrop over several square kilometres but are probably no more than 120 m thick north of Loch Gaineamhach. Palaeocurrents in the sandstones of the Allt na Béiste Member SW of Baosbheinn flowed towards the SE (Lawson 1970, fig. 7-2), exactly like those in the overlying Applecross Formation. The medium-grained red sandstones of the Allt na Béiste Member around Badachro and Loch Shieldaig, close to the basement, are only about 25 m thick. South of Loch Gaineamhach [NG 8464] the member thins over a basement hill and at Diabaig is only 22 m thick.

The Applecross Formation above the Allt na Béiste Member consists of coarse to very coarse red sandstones with abundant well-rounded pebbles of quartz, jasper, porphyry, and fine-grained quartzite (Peach *et al.* 1907, p. 274, 279–84). The pebbles are generally over 2 cm in diameter. The sandstone is typically cross-bedded and highly contorted. The base is gradational over metres or tens of metres into the underlying Allt na Béiste Member. Good sections of the contact can be seen 500 m west of Big Sand fishing station [NG 74087913], at Badachro Farm [NG 78287312], Camas na h-Airigh [NG 793734], about 1 km north of Loch Gaineamhach [NG 829681], and south of Garbh Choire [e.g. NG 870690 to NG 871688]. The Applecross Formation above the Allt na Béiste Member is about 330 m thick on Baosbheinn and 600 m on the coast between Strath and Sand. The much greater thickness of 2000 m which can be calculated from the uniformly dipping beds SE of Red Point very probably arises from fault duplication. Palaeocurrents measured by Nicholson (1993, table 1) in the lowest 500 m of the Applecross between Big Sand fishing station and Carn Dearg flowed southeastwards ($\theta = 134°$, $n = 147$).

NE trending strike faults, downthrowing to the SE, traverse the whole area. Consequently, the basement is exposed at both the southeastern and northwestern extremities of the area, roughly 30 km apart, despite a northwesterly dip in the Applecross Formation. North of Loch Gairloch, where the dip is roughly 15°, the

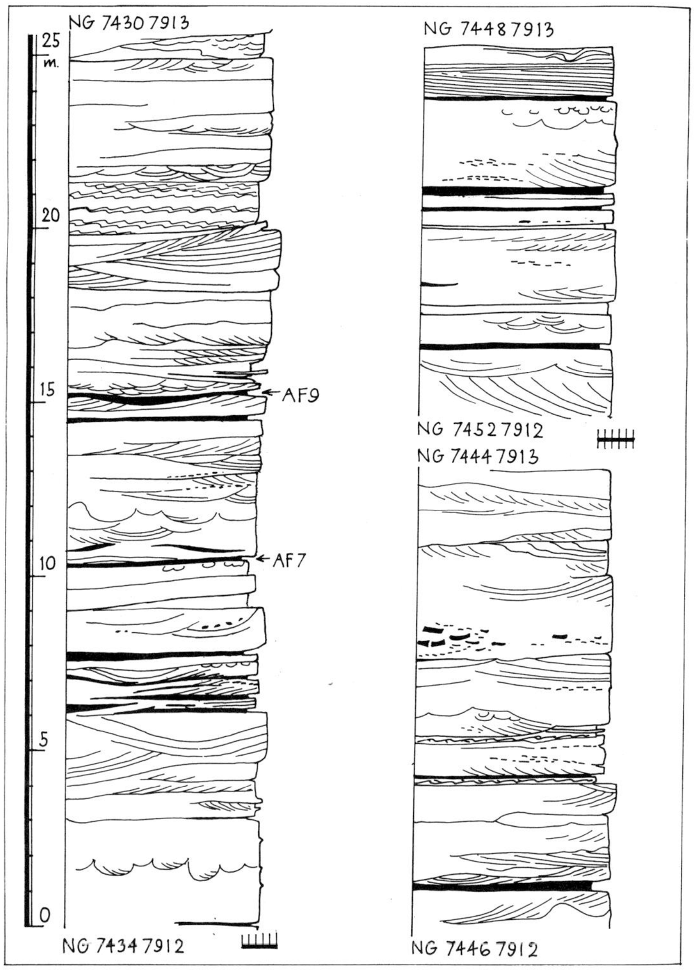

Fig. 90. Graphic logs of the Allt na Béiste Member (lowermost Applecross Formation) on the shore about 300 m west of Big Sand fishing station. Sedimentary structures are drawn to scale. The grain size scale spans +4 to $-1\ \phi$ units (0.06–2 mm). Sampling levels AF7 and AF9 of Stewart (1995*b*) are indicated.

faults must have a total movement of some 4.7 km. Of this movement the fault at Sand, that brings Triassic sandstone down to sea level, accounts for only 1.2 km. The rest must be taken up by faults within the Applecross Formation, most of which have escaped detection.

Faulting has also played a role in displacing the Diabaig–Applecross Formation boundary about 4 km between Lochan Sgeireach [NG 805635] and Beinn Bhreac [NG 848640]. An east to west section between these two points (Fig. 91) requires a fault downthrowing to the east about 100 m. This is quite likely, for there are two faults near Diabaig with a total downthrow of 280 m to the east that strike towards the section in the poorly exposed ground east of Meall an Tuim Bhuidhe. Figure 91 also shows how superposition of the Applecross Formation directly on Lewisian gneiss along part of the section can be accounted for by high palaeorelief.

The NE trending faults in the Gairloch area may be associated with the sandstone dykes that cut both the gneiss and the overlying Torridon Group. The dykes strike about 060° and outcrop at intervals for 14 km along the same trend. They have also been intercepted in the gneiss at a depth of 500 m below ground level during exploratory drilling for Cu–Zn–Au minerals (Jones *et al.* 1987). Dykes at six localities are listed by Peach *et al.* (1907, p. 193, 333):

(1) On the coast 3 km SE of Red Point [NG 751663 and NG 751661] cutting the Applecross Formation. These dykes are 2 m wide – much the thickest and also the stratigraphically highest.

(2) About 0.5 km SE of Loch Bràigh-Horrisdale [NG 804700 to NG 808702, and NG 81347038] cutting Lewisian basement.

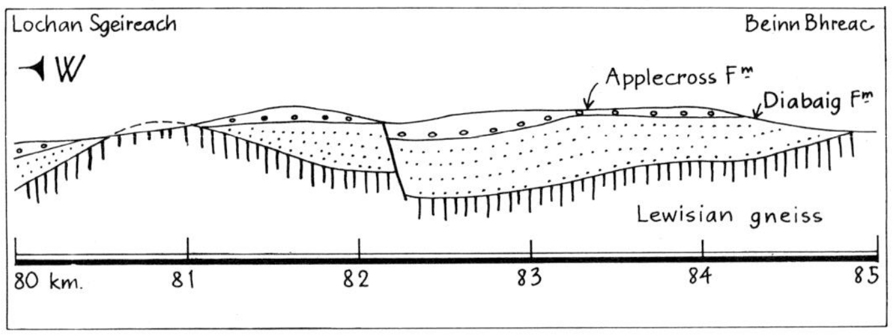

Fig. 91. Cross section showing the relationship between the Torridon Group and basement between Lochan Sgeireach [NG 805635] and Beinn Bhreac [NG 848640]. The scale bar shows National Grid lines at kilometre intervals.

(3) In and near Allt Loch na Doire Móire, about 1 km NW of Lochan Druim na Fearna [NG 820715 and NG 820713] cutting the Diabaig Formation.
(4) By the road side 0.5 km west of Loch Bad an Scalaig [NG 843721] and again about 1 km to the west [NG 834719 and NG 835719], cutting Lewisian gneiss.
(5) On the west side of Meall Aundry [NG 841725; NG 843730; NG 845732; NG 844735] cutting gneiss.
(6) On the north side of Loch na Féithe Mugaig [NG 860749] cutting gneiss.

Diabaig

The area was mapped by Horne, Clough and Hinxman for the Geological Survey in 1889–94 and designated the type area for the eponymous formation by Peach *et al.* (1907, p. 324). It was remapped by Maycock (1962), a revised version of whose map (scale 1:10 560) has been deposited with the Geological Survey in Edinburgh. The northern part of the Diabaig area appears on the 1:50 000 Gairloch map sheet, published by the British Geological Survey in 1999. The rest is shown in Figure 92. The sedimentology of the Diabaig Formation has been studied by Allen *et al.* (1960), the boron-in-illite content by Stewart & Parker (1979), the mineralogy and geochemistry by Rodd & Stewart (1992) and Stewart (1995*b*). The coast section at Diabaig is a Geological Conservation Review site (Mendum *et al.* 2003).

Basement relief was about 250 m when sedimentation started. All the sediments of the Diabaig Formation accumulated in the lower parts of the palaeovalley and laterally abut the basement gneiss that supplied the coarser detritus. The sediments nearest the gneiss are the coarsest at any given stratigraphic level, consisting usually of either massive breccias, or interbedded breccia and tabular red sandstone. This is called the breccia facies (Db1a) and is illustrated in Figure 93. Splendid exposures of the facies are also to be seen by the roadside at Ardheslaig, on the other side of Loch Torridon [NG 782563].

The clasts in the breccia facies are 1–10 cm in size and only rarely over a metre. Reddened rims are usual. Away from the unconformity clast size diminishes and the proportion of sand increases. When clast content by volume falls below 50%, which at Diabaig is always less than 400 m from the unconformity, the sediments are assigned to the tabular sandstone facies Db1b (Fig. 94).

The tabular sandstone facies owes its well-defined bedding to films of relatively fine grained sediment, which is frequently micaceous and ornamented by straight crested, symmetrical ripples. Small scours are common.

Liquefaction of the tabular sandstone facies has produced large-scale mass flow deposits at three localities in Upper Diabaig. The first locality is 200–500 m NE of Loch Roag [NG 823613], the

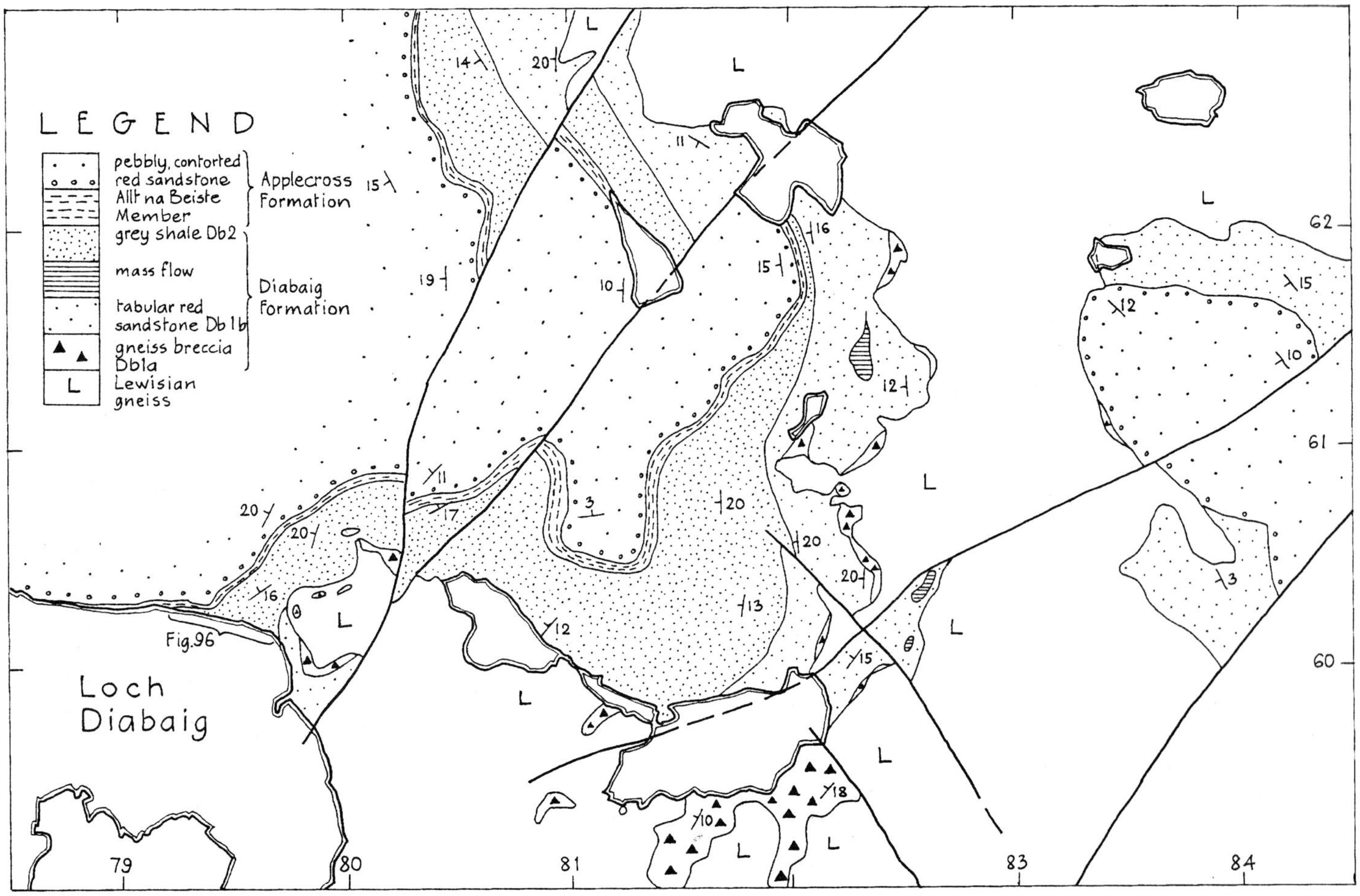

Fig. 92. Geological map of the Diabaig area, based on an outcrop map by Maycock (1962), revised and with boundaries by the author. See Plate 2 for location.

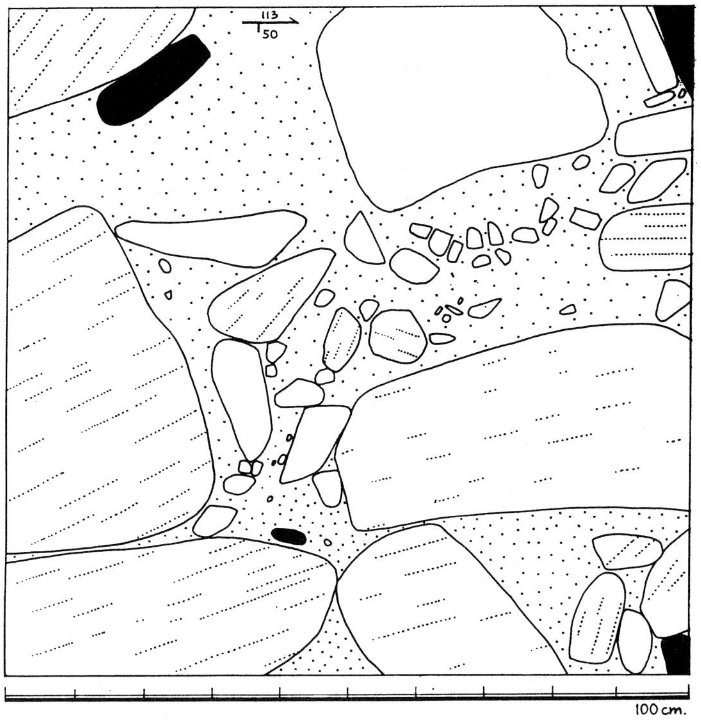

Fig. 93. The massive breccia facies (Db1a) of the Diabaig Formation at Diabaig. The figure is based on a scaled drawing of an outcrop by the roadside a kilometre east of Upper Diabaig [NG 82166012]. Basic clasts are black and the rest acid gneiss. Matrix grain size is about 1 mm.

second and third are about 1 km SE of the same loch [NG 825603 & NG 825600]. The mass flow deposit nearest the loch is 20 m thick. The basal contact is sharp, with signs of injection and disruption. Wispy folds within the flow have a symmetry indicating movement towards the SW, down the palaeoslope. At the third locality the base of the massive deposit cuts down through more than 5 m of tabular sandstone. All three flows are covered by undisturbed tabular sandstones. The mobilization of the sands may be seismic in origin, but predates the NE trending intrusive sandstone dykes that cut the gneiss nearby [NG 818590 & 828598] because this dyke suite also cuts the Applecross Formation in the Gairloch sub-area.

The tabular sandstone facies farther away from the unconformity is interbedded with micaceous fine sandstones and siltstones, sometimes red but more often grey. Those shown in Figure 95 have abundant westward migrating climbing ripples and are about 400 m down the palaeoslope from the gneiss. The persistency factor for the thinnest sandstone beds is 1000–10 000, whereas for the pebbly beds it is only 50–100. Farther still from the gneiss the sandstone fails altogether and the sediments are exclusively grey siltstones. This is defined as the grey shale facies (Db2).

The grey shale facies is splendidly exposed along the shore NW of Diabaig jetty. The jetty itself is built on gneiss, flanked to the north by some tabular beds of red sandstone containing angular gneiss clasts up to about a decimetre. There is no contact with the shale at this point. The main shale section starts on the beach about 300 m to the north where it is laterally equivalent to grey sandstone with gneiss fragments, seen by the road [NG 797601], and also to

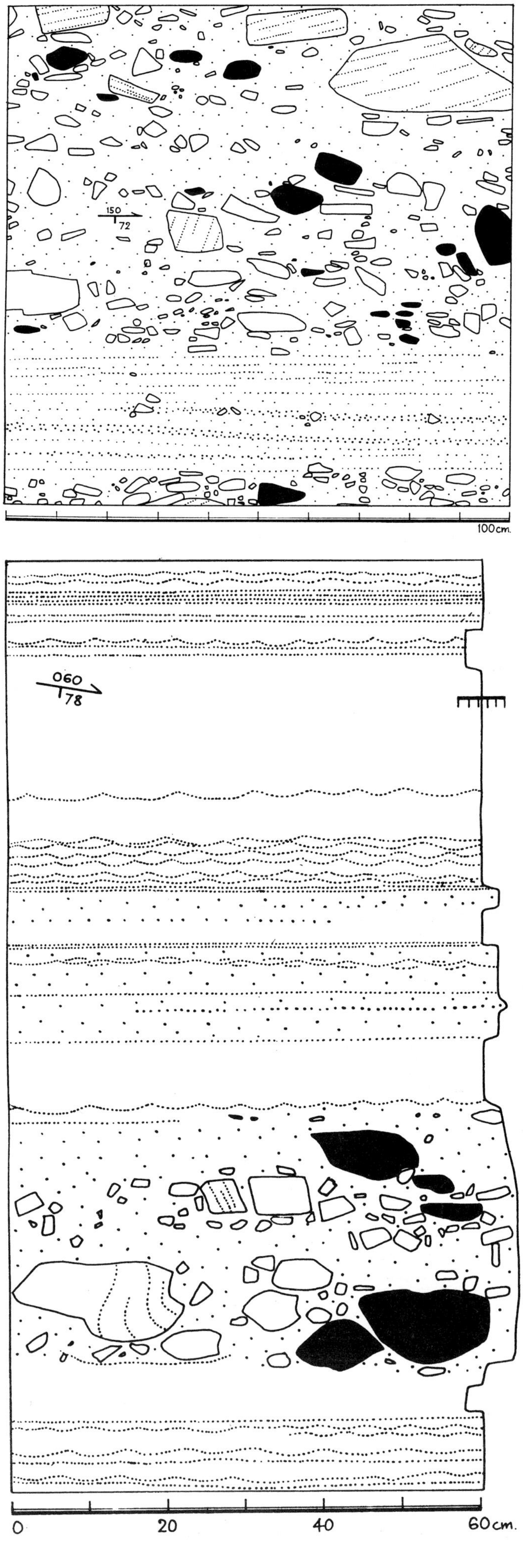

Fig. 94. Interbedded breccia and coarse red sandstone in the Diabaig Formation at Diabaig. The upper figure, which shows breccia dominant (facies Db1a), is a scaled drawing of a rock face about 400 m south of Loch Roag [NG 82156068]. In the lower figure breccia is subordinate (facies Db1b). This is a scaled drawing of a smooth rock face 500 m south of Loch Roag [NG 82096057]. The grain size scale spans +3 to -2ϕ units (0.12–4 mm). In both figures basic clasts are black, the rest acid gneiss or, rarely, quartz. All clasts over 0.5 cm in size are shown.

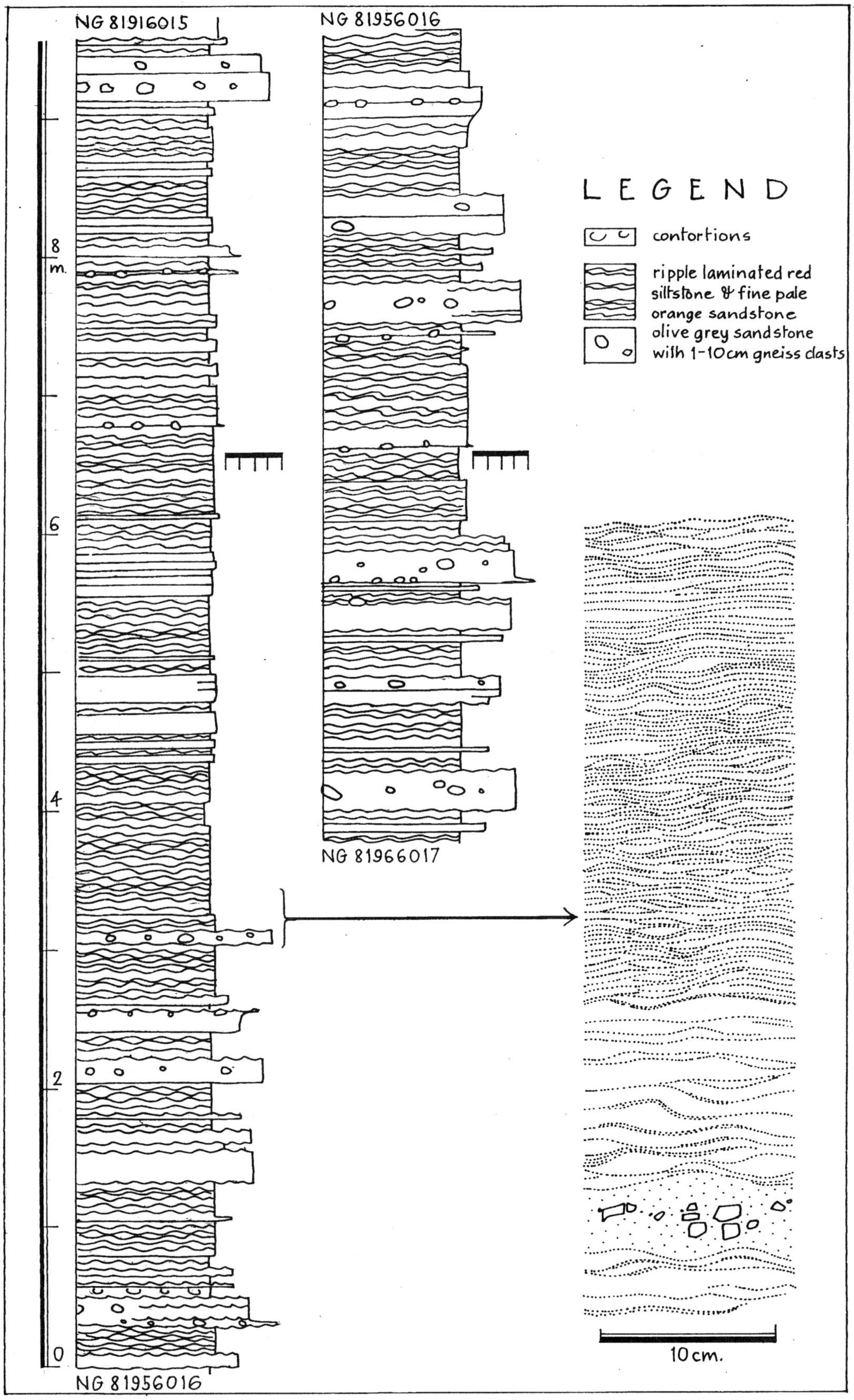

Fig. 95. Graphic log of the tabular sandstone facies of the Diabaig Formation (Db1b) by the roadside about 600 m east of Upper Diabaig [NG 81956016]. The grain size scale spans +2 to -2ϕ units (0.25–4 mm). The palaeocurrent direction from ripple drift was westwards ($\theta = 245°$).

gneiss forming the hill (An Torr) that towers above the road to the east. The distance from the shale facies to the Lewisian hillside here is only about 30 m. About 520 m to the NE, on the north side of An Torr [NG 80046057], sandy grey shale is actually seen in contact with the gneiss. The measured section through the grey shale facies (Fig. 96) follows the shore and includes 115 m of beds. It is overlain by the Allt na Biste Member of the Applecross Formation.

Three subfacies can be distinguished in the shales:

(1) Silt–mud rhythmite, with laminae averaging 0.1 mm in thickness and only rarely reaching 2 mm. They have a persistency factor of about 3000. Pale blue weathering phosphate concretions are common throughout. They are oval, up to a centimetre thick across the bedding and 5 cm along it. They are

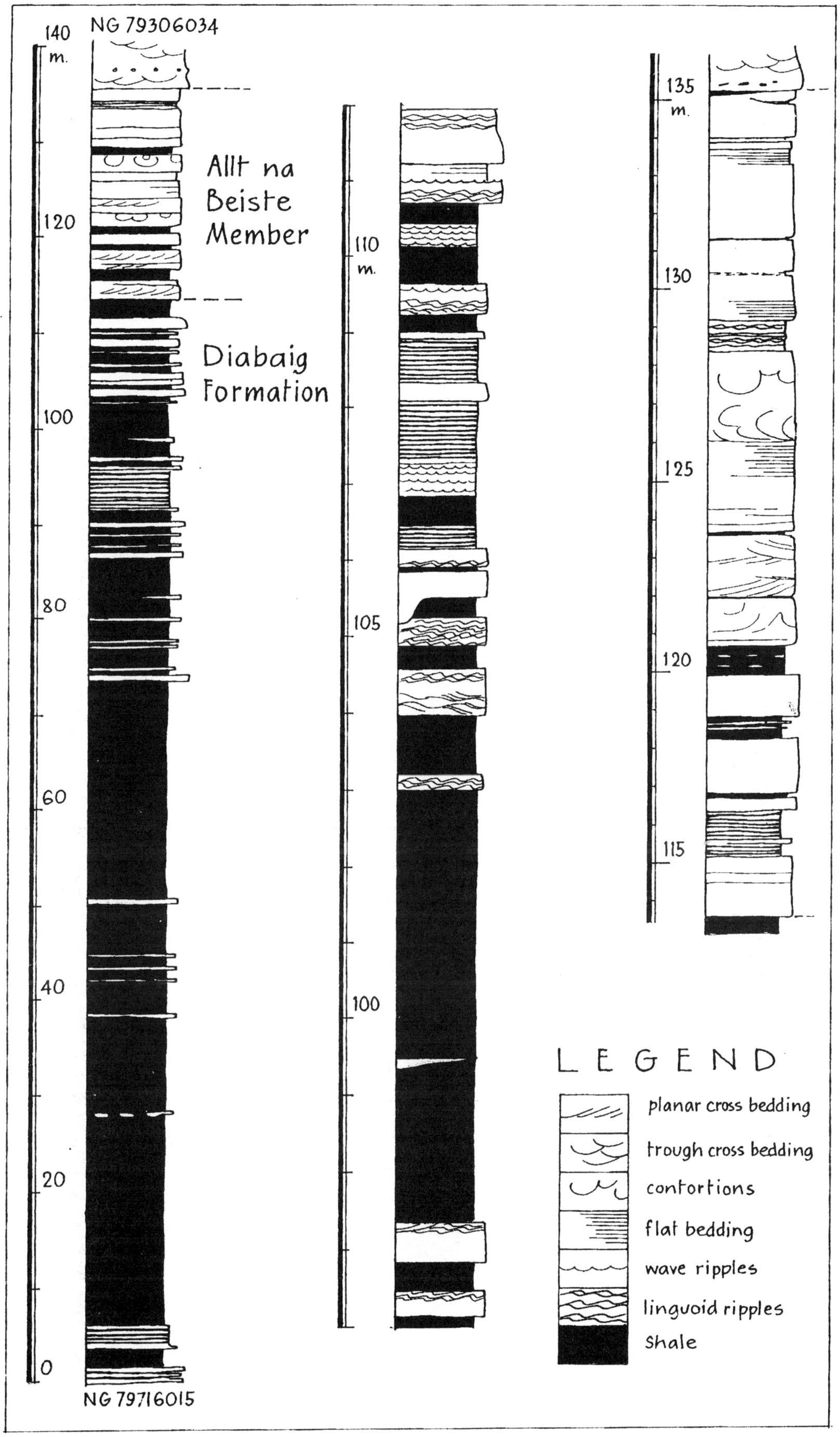

Fig. 96. Graphic logs of the Diabaig Formation and Allt na Béiste Member (lowest Applecross Formation) at Diabaig. See Figure 92 for location. The left-hand log shows the type section exposed along the shore and in the stream Allt na Béiste. The base of the section is separated from the Lewisian by a few metres of poorly exposed grey sandstones and breccias. The centre log shows the upper 16 m of the grey shale facies (Db2), which contains frequent grey sandstone beds. The right-hand log shows the Allt na Béiste Member exposed in the stream of the same name. Shales in the member above the 125 m level are red while those beneath are grey.

obviously precompactional, for the shale lamination bends round them. More extensive phosphate laminae also occur. Organic walled microfossils are common in the shales (Downie 1962; Peat & Lloyd 1974, Peat & Diver 1982; Peat 1984) but are best preserved, quite undeformed, in the phosphate (W. L. Diver, pers. comm.). What appears to be a broad channel about half a metre deep cuts the shales 75 m above the base of the section. The lowest decimetre of the channel is filled by calcareous sandstone and the rest by shale with an orientation slightly different to that beneath.

(2) Ripple-laminated sandstone beds, millimetres to centimetres thick. The ripple foresets dip in no consistent direction and although the crests are fairly constant in orientation in any given stratigraphic interval, they swing gradually from east–west at the base of the section to north–south at the top. They are not, therefore, palaeocurrent indicators and were probably formed by waves related to the emergent gneiss topography. Desiccation cracks which formed in the rhythmite were filled by sand derived from overlying ripple-laminated sandstone beds (Fig. 97 and Stewart 1991*c*, fig. 3.16). There are roughly 2000 desiccated horizons in the 115 m section. The shales also show a reticulate pattern of small ridges on upward-facing bedding planes, mistaken for rainprints by the Geological Survey (Peach *et al.* 1907, p. 325) and by Allen *et al.* (1960), but which closely resemble certain micro-load structures described by Dzulynśki & Walton (1965, fig. 140).

(3) Grey sandstone beds up to a metre thick appear in the upper part of the grey shale facies and increase in frequency and thickness towards the top (Fig. 96 and Stewart 1991*c*, fig. 3.17). The bases of these beds are sharp. The upper parts of the beds typically show ripple-drift lamination, especially clear where secondary calcification has occurred. The ripple-drift lamination shows palaeocurrent directions from the west, as in the overlying Applecross Formation. The sandstone is fine to medium-grained greywacke, with about 20% matrix and rare volcanic grains. A few beds have a quartz cement. Both the bases and the tops of the beds are often channelled, so that the persistency factor is only 300–1000. The shales in the uppermost 25 m of the section, where the grey sandstones are most abundant, contain rippled sandstones much coarser than those lower in the section. Despite this the shales are still desiccated.

Small pyrite cubes have been noted near the base of the shore section but not at higher levels. There is no trace of either primary carbonate or evaporite minerals in the formation.

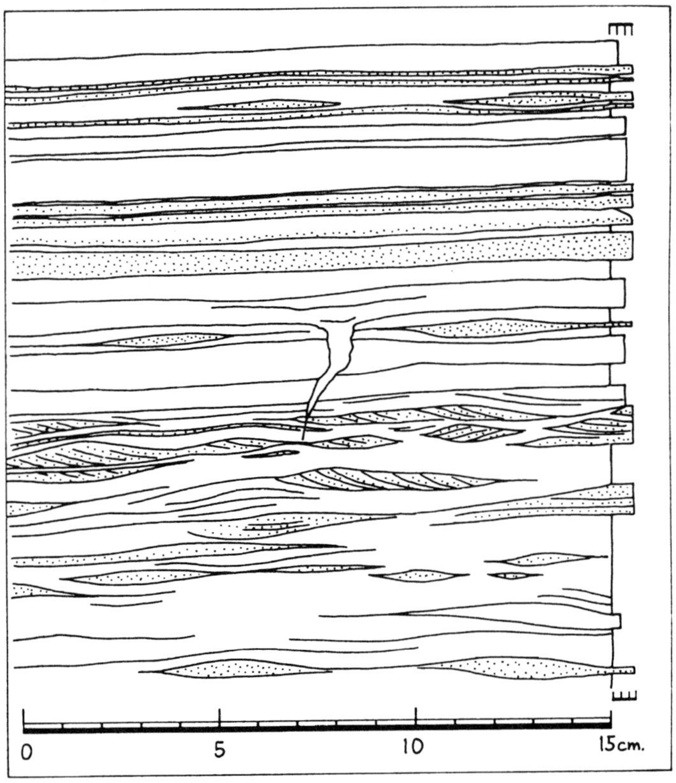

Fig. 97. Graphic log of desiccated grey shale (facies Db2) in the Diabaig Formation, based on a tracing. The grain-size scale is based on the sediment colour, which ranges from silt (N5) to fine sand (N8). Only the fine sand is stippled. Note the desiccation crack in the middle of the log, and the ripple lamination in the sand. The persistency factor (*p*) of the thinnest laminations (0.1–3 mm thick) is about 3000. The shale is 56 m above the base of the Diabaig shore section [NG 79476023].

Facies inter-relationships in the Diabaig Formation are conveniently studied along the roadside a kilometre east of Upper Diabaig. Exposures begin just above the bend in the road [NG 82146008] where massive breccia is in contact with gneiss. Stratigraphically higher and progressively finer grained beds crop out to the west, the last exposed being the red, micaceous ripple-laminated beds shown in Figure 95 [NG 81876008]. The up-dip equivalents of the micaceous beds, which are breccias, can be seen to the NNE above the recent screes, with the Lewisian basement beyond. In other words, the Diabaig sediments get finer stratigraphically upwards and also with increasing distance from the basement forming the palaeovalley sides.

The grey shale facies is thought from the boron content of the illite fraction to be lacustrine (Stewart & Parker 1979). The geochemistry of the shales shows that the lakes were initially supplied with detritus from local hills but were later invaded by a large river system that contributed non-Lewisian detritus (Stewart 1995*b*). The grey sandstones within the shale facies record turbidity flows across the lake bottom stemming from the river system. Sediment from these rivers ultimately filled the lake basins and buried the remaining hills with sand (i.e. the Applecross Formation).

The contact between the grey shale facies of the Diabaig Formation and the overlying Applecross Formation is defined at the west end of the shore section by the abrupt appearance of trough cross-bedded red sandstone. The lowest 20 m of the Applecross Formation form a distinctive, mappable member, all exposed in the stream Allt na Béiste where it dashes down through the wooded scarp to the sea (Fig. 96). This is the type section of the Allt na Béiste Member. The unit was described by the Geological Survey as 'massive bright-red sandstones with shale partings' (Peach *et al.* 1907, p. 325) and included by them in the Diabaig Formation, doubtless because it lacked large durable pebbles. However, the sandstones of the Allt na Béiste Member, though finer grained than much of the Applecross, have the same trough cross-bedding, locally contorted. The modal mineralogy is very similar to that of the Applecross Formation, but unlike that of the sandstones in the Diabaig Formation (see below). In addition, the Allt na Béiste Member contains small pebbles of porphyry and black chert. These can be found on the low cliffs about 500 m west of the type section [NG 78776029] where the exposures are fresher than in Allt na Béiste. For these reasons it is thought desirable to redefine the base of the Applecross Formation by moving it down to the base of the Allt na Béiste Member.

The sandstone of the Allt na Béiste Member weathers pale reddish-brown, but when fresh is pale red to greyish red. These colours are largely due to the feldspar that forms about 25% of the rock and is mostly potassic. The tabular sandstones of the Diabaig Formation, by contrast, contain about 40% detrital feldspar, mostly plagioclase derived from the local basement gneiss. The shales noted in the Allt na Béiste Member by the Geological Survey are grey, except at the top where they are red. Palaeocurrents flowed eastwards as in the beds above and below.

The Allt na Béiste Member is 20 m thick where it overlies the Diabaig shale facies, but is completely absent at Ruadh Mheallan, only 5 km NE of the type section at Loch Diabaig. Basement relief here protrudes through the Diabaig Formation so that pebbly Applecross Formation directly overlies gneiss. The top of the Allt na Béiste Member in the type section is marked by the highest red shale, above which the sandstone is noticeably different to that below. The colour is reddish purple and the grain size is coarse to very coarse – both features that are contained in the Geological Survey's description of the boundary. About 50% of the beds above this boundary are contorted as compared with about 5% in the Allt na Béiste Member. Half-centimetre sized durable pebbles appear about 2 m above the top of the Allt na Béiste Member and pebbles over a centimetre become common about 10 m higher. A huge exposure of these coarse, contorted Applecross beds (Fig. 98), representative of the formation, can be studied about 200 m west of the Diabaig township wall, 45 m above sea level [NG 786603].

The definitive features of the Diabaig and Applecross Formations can now be stated. The Diabaig Formation is formed of red

Fig. 98. Contorted bedding in part of the type exposure of the Applecross Formation at Diabaig [NG 786603]. The photograph shows medium grained red sandstones in which several cosets have been contorted together. They are erosively truncated at the level of the hammer head. The hammer handle is 0.5 m long. Elsewhere in the outcrop coarser, pebbly sandstone are exposed, about half of them contorted.

breccias and sandstones contained within palaeovalleys and derived from the unconformably adjacent basement. Lateral passages from sandstone to grey shale are common. The Applecross Formation consists of red sandstone with a maximum grain size over 0.5 mm and siliceous pebbles, including porphyry, scattered through the rock or arranged in seams. About 50% of the beds are strongly contorted.

Alligin to Liathach

This strip of country flanking the north side of Upper Loch Torridon was mapped by Clough and Hinxman for the Geological Survey in 1889–91. The geology has also been studied by Maycock (1962) and Rodd (1983). Irving started the earliest palaeomagnetic work on the Precambrian of Britain near Alligin in 1952 (Irving & Runcorn 1957). The main features of the stratigraphy and structure are shown in the map, Figure 99 and section Figure 100. The stratigraphy shown on the currently available one-inch to the mile geological map of 1896 (Applecross, sheet 81) is out of date.

Despite the generally westerly dip, faults bring the oldest rocks to the surface at the western end of the section where a steep contact between the Diabaig Formation and the basement gneiss is well exposed in a stream due north of Canapress [NG 82845793]. Basement gneiss is probably present not far beneath sea level all along the section for only a kilometre from the section line, north of Sgorr a' Chadail, it reaches the surface at an altitude of 530 m. Along much of the south side of Upper Loch Torridon the basement is near sea level. Basement relief in Diabaig times must, therefore, have been about 500 m.

The Diabaig Formation is represented by pale red tabular sandstones (facies Db1b). There are no grey siltstones and hardly any red ones. The sandstones have an overall thickness of about 120 m but their true base is below sea level. Down dip to the south and SE they probably interfinger with the grey shale facies seen

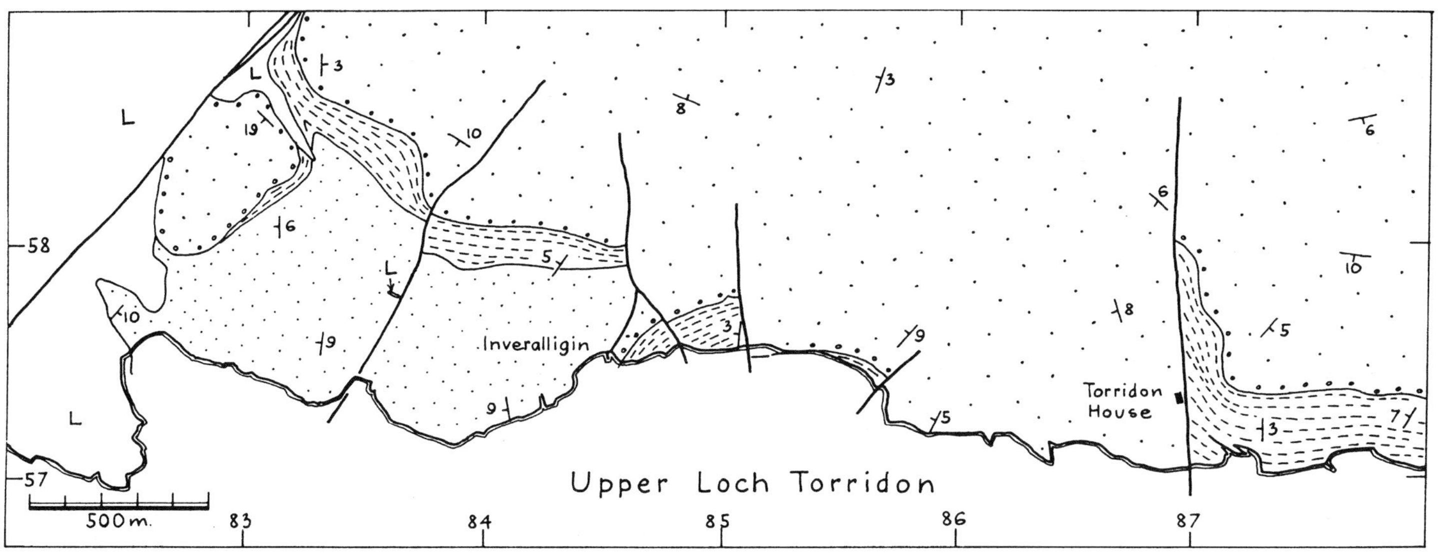

Fig. 99. Geological map of the Alligin area, based on 1: 10 560 mapping by the author. For location see Plate 2. The units present are the Lewisian gneiss complex (L), Diabaig Formation (facies Db1b), Allt na Béiste Member, and pebbly Applecross Formation, all ornamented as in Figure 92.

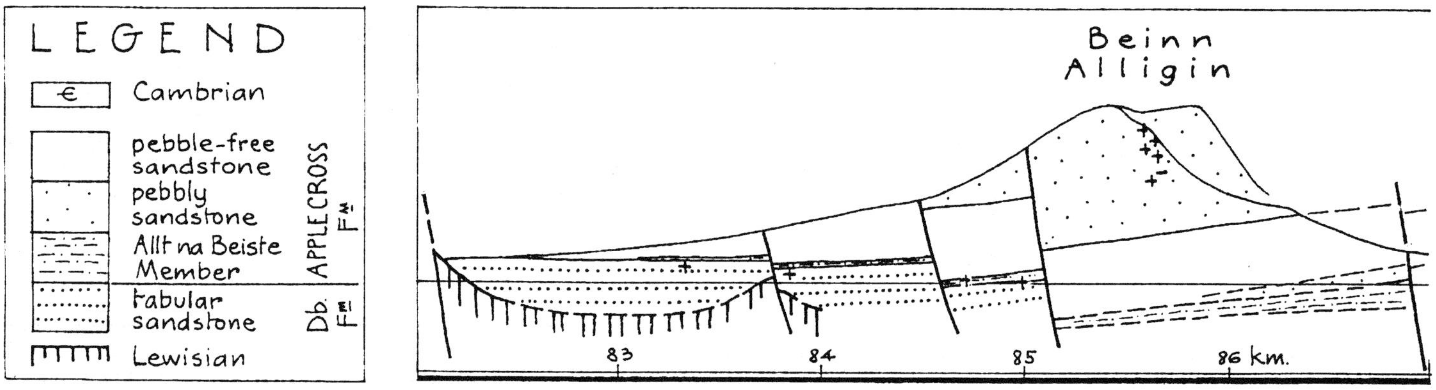

Fig. 100. Section from Alligin to Liathach, incorporating palaeomagnetic data from Irving & Runcorn (1957, fig. 5) and E. A. McClelland (pers. comm.).

cropping out on the south side of Upper Loch Torridon. The very existence of the loch may be due to the excavation of these shales during Pleistocene glaciation.

The tabular sandstones are mainly parallel laminated, sometimes flat bedded, but never contorted except at the very top of the formation. Ripple lamination, probably wave formed, is quite common. Trough cross-bedding is locally abundant, for example on the coast near Òb a' Bhrighe [NG 83005745 to 83105742]. Troughs are 5–10 cm deep and up to 70 cm wide, their bases cutting the flat-lying laminations mentioned above. The southeasterly palaeocurrent direction deduced from these troughs agrees with that from primary current lineation and from eddy scours around gneiss pebbles and cobbles. A good example of the latter, due to an isolated block of acid gneiss, can be seen on an ice-scoured bluff immediately west of the road bridge over the Alligin River [NG 83325791]. In addition to the small-scale structures listed above the tabular sandstones also contain large scours up to a metre deep and 20 m wide, well exposed on the low cliffs 450 m west of the Alligin River mouth [NG 83785718]. These are sections through sinuous channels filled by lateral accretion deposits. Individual foresets are gently curved in plan as if they formed part of oblique bars. The dip directions of the foresets are bimodal, with one mode directed to the east and the other to the west. Channels of this kind often cut the tabular sheet flood deposits on the middle and upper parts of recent alluvial fans.

The uppermost 30 m of the Diabaig Formation is formed by alternations of trough cross-bedded sandstone and tabular red sandstone, each about 5 m thick. They are exposed on the crags beneath the road north of Inveralligin [NG 844578]. The trough cross-bedded sandstones have erosive bases and are locally contorted. They probably represent channel deposits, precursors of those seen in the overlying Allt na Béiste Member, here cutting Diabaig sheet flood deposits.

The Diabaig Formation is overlain by beds that closely resemble the Allt na Béiste Member of the Applecross Formation at Diabaig. The sections exposed by the roadside 500 m east of Inveralligin jetty (Fig. 101) are representative. Fining-upward cycles 2–8 m thick are discernible – good examples overlie erosion surfaces **a**, **c** and **d** shown in Figure 101. The erosion surfaces extend 80–100 m, much farther than in the tabular sandstones of the Diabaig Formation. Desiccation cracks are common in the dark red siltstones. The trough cross-bedded sandstones are probably alluvial channel bar sediments. The silty sandstones and siltstones seem to be flood plain deposits rather than abandoned channel fill, for in sections 4 & 5 of Figure 101 they overlie erosion surface **b**.

The Allt na Béiste Member is only 30 m thick at Alligin but when it reappears east of Torridon House it is thicker and coarser. It reaches about 180 m near Torridon village. A large bedding plane behind the trees at Torridon jetty [NG 89465658] is strewn with durable pebbles up to 2 cm across, including white vein quartz and quartzite, jasper, quartz-porphyry, quartzo-feldspathic gneiss and feldspar. These beds may be at the base of the Member for thin bands of grey siltstone crop out not far beneath, at the roadside [NG 89405658].

The top of the Allt na Béiste Member is marked by a rapid increase in grain size, a change in colour from pale red to purplish red, and the appearance of scattered durable pebbles. Contorted bedding is also much more common in the overlying beds. The base of these coarse, pebbly sandstones is erosional where it overlies the Allt na Béiste Member 500 m ENE of Inveralligin jetty [NG 849577] and also 300 m north of Canapress [NG 82755796]. Farther east, however, it is gradational over 10 or 20 m.

The Applecross Formation east of the Fasag fault cannot be correlated with that to the west (see Fig. 100), implying an easterly downthrow of at least 1000 m. The palaeomagnetic data in Figure 100 show three reversals of polarity in the upper slopes of Liathach, roughly in correspondence with an intercalation of fine grey sandstones and laminated siltstones. These grey beds, which are about 6 m thick, crop out on the western side of Coire Liath Mhór [NG 93515781] and are labelled as shale on Figure 100. A similar association of a grey siltstone unit and rapid reversals of palaeomagnetic polarity occurs at Toscaig and in the section east of Isle Ristol, in both cases close to the boundary between the Applecross and Aultbea Formations. Hinxman recorded green shale beneath the Cambrian on Mullach an Rathain which, he suggested, might represent the base of the Aultbea Formation (Peach *et al.* 1907, p. 324). This is about 400 m stratigraphically higher than the grey siltstone horizon in Coire Liath Mhór. If the Applecross–Aultbea boundary is actually present at the summit of Liathach then the easterly downthrow on the Fasag fault must be near 2000 m.

Details of the sedimentary structures in the Applecross Formation at Torridon [NG 906561], including a graphic log of a 97 m thick section of strata and lateral correlations over 700 m have been published by Owen (1995).

Upper Loch Torridon (south side)

The southern shore of Upper Loch Torridon intersects the Diabaig and Applecross Formations which infill 250 metre-deep palaeovalleys in the gneiss. The sea has invaded the sediments preferentially so that the bays coincide with the ancient valleys. From west to east these are Loch Shieldaig, Ob Mheallaidh, Balgy River, Ob Gorm Beag and Ob Gorm Mór, as shown on the map, Figure 102. The currently available Geological Survey one-inch to the mile map of 1896 (Applecross, sheet 81) is out of date. The area is briefly described by Peach *et al.* (1907, p. 325–60). The coast section is a Geological Conservation Review site (Mendum *et al.* 2003).

The palaeovalley presently occupied by Loch Shieldaig is filled by red sandstones of the Applecross Formation at the present level of erosion. There are good exposures along the road that follows the eastern margin of the Loch.

Ob Mheallaidh is surrounded by well-exposed beds of the Diabaig Formation dipping gently off the gneiss. Exposures at high water mark along the south side of the bay show the grey shale facies (Db2) typical of the formation, but also some red micaceous sandstones. Wave ripples in the shales trend roughly SE. The road

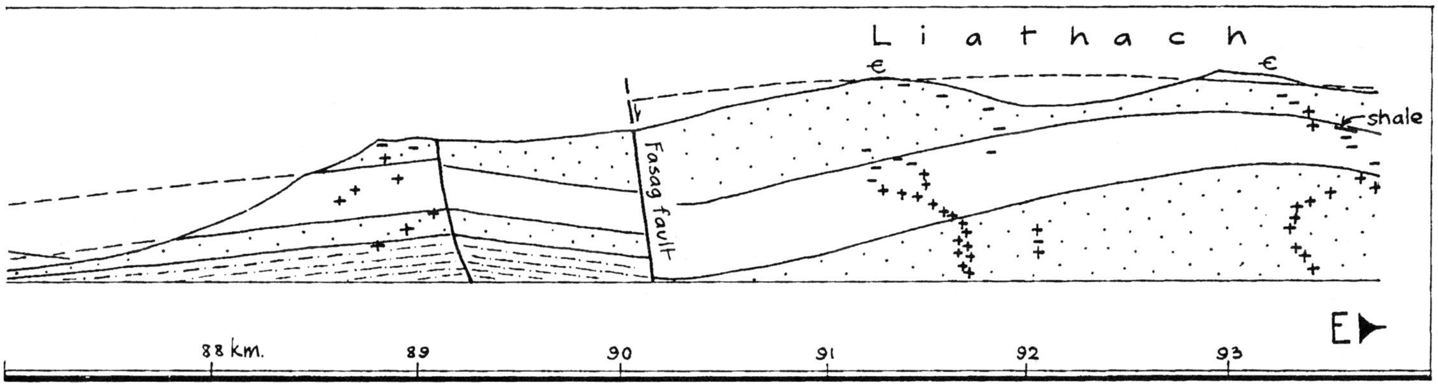

section about 15 m above sea level exposes coarse and sometimes pebbly red sandstones and red siltstones. The beds are quite similar to those near Upper Diabaig (Fig. 95) and have similar lateral persistency of 300–1000. Ripple-drift lamination is not developed. The ripples are all symmetrical and trend SSE. They were probably wave induced like those in the shale exposed along high water mark. On the east side of the bay the contact between the Diabaig Formation and the Lewisian basement is visible at several points. Red sandstone (Db1b) rests unconformably upon a gneiss crag by the roadside 87 m NE of the stream flowing into the SE corner of the bay [NG 83385364]. Tabular bedded red sandstone (Db1b) and up to half a metre of massive breccia (Db1) coat the hummocky gneiss surface near high water mark [NG 840551], passing rapidly upwards into grey shale. Compactional dips are particularly noticeable here. The transition from red sandstone to grey shale is seen on the coast due west of the sharp bend in the main road [NG 838548], and again on the coast about 700 m to the SW [NG 833543]. The Applecross Formation overlies the shales with an erosive contact well exposed in the wooded bluff overlooking Camas a' Chlarsair [NG 835545] and again about 300 m to the NW, at low tide. The grey shales closely resemble those in the lower part of the type section at Diabaig, and like them contain phosphatic laminae. They differ, however, in lacking the grey sandstone beds found in the upper part of the type section.

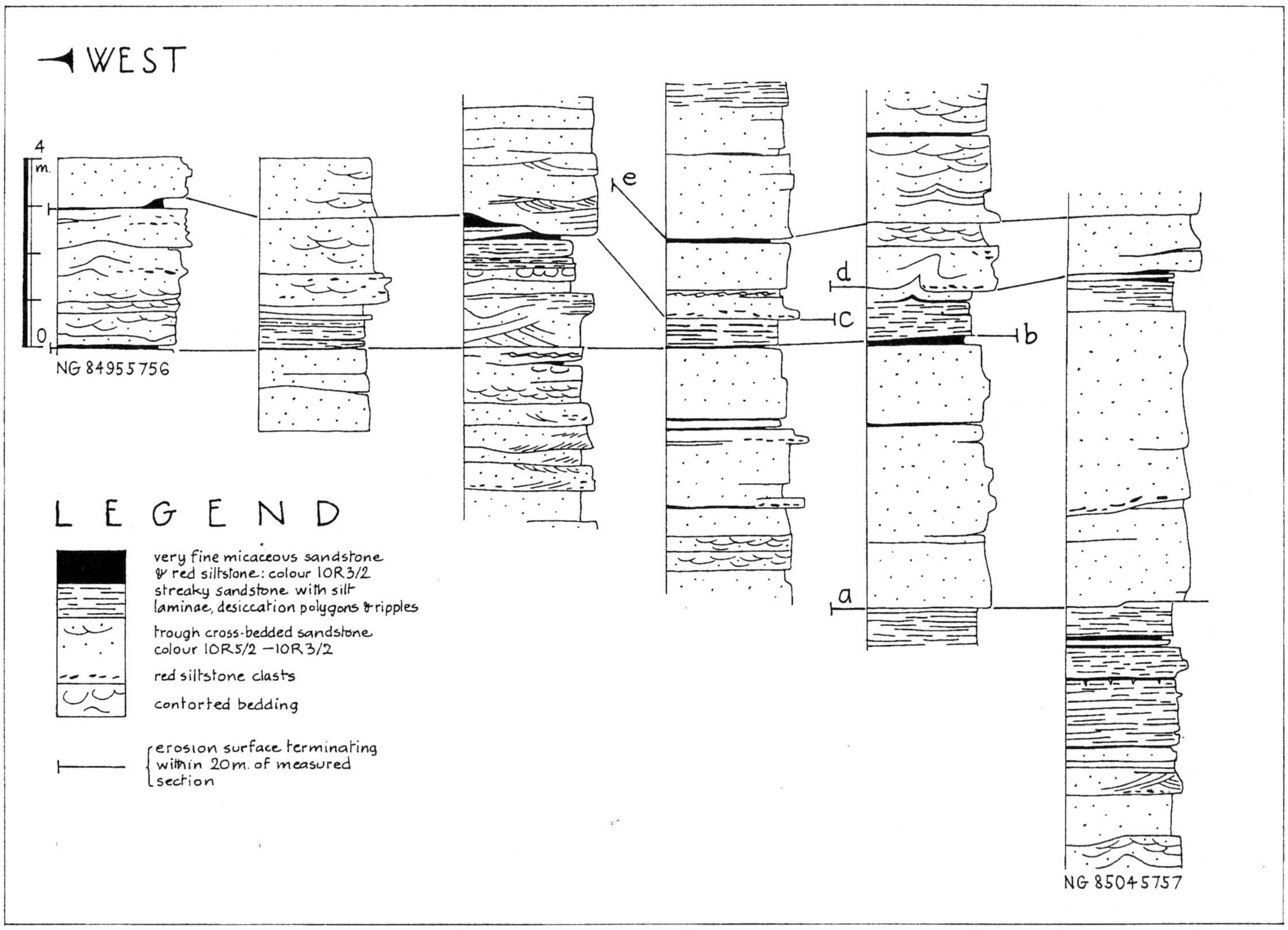

Fig. 101. Graphic logs of the Allt na Béiste Member (lowest Applecross Formation) at Alligin. Sedimentary structures are drawn to scale. The logs were compiled at 20 m intervals along a continuous, fresh road cut in 1975.

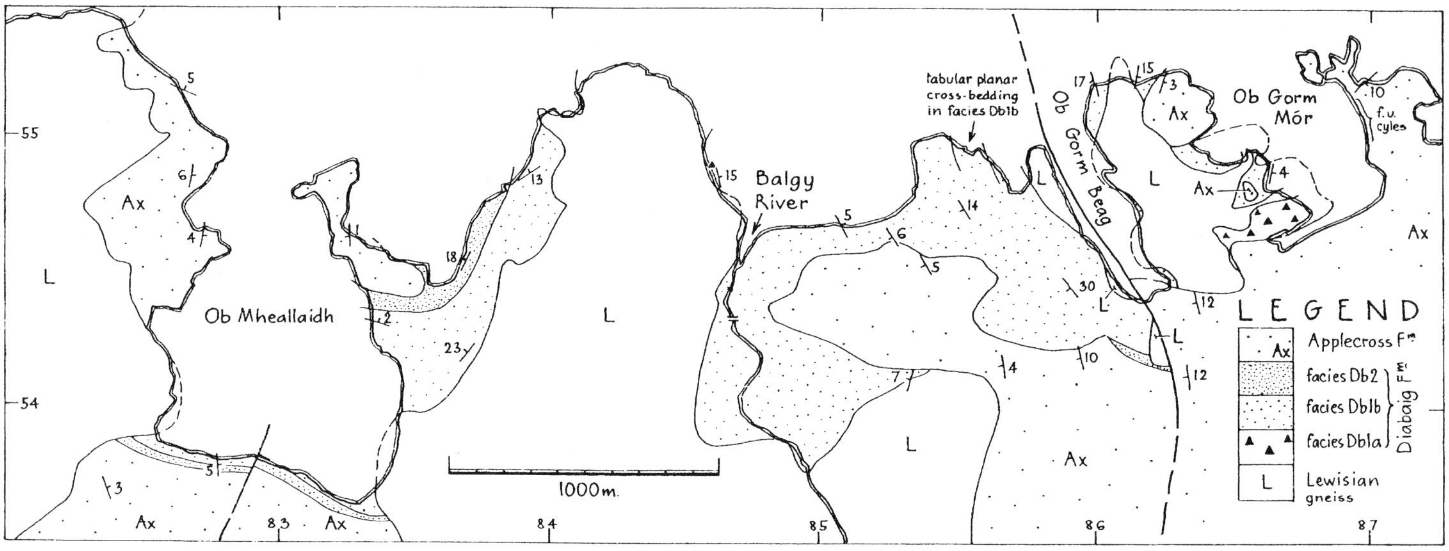

Fig. 102. Geological map of the south side of Upper Loch Torridon, based on 1:10000 mapping by the author. See Plate 2 for location.

Diabaig breccias (Db1a) are well exposed on the coast 300 m north of Balgy River mouth [NG 84N5549]. The clasts derive from the immediately adjacent gneiss and are predominantly amphibolites, with hematitic rims. Rounding is good. The blocks average about 10 cm in size and may reach a metre. Their magnetization has been studied by Poppleton & Piper (1989) in order to establish that the basement remanence is Laxfordian. Conglomeratic units 0.5–1 m thick alternate with grey feldspathic sandstone, the proportion of which increases rapidly away from the gneiss. Sandstones overlying the breccia are exposed at low tide. Extensive epidotization of both the matrix and detrital plagioclase grains in the sandstone at this locality has been described and figured by Maycock (1962, p. 120–1). A few metres of breccia are also seen on the eastern edge of the palaeovalley at Camas na Nighinn [NG 857548]. Elsewhere the gneiss is overlain by red sandstone with scattered gneiss fragments.

Stratigraphically higher sediments near Balgy are mainly tabular bedded grey sandstones (facies Db1b) with films of greenish-grey siltstone and abundant ripples. Ripple trends are uniformly SE and though frequently asymmetrical were probably wave induced (Stewart 1988*b*, fig. 9.6). The grey sandstones are separated from the tabular red sandstones (the same facies) on the eastern margin of the palaeovalley by a peculiar subfacies not found elsewhere in the formation, except perhaps at Annat Bay, near Scoraig. It consists of red sandstone with planar cross-bedding in sets about a metre thick. The average grain size is 0.5–1 mm with pebbles (of quartz and feldspar) up to a centimetre in diameter concentrated along set boundaries. In thin section volcanic lithic grains are detectable. Similar cross-bedding is seen in sandstones directly in contact with gneiss on the NW shore of Aird Mhor [NG 860552]. At both localities the boundary with the overlying tabular bedded sandstones (facies Db1b) is conformable.

In Ob Gorm Mór, the Diabaig Formation consists of gneiss breccias and red sandstones overlying irregular basement topography. Ripple marked sandstone envelops gneiss blocks at two localities (cf. Rubha Dunan) – on the north shore of Aird Mhor [NG 86125521], and on the shore in Ob na Glaic Ruaidh [NG 86475491].

An intercalation of interlaminated red siltstone and pale grey sandstone 65 cm thick can be seen in the massive breccia in the SE corner of Ob Gorm Mór [NG 867547]. The intercalation can be traced about 100 m along the low cliff. A photograph of it has been published by Bull (1972, fig. 15) and attributed to playa lake deposition within a fanglomerate.

The contact between the Applecross and Diabaig Formations along the south side of Upper Loch Torridon is sharp and locally erosive. It cuts down over a metre into breccia on the west side of Ob Gorm Mór [NG 86635492] suggesting a degree of pre-Applecross cementation. Sharp, planar contacts are seen on the north shore of the peninsula Aird Mhor [NG 86205523] and can be inferred along the south side of the estate road immediately east of Balgy.

The Applecross Formation is entirely composed of medium to coarse grained sandstone with trough cross-bedding. The only silty intercalations crop out on the east side of Ob Gorm Mór, forming the upper parts of fining-upward cycles about 10 m thick. The main part of each cycle is normal Applecross lithology, with an erosional base. The upper part is formed of 1–10 cm thick tabular red sandstone beds with rippled surfaces and interbeds of red siltstone and pale grey sandstone. The siltstone dominates the cycle tops if not destroyed by erosion. Palaeocurrents deduced from trough axes flowed southeastwards. Whether the silty beds at Ob Gorm Mór should be regarded as part of the Allt na Béiste Member is uncertain.

Pebbles up to about a centimetre in size are only rarely found in the lower part of the Applecross Formation in this area. Where the Applecross rests on gneiss it contains strings of decimetre-sized angular gneiss blocks for 10–20 m laterally from the contact. Such breccias are well exposed in the bed of the stream that falls into the SE corner of Ob Mheallaidh. About 350 m above sea level on nearby Beinn Shieldaig [NG 829530] and Sgurr na Bana Mhoraire [NG 870526] durable pebble size and abundance increase markedly. Above this level there are thick seams of pebbles 2–3 cm in size.

Applecross

The stratigraphy of the red sandstones forming the mountains between Loch Torridon in the north and Lochs Kishorn and Carron in the south was first investigated by Sedgwick and Murchison (1835). They described the strata cropping out near the only road crossing the area, from Tornapress to Creag Ghorm, but provided no stratigraphic thicknesses. The area was mapped by Horne and Peach for the Geological Survey in 1892–93. They thought that most of the sandstones belonged to the Applecross Formation, with the Aultbea Formation confined to the Crowlin Islands and the area around Toscaig. The total stratigraphic thickness was stated to be about 2400 m of which the Applecross Formation accounted for 1800 m and the Aultbea Formation 600 m (Peach *et al.* 1907, p. 338–9). An outline geological map of southern Applecross is shown in Figure 103. The currently available geological map (Applecross, sheet 81, published in 1896) is out of date.

The lowest 500 m of the Applecross Formation exposed along the northern coast is mainly composed of fine to medium-grained,

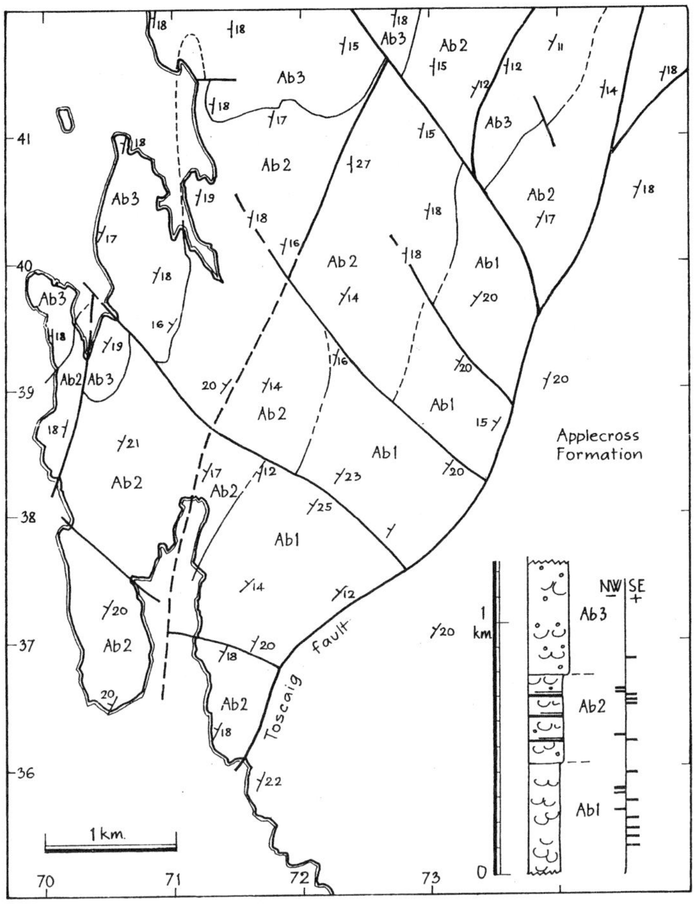

Fig. 103. Geological map of the Aultbea Formation around Toscaig in southern Applecross. See Plate 2 for location.

pale red sandstones with scattered durable pebbles up to 3 cm in size. Coarse sandstones form less than 20% of the sequence. However, pebbly bands occur near the abandoned village of Fearnmore [NG 729600; NG 724610; NG 722614]. A graphic log of a 20 m thick sandstone sequence in this area has been published by Nicholson (1993, fig. 3), who measured palaeocurrents here flowing southeastwards ($\theta = 140^\circ$, $n = 246$). Pebbly sandstone units are found sporadically to the south of Fearnmore, for example by the roadside east of Ob Chuaig [NG 711589] and on the coast west of Cuaig [NG 699580].

The road section from the old pony stables at Tornapress [NG 833422] to Creag Ghorm [NG 765431] exposes 2060 m of fine to medium grained pale red sandstone that hardly ever contains pebbles. Sparsely pebbly sandstones of medium to coarse grain, typical of the Applecross Formation, outcrop only at the head of Coire na Bà [NG 789411]. Palaeocurrents at this point flow southeastwards ($\theta = 146^\circ$, $n = 137$: Nicholson 1993, table 1). Two metre-thick laminated siltstone beds crop out near the road at Creag Ghorm [NG 770430; NG 772429]. Stratigraphically, they are about 480 m from the top of the section. The finest laminae are dark grey (N3–N4). Grey siltstone about 500 m stratigraphically lower was mapped by the Geological Survey along the eastern escarpment of Carn Dearg, about 2 km to the NE. The sequence of red sandstones overlying the shales at Creag Ghorm is, if anything, slightly coarser than that below. The highest sandstones are truncated by the Toscaig fault about a kilometre NW of Creag Ghorm but are preserved farther SW. About 500 m of coarse sandstone are exposed along the southern coast of Applecross, immediately east of the Toscaig fault. The sandstones commonly contain pebble bands up to 10 cm thick, for example in the kilometre grid square NE of Uags [NG 7235].

The stratigraphic level of the base of the sequence, at Tornapress, is probably several hundred metres above the base of the Applecross Formation. A section drawn from the base of the Applecross at Shieldaig to the Toscaig fault near Carn Breac using Geological Survey dips shows about 2200 m of strata.

The Aultbea Formation at Toscaig crops out in a triangle bounded to the north by the Applecross fault, and to the SE by a branch of the Applecross fault that reaches the sea about 2.5 km south of Toscaig village, and here called the Toscaig fault. The area is shown in Figure 103. The rocks can be informally divided into three members.

The lowest member (Ab1), which is at least 450 m thick, is composed of pebble-free, medium grained, contorted pale red sandstone.

The next member (Ab2), is sparsely pebbly, pale red or light brown medium-grained sandstone at least 340 m thick. Beds of fine-grained, flat-bedded and rippled sandstone, that may be associated with greenish grey or grey siltstones, form the bases of coarsening-upward cyclothems averaging 10 m in thickness (Rayner 1981, p. 33). Laminae rich in heavy minerals are common in the flat bedded sandstones. The flat bedding lacks current lineation and is frequently gently inclined to bedding, suggesting that it represents the toes of planar cross beds. The siltstones are ripple laminated but not desiccated. Some representative sections are shown in Figure 104. Large scale contorted bedding, and drop structures in heavy mineral bands in this member have been studied by Stewart (1963) and Irving (1964, fig. 5.10).

The uppermost member (Ab3) is a pale red, coarse to very coarse sandstone with abundant pebbles of quartz, quartzite, porphyry and jasper up to 3 cm in diameter. The member is at least 450 m thick. It has all the characteristics of the Applecross Formation of which it may, in fact, be a part. Its inclusion in the Aultbea by the Geological Survey merely emphasizes the unsatisfactory nature of the existing lithostratigraphic nomenclature.

The contact between Ab1 and Ab2 exposed on the NE side of Loch Toscaig is gradational over about 30 m. The contact between Ab2 and Ab3 is gradational over 10–30 m west of Culduie [NG 710401] but erosional at Ard Ban [NG 702394]. The base of Ab3 is conveniently defined at the base of the lowest metre-thick interval with an average grain size of 0.75 mm *and* durable pebbles greater than a centimetre. The base of the member thus mapped strikes SSW to touch the northern edge of Eilean Beag [NG 682361] in the Crowlin Islands. The remainder of the Crowlins is lithologically like member Ab2 and so, too, is the island of Longay 4 km to the SW (see p. 109). The top of member Ab3 is concealed beneath the sea. To the north it disappears beneath the Triassic outlier at Applecross village.

Palaeomagnetic reversals are present in all three members of the Aultbea Formation at Toscaig (see Fig. 103) and in the Crowlin

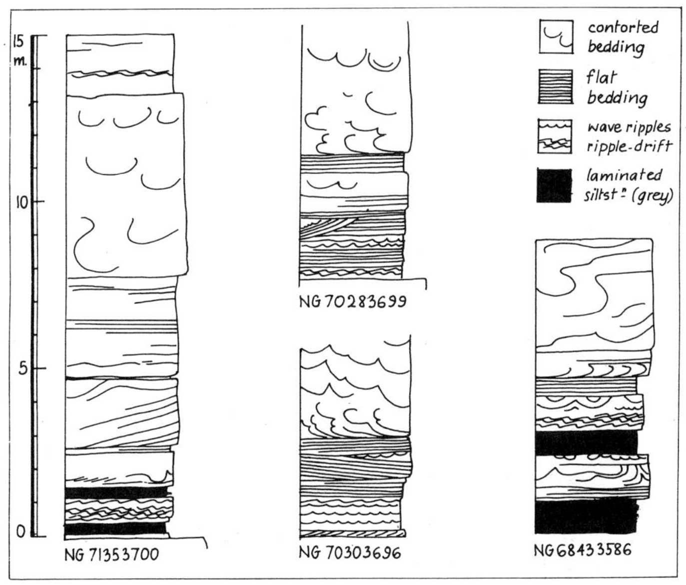

Fig. 104. Graphic logs of coasening-upward cycles in the Aultbea Formation near Toscaig and in the Crowlin Islands.

Islands (Irving & Runcorn 1957; Smith *et al.* 1983). Elsewhere in the Torridon Group reversals occur only high in the Applecross Formation and in the basal Aultbea. In the section east of Isle Ristol, for example, where the boundary is well defined, reversals are found in the lowest 150 m of the Aultbea Formation. This suggests contemporaneity between the strata at Toscaig and the Aultbea Formation farther north.

The highest member of the Aultbea Formation may once have been covered by Cambrian rocks, for blocks of Durness Limestone are abundant in the Triassic basal conglomerate near Applecross village (Lee 1920, p. 5).

Neither the direction nor amount of movement on the Toscaig fault (Fig. 103) are known, and correlation across it difficult. The grey siltstones at Creag Ghorm are not sufficiently distinctive to be correlated with those in member Ab2 at Toscaig, or with those near the Applecross–Aultbea boundary at Aultbea and the Summer Isles (*q.v.*). There are, unfortunately, no palaeomagnetic data for the road section. In constructing Figure 23 the strata west of the Toscaig fault have been assumed to be stratigraphically higher than those to the east.

Occurrences of copper ore in Applecross were reported to MacCulloch when he traversed the area in the early nineteenth century, but he was unable to ascertain the locality (MacCulloch 1836, p. 137).

Raasay and Fladday

These islands were mapped for the Geological Survey by Hinxman in 1896 (Peach *et al.* 1907, p. 340–41) but the account that follows is mainly based on the work of Selley (1963, 1965*a*, 1965*b*, 1966, 1969) who remapped them in 1959–62. A 1: 50 000 geological map of the area incorporating Selley's mapping (Raasay sheet 81W) is in the course of preparation by the British Geological Survey. Only the Torridon Group is present on the islands; a synopsis of the stratigraphy is shown in Figure 105. Selley's stratigraphic units are not redefined, for most of them are redundant. About 2 km of sediment are exposed on the islands but offshore, to the NW, seismic and gravity data show that the Torridon Group thickens to 5 km just east of the Minch fault (O'Neill & England 1994). Pre-Torridon Group relief was over 420 m, easily appreciated from the stratigraphic profile in Figure 106.

The sediments infilling topographic depressions are locally derived gneiss breccias and very coarse red sandstones, the most extensive outcrop constituting the Torran Member (part of the Diabaig Formation) with its type section on the east side of Loch Arnish [NG 592490 to NG 592482]. The member is 60 m thick but the top is not exposed. The lower half of the sequence here has an average grain size of about 2 mm, with gneiss clasts frequently reaching 10 cm. Much larger gneiss blocks are present next to the basement gneiss. Bedding is planar, with low lateral persistency. Stratigraphically upwards the member becomes finer. In the upper half of the sequence the sandstones have an average grain size less than a millimetre and show low angle cross-bedding and shallow scours. Authigenic epidote is common. No palaeocurrent data are available and the suggestion by Selley (1965*a*, p. 368 & fig. 5) that the sediments were deposited as three alluvial fans is based mainly on the distribution of dips. It is, however, difficult to say to what extent these dips are truly depositional rather than the result of differential compaction over irregular gneiss topography. Most of the dips are about 20° after correction for regional tilt, which is far too high for fan deposits. The sedimentary structures do not suggest that any part of the Torran Member was deposited in talus fans.

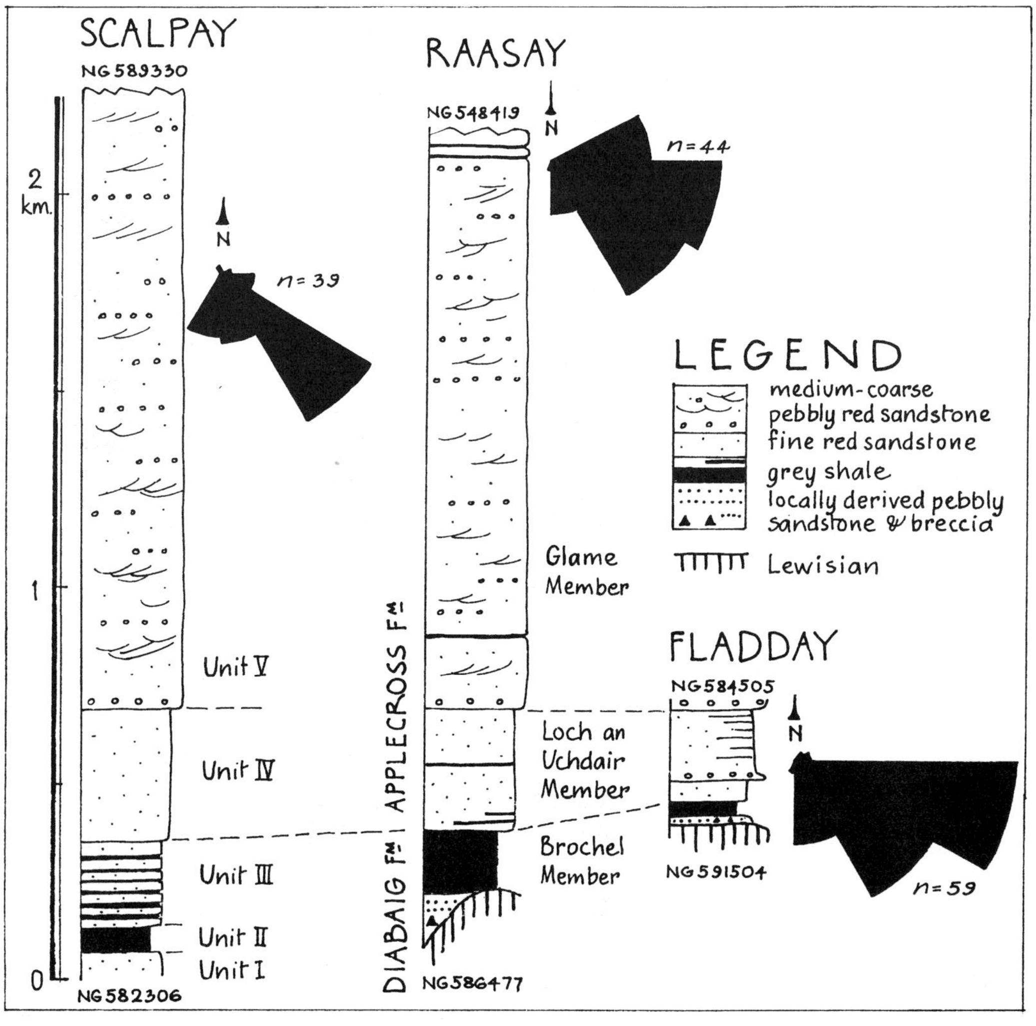

Fig. 105. Torridon Group stratigraphy and palaeocurrents in the isles of Scalpay, Raasay and Fladday.

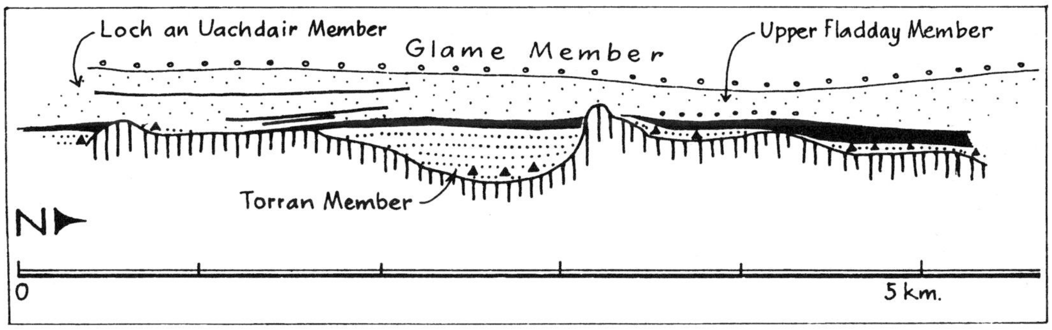

Fig. 106. True-scale stratigraphic profile of the lowermost Torridon Group in Raasay, showing palaeorelief.

The main sequence in Raasay, which is over 1500 m thick, starts on the east side of the island with the breccias of the Brochel Member (part of the Diabaig Formation). Locally derived gneiss blocks up to a metre across lie in a matrix of coarse red or grey micaceous sandstone about 400 m NE of the ruins of Brochel Castle [NG 587464]. These were interpreted by Selley as 'screes and fanglomerates'. The unconformable contact between breccia and basement forms the base of the type section. Beds laterally equivalent to the breccia, probably grey shales and sandstones, are concealed beneath the shingle beach NE of the castle. Stratigraphically higher grey beds within the Brochel Member are, however, well exposed along the coast south from the castle (Fig. 107). They are about 130 m thick. Three subfacies can be recognized, different to those in the Diabaig Formation at Diabaig:

(1) Grey mudstone, micaceous siltstone and fine sandstone. The sandstone forms millimetre to centimetre-thick bands, with wave ripples. Polygonal desiccation cracks in the mudstone are filled by the sandstone. Microfossils have been extracted by maceration of these shales by Naumova & Pavlovsky (1961). They are mainly unicellular forms with diameters of 3–8 μm. Most were assigned to the group *Psophosphaera* Naum. 1937. In addition, there are two genera that also occur in the lower Palaeozoic; *Archaeodiscina* Naum. and *Archaeosacculina* Naum. Three species belonging to the group Triletes were noted; *Pavlovskaya augenia* Naum. gen. & sp. nov., *Minutissima prima* Naum. gen. & sp. nov., and *Minutissima atava* Naum. Downie (1962) and Sutton (1962) have commented on these microfossils.

(2) Decimetre-thick beds of fine-grained greywacke. Matrix forms about 20% of the rock, and feldspar, mainly plagioclase, 30–45%. Current ripple lamination is common in the upper parts of the beds; coarse grains (including volcanic clasts) and shale fragments are commonly concentrated at the base. Their lateral persistency (p) is 500–1000.

(3) Medium to coarse-grained grey or reddish-grey sandstone in beds up to 8 m thick. The bases of the beds are erosive with relief of up to a metre. One of the sandstones has a coarse base including gneiss pebbles up to 6 cm in size. Common sedimentary structures in these sandstones include planar and trough cross-bedding in sets 1–1.5 m thick. Ripple bedding, flat bedding and large-scale contorted bedding also occur. The top of one of the sandstones has trough-shaped channels 4 m wide and 0.7 m deep filled concordantly by apparently flat bedded sandstone that passes up into ripple bedding (cf. the Beinn Bhreac Member in Soay).

The grey shales and greywackes constituting the first two facies are similar to the shale facies (Db2) of the Diabaig Formation at Diabaig, whereas the reddish-grey sandstones, that occur throughout the shales, closely resemble both the overlying Loch an Uachdair Member in Raasay, and the Allt na Béiste Member at Diabaig.

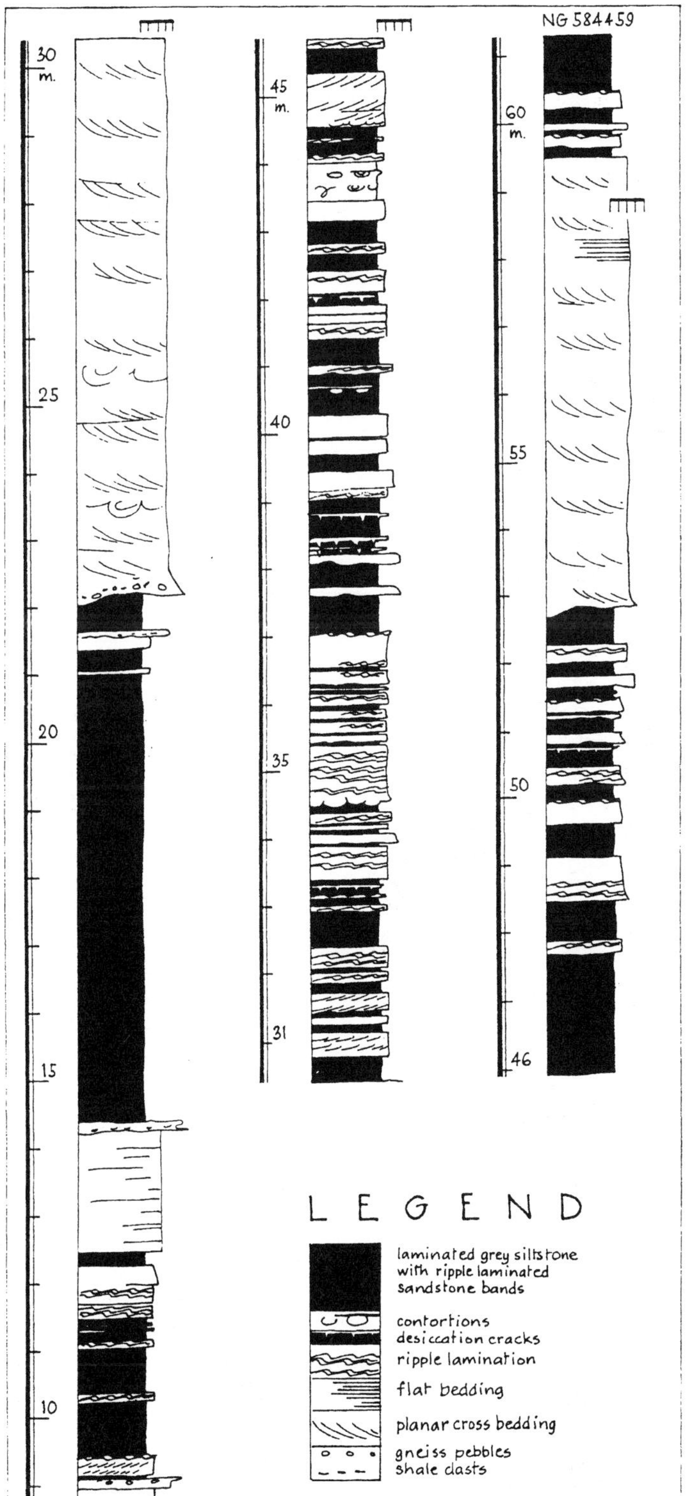

Fig. 107. Graphic log of part of the Brochel Member (Diabaig Formation) near Brochel, in Raasay. The grain size scale spans 4–0 ϕ units (0.06–1 mm). From Selley (1996, fig. 5.6) © with permission.

Selley concluded that the grey facies represented lacustrine conditions, with the grey greywackes deposited by turbidity currents.

The Loch an Uachdair Member (300 m thick) consists of fine-grained red sandstones lacking pebbles. It was included in the Diabaig Formation by Selley (1965*a* & *b*) but is here placed in the Applecross Formation in accordance with the revised definition of the boundary at Diabaig. Cross-bedding and large-scale contorted bedding are typical of the Loch an Uachdair Member. Petrographically it resembles the overlying Glame Member, described below. The base of the member immediately SE of Loch an Uachdair is conformable on red breccias that here fringe the gneiss basement. To the north, on the shore of Loch Arnish [NG 585477] it erosively overlies about 10 m of grey shales, probably part of the Brochel Member. The contact with the Brochel Member is not exposed on the coast near Brochel, but may be seen a short distance inland, about 450 m SW of the castle [NG 583458]. For about 10 m above the contact grey shaly intercalations are present in the Loch an Uachdair Member. Four mappable grey units, similar to the fine-grained facies of the Brochel Member, occur in the lower half of the member north of Loch an Uachdair.

The Glame Member (Applecross Formation) is composed of medium to coarse grained pebbly red sandstone. The base of the member is defined about half a kilometre west of Loch an Uachdair [NG 575467] by the abrupt appearance of durable pebbles. Sedimentary structures in the member are mainly planar cross-bedding (44% of the section), flat bedding (26%) and apparently massive bedding (18%). Trough cross-bedding is present in only 10% of the section. Roughly half the sediments show contortions. There is no statistical evidence for fining upwards sandstone sequences in either the Glame or Loch an Uachdair Members (Selley 1970). Petrographically the sandstone of the Glame Member is arkose, with about 30% feldspar. The pebbles are mainly vein quartz (47%) volcanics (36%) quartzite (12%) chert and jasper (4%). Palaeocurrents in the upper half of the Glame Member measured by Nicholson (1993, table 1) flowed towards the SE ($\theta = 141°$, $n = 180$).

The highest stratigraphic level preserved in the Glame Member crops out on the coast of Raasay at Rubha an Inbhire [NG 548417]. Some 40–80 m stratigraphically below this level, and about 1730 m above the base of the Applecross Formation, the Glame Member contains two intercalations of shales like those in the Diabaig Formation. They reach the coast just south of Rubha an Inbhire [NG 550412]. There is another about a kilometre to the north [NG 575467].

The island of Fladday duplicates the lower part of the Raasay succession. The lowest sediments (Caolas Fladday Member, part of the Diabaig Formation) consist of grey chloritic tabular-bedded sandstones containing blocks of gneiss up to 2 m in size, overlying Lewisian basement. These grey beds, like similar sediments at Diabaig (Db1), pass laterally into the well-sorted, cross-bedded epidotic sandstones seen 200 m north of the causeway linking Raasay and Fladday at low tide. A short distance to the west, on the island of Fladday, 37 m of grey silty beds like those in the Brochel Member outcrop [NG 591505 to 589504]. They were called the Caolas Fladday Member by Selley and also form part of the Diabaig Formation (Db2). A cross-section, assuming no strike fault, shows that these beds are lateral equivalents to the breccias and sandstones that directly overlie the gneiss basement. The section also reveals at least 60 m of pre-Diabaig relief (Selley 1965*a*, fig. 7). The grey beds were formerly quarried for roofing slate. They contain red hematite nodules up to a centimetre across and traces of malachite [NG 591504]. Graded sandstone beds with volcanic lithic grains are also present in the shales.

The red sandstones of the Lower Fladday Member (Applecross Formation) that abruptly follow the Caolas Fladday Member are well exposed on the cliff flanking the west side of the strait [NG 589504]. This is the type section. The Member is about 50 m thick and sedimentologically similar to the Loch an Uachdair Member. It is overlain by coarse red sandstones with durable pebbles up to 2 cm in size [NG 589504]. One red sandstone pebble was noted by Selley at this locality. These sandstones were called the Upper Fladday Member by Selley, and correlated with the Glame Member of Raasay, but pebbles are only present at the base – the stratigraphically higher medium to fine-grained sandstones lack them and a correlation with the Loch an Uachdair Member (300 m thick) accords better with both the strike evidence and the total stratigraphic thickness of the Lower and Upper Fladday Members (240 m). Pebbles reappear on the west coast of the island, reaching about a centimetre in size [NG 584505], and are also present on the islets of Griana-sgeir and the western part of Glas Eilean. These pebbles are taken to mark the base of the Glame Member (Fig. 106).

Intercalations 1–2 m thick of fine red sandstone or siltstone with ripple bedding and flat bedding occur throughout the Applecross sandstones on Raasay and Fladday. Selley (1969) recognized two types. The first is an upward-fining sequence cut by an erosion surface; the second is confined between two erosion surfaces. He interpreted them as overbank and abandoned channel deposits, respectively, within the Applecross low sinuosity braided river environment.

From Selley's measurements the palaeocurrents in all the red sandstones of the Applecross Formation flowed towards the SE as shown in the rose diagram in Figure 105. Similar results were obtained by Nicholson (1993) from the Glame Member of the Applecross Formation (see above). Ripple cross-lamination in the Caolas Fladday Member (Diabaig Formation) also gives a very similar direction.

Scalpay, Longay and adjacent parts of Skye

Scalpay was mapped by Harker for the Geological Survey in 1900 (Peach *et al.* 1910, pp. 63–64) but the following account is based mainly on the resurvey of Scalpay by Selley between 1959 and 1962 (Selley 1965*a*, 1965*b*). A synopsis of the stratigraphy with Selley's member names (designated by Roman numerals owing to the lack of suitable place names) is shown in Figure 105. All the beds belong to the Torridon Group. The regional dip is westwards but the stratigraphically lowest beds crop out on the west side of the island, forced up by a Cenozoic granite.

Member I consists of fine to medium-grained sandstones intruded and bleached by a Cenozoic granite. The base is not seen. The type section is 200 m SW of Allt Reireag [NG 582306].

Member II (grey shale), Member III (interbedded grey shale and red sandstone, Member IV (pebble-free red sandstone) and the lower part of Member V (pebbly red sandstone) have their type sections on the coast from Rubha Reireag [NG 577309] to Rubh' a' Chonnaidh [NG 582327]. The NW dipping pebbly sandstones at the south end of Raasay may be an upward continuation of this section. Members I–V were recognized by Harker but not named or definitely correlated with the mainland succession. Members I–III closely resemble the Brochel Member of Raasay, i.e. the Diabaig Formation, and Members IV and V resemble the Loch an Uachdair and Glame Members in Raasay, i.e. the Applecross Formation. The only significant difference is in the coarseness of the Scalpay succession. Member V is 70% pebbly and has no siltstone bands, whereas in Raasay only 10% of the equivalent Glame Member is pebbly and 2% is silty.

The palaeocurrents in Members IV and V flowed towards the SE, as in Raasay and Fladday. The directions are shown in Figure 105.

Grey and red sandstones that outcrop on the coast of Skye only 2 km west of Scalpay probably belong to the Applecross Formation. According to Harker (in Peach *et al.* 1910, p. 64) they are medium grained east of Sconser [NG 530320] and pebbly to the west. Rare-earth element data for the sandstones near Sconser have been published by Thorpe *et al.* (1977).

Creag Strollamus in Skye [NG 606260] has beds belonging to the lower part of the Scalpay succession (i.e. Diabaig Formation), according to Harker (*in* Peach *et al.* 1910, p. 63). Bailey (1954) described them as 'thinly bedded fine-grained feldspathic sandstones, only occasionally gritty' and about 450 m thick. They are enveloped by Cenozoic intrusions. The best exposures are south of the summit of the Creag.

The islet of Longay is composed of fine to medium-grained red sandstones of the Aultbea Formation. Beds of red shale and flat bedded fine-grained sandstone with current lineation occur at several horizons. The strata on Longay are probably a southwestward continuation of those seen in the Crowlin Islands, separated from the Applecross Formation in Scalpay by a major fault.

The Sleat of Skye

The Sleat of Skye was mapped by the Geological Survey between 1892 and 1895 and appears on the one-inch to the mile sheet 71 (Glenelg) published in 1909. The west half only of this map was reprinted on 1:50 000 scale in 1976. Some 4500 m of clastic sediments were found by the Survey, the uppermost kilometre of which was identified as the Applecross Formation. The rest was divided into four lithostratigraphic units, thought to equate with the Diabaig Formation, but transferred to the newly created Sleat Group by Stewart (1969). There is a good general description of the rocks by Clough in the NW highlands memoir (Peach *et al.* 1907, p. 348–362). The first sedimentological studies were by Sutton & Watson (1960, 1963, 1964). The palaeocurrent directions cited below are from this source. The palaeomagnetism has been examined by Potts (1990) and the geochemistry by Stewart (1991*b*). A field guide to the Torridonian rocks of the Sleat of Skye was published by Hambrey *et al.* (1991, p. 86–92). Three Geological Conservation Review sites in the Sleat Group are noted below. The stratigraphy, composition and palaeocurrents are summarized in Figure 20. The rocks are described in stratigraphic order in the following paragraphs, starting with the oldest.

The Rubha Guail Formation

The formation was called the Epidotic Grit and Conglomerate by Clough but given its present name by Stewart (1975). The characteristic lithology is trough cross-bedded coarse or very coarse sandstone (average grain size about 1.3 mm) coloured shades of green according to the relative proportions of epidote and chlorite in the matrix. Pebbles frequently reach a centimetre, but the 'quartz-felsite, purple felstone and quartzite' pebbles recorded by Clough from the top of the formation at Port Aslaig [NG 76901777] are very rare. The Lewisian basement provided the great bulk of the material comprising the formation, a fact especially evident farther south where gneiss blocks over 30 cm in size, together with detrital hornblende, have been recorded (Peach *et al.* 1907, p. 352–353; Bailey 1955, fig. 8) in a tectonic slice severed from the Lewisian basement about 5 km west of Armadale [NG 588047].

Basement relief was substantial when sedimentation started, for at Loch Carron several hundred metres of strata are cut out against the basal unconformity, including the whole of the Rubha Guail Formation and the overlying Loch na Dal Formation (Peach *et al.* 1907, p. 343). The base of the Rubha Guail is well exposed at Fernaig, 10 km NE of Kyle of Lochalsh [NG 842336], but not in Skye.

The type section of the formation extends from the mouth of Allt Coire Gasgain [NG 73831595] SW along the coast to the mouth of Allt an Teanga Odhair [NG 72201504]. The section is cut by faults so that its true thickness is hard to establish. The best continuous section, shown in Figure 108, has the lowest 30 m composed entirely of trough cross-bedded very coarse sandstone. Over the next 70 m medium-grained green sandstones are interbedded with striped greenish-grey siltstone and pale grey mudstone that locally show desiccation polygons on bedding surfaces (Stewart 1962, figs 11 & 12). The bases of the sandstone beds are strongly erosional. Above this level the formation is dominated by grey siltstone, mostly millimetre laminated, with only occasional sandstone beds. The section exposed is 270 m thick. The laminated grey

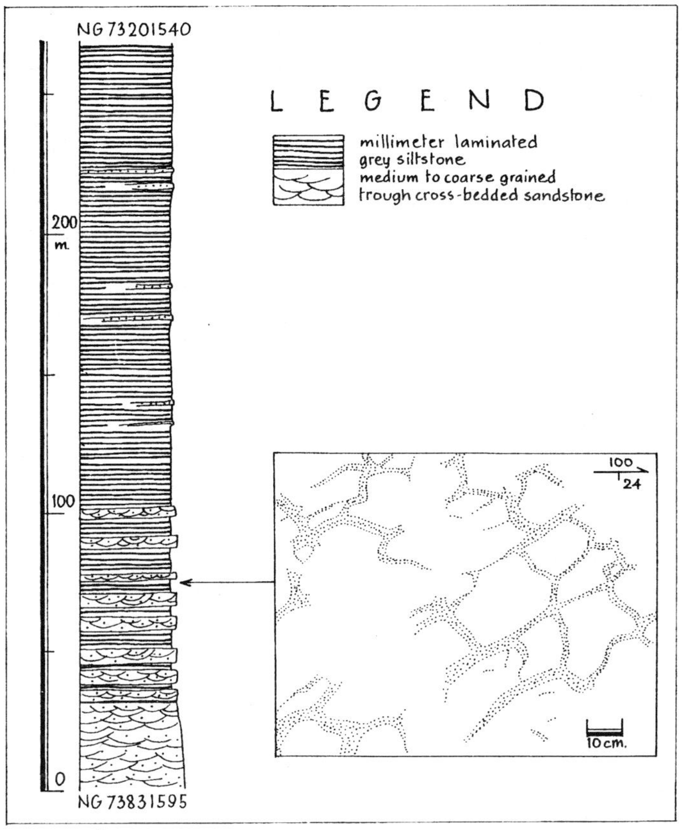

Fig. 108. Graphic log of the Rubha Guail Formation on the type section. The base of the section is 400 m NE of Rubha Guail. The top is truncated by a fault 220 m SW of Rubha Guail. The desiccation polygons figured are exposed, with others, on outcrops below high water mark about 200 m NE of Rubha Guail [NG 73421565]. See also Stewart (1962, fig. 11) and Sutton & Watson (1960, fig. 7) for desiccation cracks at this locality. The polygons have been shortened perpendicular to the axial plane of the Lochalsh fold during Caledonian deformation.

siltstones are like those in the overlying Loch na Dal Formation from which they were excluded by Clough only because some of them are tinted green.

Troughs in the very coarse sandstones are 10–20 cm deep and several metres wide. Palaeocurrents flowed towards the east ($\theta = 073°$, $n = 49$), as shown in Figure 20, the low dispersion of directions suggesting a braided fluvial environment with a relatively steep palaeoslope. Soft sediment contortions are absent from the very coarse sandstones but appear commonly in the medium-grained sandstones that form tabular cross-bedded sheets ($p \approx 100$) up to 0.5 m thick. The contortions often take the form of domes and basins that die out towards the top and bottom of a bed. Some, however, intrude the bed above. In plan they are elongated parallel to the axis of the Lochalsh fold, due to strain suffered by the beds during the Caledonian orogeny.

The basal breccias and trough cross-bedded very coarse sandstones are interpreted as having been deposited on small alluvial fans that derived their sediment from local gneiss hills. This conclusion is supported both by the basal breccia noted above but also by the mineralogy of the Rubha Guail sandstones which is similar to that of the basic Scourian gneisses and pegmatites in the Lewisian nearby (see pp. 25–26). The fans graded upwards and laterally into lacustrine or shallow marine deposits, i.e. the millimetre-laminated siltstones forming the top of the formation.

The Loch na Dal Formation

The formation is about 800 m thick and consists mainly of interbedded dark grey mudstones and slightly calcareous medium-grained grey sandstones. The contact with the underlying Rubha

Guail Formation is concealed by a shingle beach at the mouth of Allt na Teanga Odhair [NG 72201504]. Some of the mudstones appear to be massive but most show millimetre-thick laminations formed by silt–mud couplets. Many couplets are graded (Bailey 1955, p. 123). Some laminae are phosphatic and are reported to contain cryptarchs (W. L. Diver, pers. comm.) but no details have been published.

The sandstones of the Loch na Dal Formation weather to a yellowish-grey colour. Although they are on average medium-grained (0.3 mm), many are coarse, with seams of quartz and K-feldspar grains as much as 0.5 cm in size. The beds range in thickness from millimetres up to decimetres and are laterally very persistent ($p \approx 10\,000$). They all have more or less erosional bases and some have gradational tops. Ripple cross-lamination is a common structure in the sandstones. A typical two metre section of the formation is shown in Figure 109 and a drawing of the characteristic interbedded siltstones and very coarse sandstones in Figure 110. The interbedding is like that seen in the Diabaig Formation where coarse material from basement hills has been washed across the floor of a shallow lake (cf. Fig. 95).

The uppermost 200 m of the Loch na Dal Formation contains coarser grained sandstone and less mudstone than the rest of the

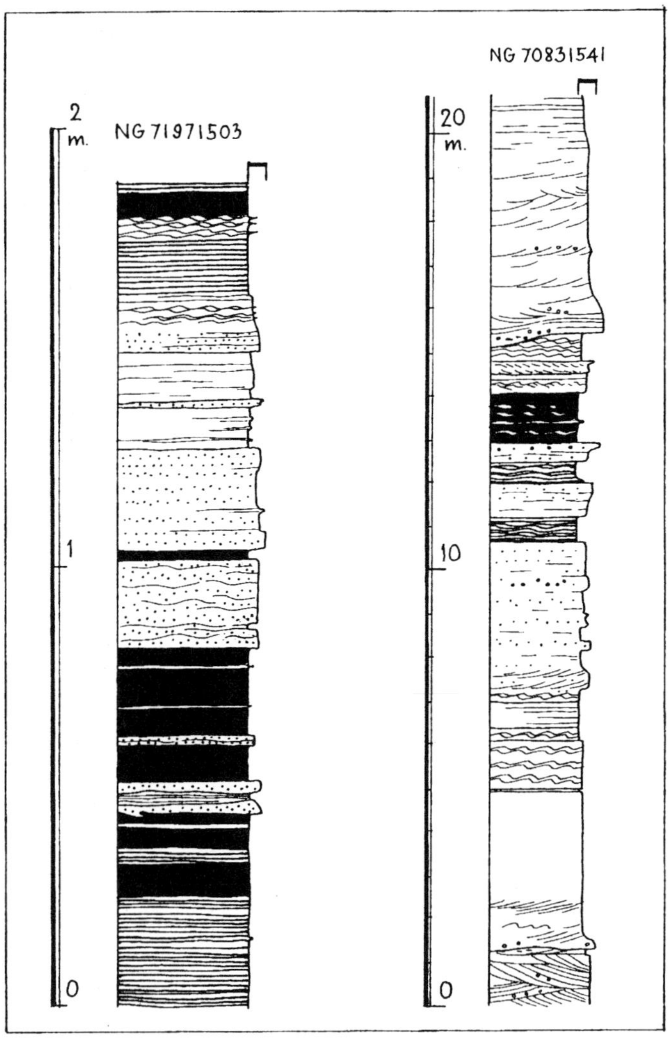

Fig. 109. Graphic logs of the Loch na Dal Formation at Loch na Dal. Lithologies figured are massive dark grey mudstone (black), millimetre-laminated dark grey mudstone and siltstone (lined), and yellowish-grey weathering slightly calcareous sandstone (stippled). Sedimentary structures are shown as seen. The left-hand log shows the interbedding of coarse and fine sediment characteristic of the lower part of the formation. The grain size scale spans 4–0 ϕ units (0.06–1 mm). The right-hand log is typical of the top of the formation, and has a grain size scale spanning +4 to $-2\,\phi$ units (0.06–4 mm). All pebbles over 0.5 cm are shown.

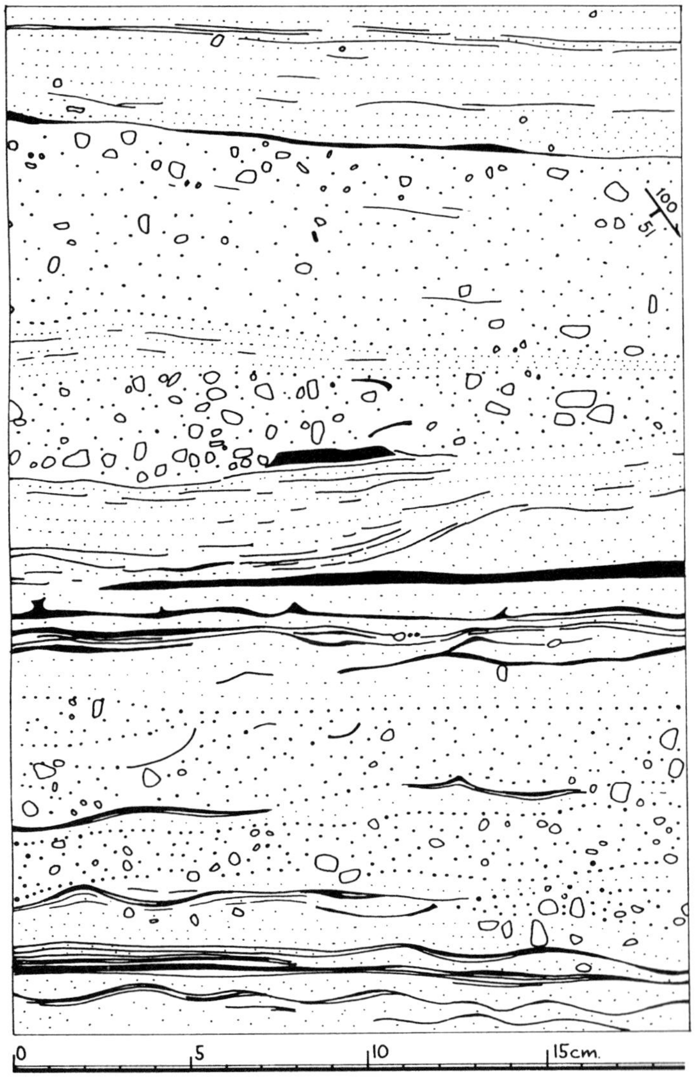

Fig. 110. Interbedded siltstones and sandstones in the Loch na Dal Formation at Loch na Dal [NG 70831541]. Siltstone in the tracing is shown black, fine sandstone by fine stipple and coarse sandstones by coarse stipple. The fine and coarse sandstones have average grain sizes of 0.2 and 0.5 mm, respectively. All grains over 2.5 mm diameter are outlined.

formation. A 20 m graphic log from this part of the section is shown in Figure 109. The overlying Beinn na Seamraig Formation is not well exposed on the coast section, and the contact with the Loch na Dal Formation is concealed.

The coarse-grained sandstones in the Loch na Dal Formation suggest a nearby source. Some may have been deposited by turbid underflows close to a fan-delta that was building out into a lake or shallow sea. Eventually the delta filled the lake so that the top of the Loch na Dal Formation is dominated by channel sands. Sandstones at this stratigraphic level contain a potassic component that is unlikely to have come from the nearby basement, and which becomes progressively more important upwards through the Sleat Group. These sands must have been contributed by a major fluvial system with a relatively distant source (Stewart 1991*b*). Palaeocurrent directions from cross-bedding, corrected for tilt, were easterly ($\theta = 093°$, $n = 26$).

The type sections of the Rubha Guail and Loch na Dal Formations at Loch na Dal constitute a Geological Conservation Review site (Mendum *et al.* 2003).

The Beinn na Seamraig Formation

The Beinn na Seamraig (1100 m thick) is composed almost entirely of fine grey sandstones (average grain size about 0.15 mm), about

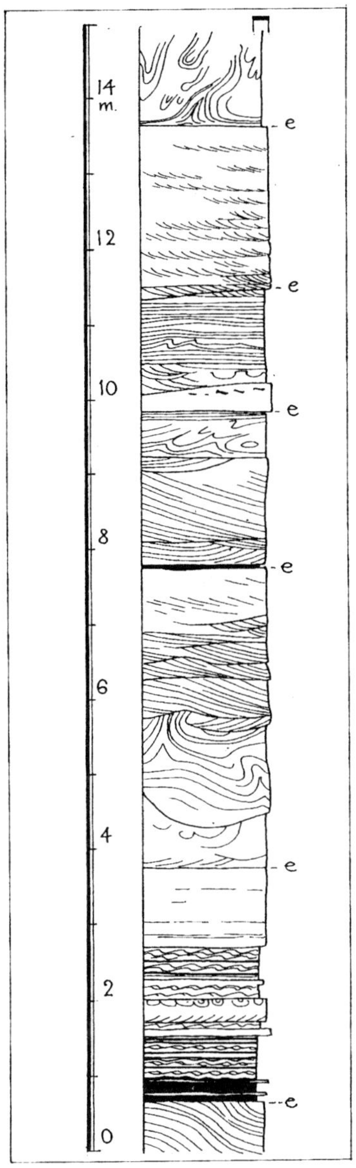

Fig. 111. Graphic log of part of the Beinn na Seamraig Formation in Glen Arroch [NG 75362087]. Laterally extensive erosion surfaces are indicated (e). The grain size scale spans 4–0 ϕ units (0.06–1 mm). Sedimentary structures are drawn as seen. The sandstones are all greenish-grey in colour, the shales and siltstones grey (N4–N5).

a half of which are strongly contorted. A typical section chosen from the weather-etched outcrops near the road up Glen Arroch is shown in Figure 111. Both trough and planar cross-bedding are common. The bases of the sandstones are commonly erosive, whereas the tops frequently show ripple-drift lamination due to waning flow. Finer grained beds are comparatively rare in the formation, though some have been mapped by Sutton & Watson (1964, fig. 2) and one such bed is included in the measured section. It contains laminated siltstones and current rippled fine-grained sandstones similar to those in the upper part of the Loch na Dal Formation. The Glen Arroch exposure, shown in Figure 111, is a Geological Conservation Review site (Mendum *et al.* 2003).

The base of the formation, marked by the appearance of contorted bedding, lacking from the underlying Loch na Dal Formation, can be located about 260 m SSW of Beinn Bhreac [NG 71771618]. The type section extends from here northwestwards to Allt a' Choin.

The coarser sandstones are thought to have been deposited in channels on a braided alluvial plain. Palaeocurrent directions from cross-beds, corrected for tilt, were southerly ($\theta = 175°$, $n = 60$). The finer beds may represent the temporary advance of a lake margin across the area.

The Kinloch Formation

The formation (1100 m thick) is mainly fine or very fine-grained grey sandstone (average grain size about 0.1 mm), with subordinate grey shales. It is best studied at Loch Eishort [NG 6716], for exposures are poor around Kinloch. At Loch Eishort the formation is built of fining-upward cycles each 25–35 m thick. The shaly tops of the cycles have been preferentially ground down by glaciation and are now covered by shingle. Graphic logs of the sandstone forming the lower and middle parts of a cycle are shown in Figure 112 and can also be taken as representative of the sandstones in the rest of the formation. Large-scale trough cross-bedding can be identified, but as can be seen from Figure 112 ripple-drift lamination is much more common. Palaeocurrent directions from cross-beds, corrected for tilt, were easterly ($\theta = 079°$, $n = 38$). The shales, that are here concealed, can be seen 500 m SW of Ob Gauscavaig

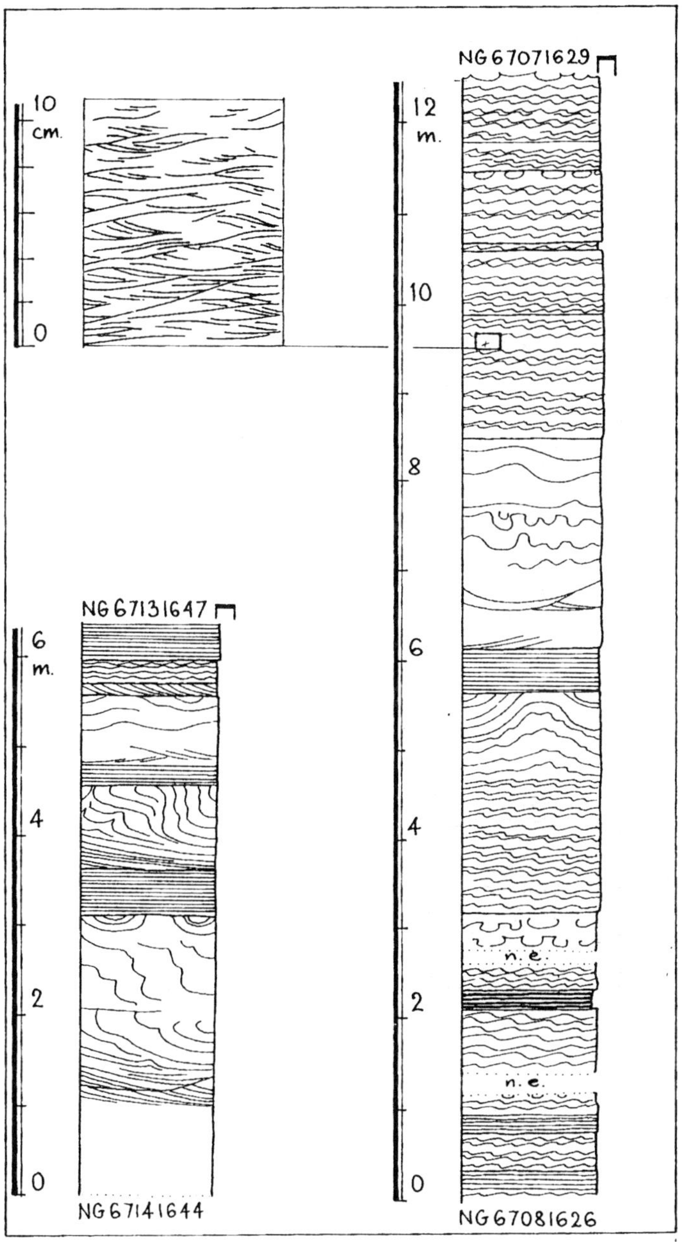

Fig. 112. Graphic logs of sandstones belonging to the Kinloch Formation on the shore of Loch Eishort in Skye. The left-hand log is from the base of a fining-upward unit. The right-hand log is from the top of a fining-upward unit and is followed by a shale sequence. Sedimentary structures are drawn as seen; blank areas are apparently structureless. The detail showing ripple-drift lamination is based on a tracing of the rock surface. The grain size scales span 4–0 ϕ units (0.06–1 mm). The sandstones are medium grey (N5) and the silts dark grey (N5). n.e. = not exposed.

[NG 591115] at the top of the Kinloch Formation where they form the tops of fining-upward cycles about 10 m thick with erosional bases (Stewart 1966*a*). The sandstones forming the lower part of each cycle are pale red, with contorted bedding like that common in the overlying Applecross Formation. The shales consist of millimetre and centimetre-thick beds of grey siltstone with ripple lamination. These are the 'carbonaceous shales' searched unsuccessfully for fossils by the Geological Survey (Geikie 1900). Similar beds are seen at the same stratigraphic level at Rubha Ard Treshnish [NG 761258], about a kilometre east of Kyleakin.

The thick cycles in the Kinloch Formation are interpreted as alluvial fans or wedges interfingering with lake or shallow marine sediments. They are analogous to the Rubha Dubha Ard and Achduart Members which occur at the base of the overlying Applecross Formation at Achduart (see Achiltibuie).

The sections of the Kinloch Formation at Loch Eishort constitute a Geological Conservation Review site (Mendum *et al.* 2003).

The Applecross Formation

The formation is about 1000 m thick (1500 m according to Peach *et al.* 1907, p. 361), all composed of fine reddish-brown sandstone (average grain size about 0.15 mm) that only rarely contains pebbles. Pebbly sandstone was, however, noted by Clough (in Peach *et al.* 1907, p. 360) on the coast NW of Sgeir Gormul [NG 631158]. Pebble lithologies listed for the Applecross Formation include pegmatite, red felstone (i.e. felsite), red porphyrite (i.e. porphyry), arkose, vein-quartz, jasper, and pink and purple quartzite.

A typical section of the beds about 700 m above the base of the formation is shown in Figure 113, in which 60% of the beds are contorted. According to Selley *et al.* (1963) contorted bedding is present in about a third to a half the beds at this horizon. About a third of the beds show flat bedding, though in many cases this may represent the toes of planar cross beds. Coarsening-upward cycles are a prominent feature of the section, resembling those in the Aultbea Formation (member Ab2) at Toscaig. Such cycles are common between Heast and An Tòrr, in a zone 500 m thick (Selley *et al.* 1963). Palaeocurrent directions, corrected for tilt, were easterly ($\theta = 106^{\circ}$, $n = 56$). The direction of magnetization accords with that for the rest of the Torridon Group (Potts 1990).

On lithological grounds the sandstones would have been better assigned to the Aultbea Formation but, presumably, this was rejected by the Geological Survey because the Aultbea was supposed to overlie the Applecross Formation. The Applecross Formation is unconformably overlain by the Cambrian basal quartzite (Eriboll Formation) near Ob Gauscavaig [NG 594121] and again about 2.5 km SW of Heast [NG 627161]. The difference in dip between the Applecross sediments and the Cambrian rocks is slight (Peach *et al.* 1907, p. 419).

The boundary between the Kinloch Formation and the overlying Applecross Formation appears to be conformable. According to Sutton & Watson (1964, p. 266) there may, indeed, be a lateral passage from lowermost Applecross into uppermost Kinloch towards the SW, but poor exposure along the boundary makes this hard to verify.

The red colour of the Applecross Formation seems to depend on the degree of alteration of the feldspars and the oxidation state of the iron. Thin sections of the sandstones below the middle of the Kinloch Formation show feldspars that are clear, but above this level untwinned detrital grains are dusted with reddish-brown iron oxide. At first the dusting follows a skeletal pattern but near the contact with the Applecross Formation it comes becomes pervasive. These altered grains may be calcic plagioclase which has been albitized – a transformation common in the Applecross Formation but less so in the underlying Sleat Group (Stewart 1991*b*).

The dominance of hematite over magnetite in the Applecross Formation of the mainland is shown by the ratio $Fe_2O_3/FeO \approx 6$ (Van de Kamp & Leake 1997, table 2) or ≈ 19 (Williams & Schmidt 1997), thermo-magnetic analysis (Torsvik & Sturt 1987; Potts 1990) and ore microscopy (Stewart & Irving 1974). In the Sleat Group,

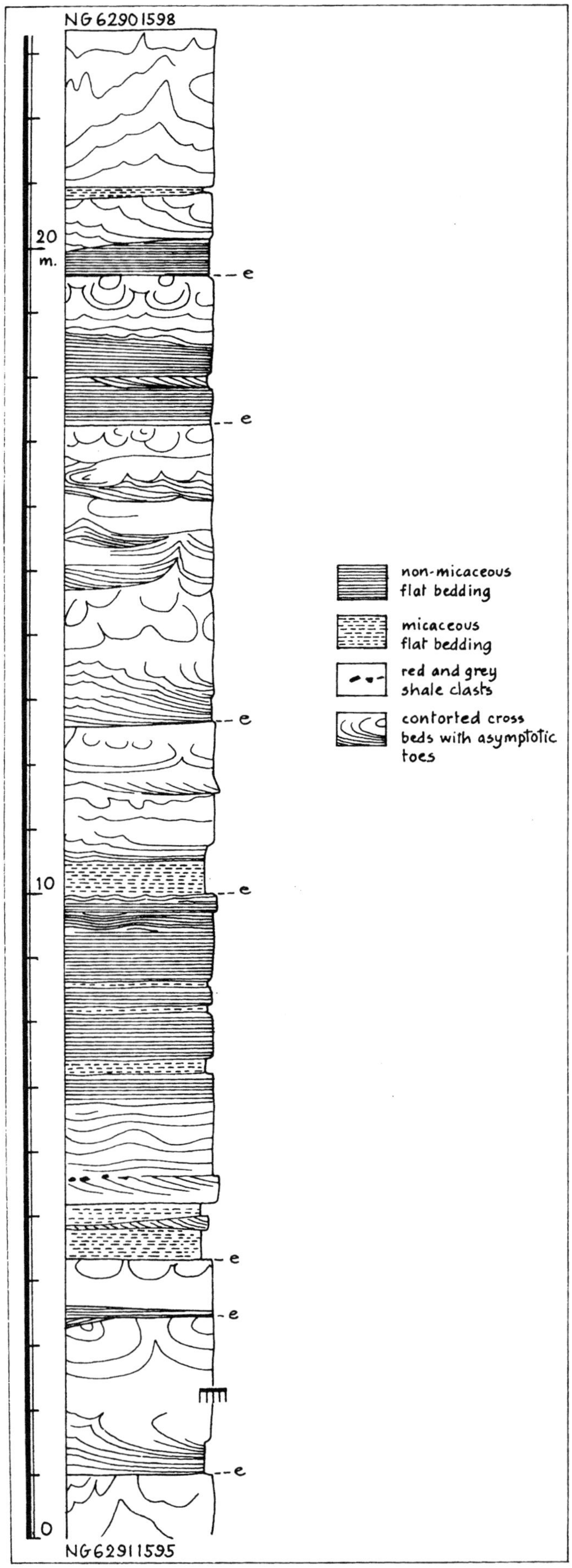

Fig. 113. Graphic log of the Applecross Formation at An Tòrr, about 2 km west of Heast in Skye. The extensive erosion surfaces indicated by the letter **e** mark the bases of upward-coarsening cycles. Sedimentary structures are drawn as seen. The grain size scale spans 4–0 ϕ units (0.06–1 mm).

however, the ratio Fe_2O_3/FeO is only about 0.5 (Kennedy 1951, table 1). Furthermore, according to the Geological Survey, the ore minerals in the Sleat Group are mainly magnetite, whereas the red Applecross of Skye contains hematite and ilmenite (Peach *et al.* 1907, p. 286, 346 & 358–360). The difference in oxidation state is probably partly due to low-grade Caledonian metamorphism that in the Sleat of Skye increases towards the SE, so that boundaries defined by the first appearance of pressure solution stripes, penetrative cleavage, the growth of chlorite, and presumably the reduction of iron minerals are all parallel to the Loch Alsh fold hinge (Bailey 1955; Coward & Whalley 1979). Note, however, that the Sleat/Torridon Group boundary at the present level of erosion is oblique to this hinge.

Potts (1990) has recorded pre-Caledonian palaeomagnetic directions from seven sites in the Applecross Formation and one in the Kinloch Formation. The latter is in the cyclic beds at Ob Gauscavaig, described above, that may be transitional between the Kinloch and Applecross Formations. The palaeomagnetic directions, though similar to those in the Applecross Formation elsewhere, have declinations mostly slightly clockwise from those found on the mainland to the north, but slightly anticlockwise from those measured by Robinson & McClelland (1987) in Rum. Bearing in mind the precision of these directions the suggestion by Potts that the Kishorn nappe has been rotated 26° clockwise is unwarranted. Indeed, a comparison of *palaeocurrent* directions in the Applecross Formation of Rum and Skye would suggest a small relative anticlockwise rotation of the latter.

Camusunary

A small area of sandstones and siltstones probably forming part of the Diabaig Formation was mapped in this area by Wedd for the Geological Survey (Peach *et al.* 1910, p. 65). The beds all lie west of the Camusunary fault. They are shown on the 1:50 000 geological map of Broadford (sheet 71W) published by the British Geological Survey in 1976. The stratigraphy is shown in Figure 114.

East and north of the house there are intermittent exposures of coarse and very coarse grey sandstone with crudely tabular bedding, similar to that in the lowermost Diabaig Formation (Db1b). Rare-earth element data for these beds have been published by Thorpe *et al.* (1977). Stratigraphically higher beds, mainly coarse siltstone, are well exposed along the coast west of Camas Fhionnairidh. The lowest beds crop out about 300 m south of the bridge over Abhainn Camas Fhionnairidh [NG 509186] whereas the highest, 70 m stratigraphically above, are 200 m beyond the headland Rubha Bàn [NG 503181]. These silty beds, despite contact alteration from the enveloping Cenozoic intrusions, have their sedimentary structures excellently preserved. They are accompanied by millimetre to decimetre-thick beds of fine sandstone with ripple lamination and small contortions. The bases of the thicker sandy beds are frequently sharp, and the tops gradational, like the grey sandstones in the shale facies (Db2) at Diabaig. Sedimentary veins are common in the siltstones but bedding surfaces are rarely exposed and complete polygons have not been seen.

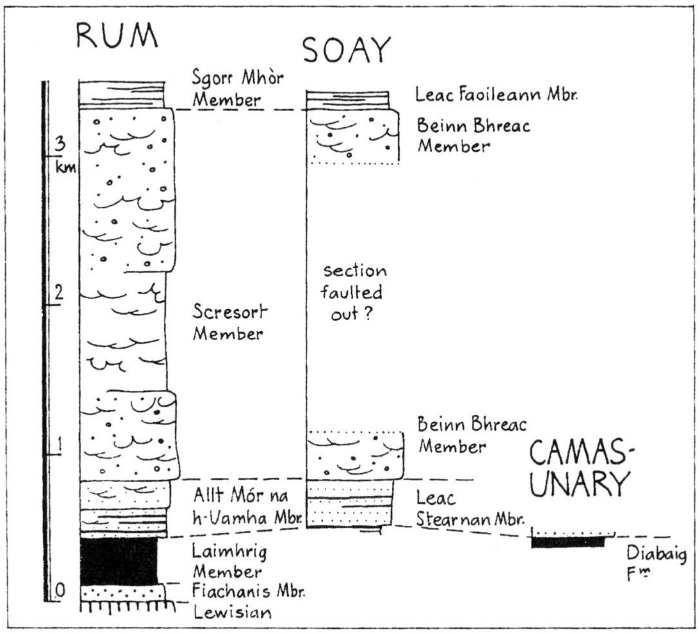

Fig. 114. Torridon Group correlation between Rum, Soay and Camusunary.

Near the top of the sequence just described thick beds of fine sandstone appear. The first bed, which is 7 m thick, shows large-scale contortions and has a markedly erosional base. Roughly 8 m stratigraphically higher another similar bed appears, but at this point the sequence is truncated by a major intrusion. These sandstones probably represent the base of the Leac Stearnan Member (basal Applecross Formation), for which the stratotype is a few kilometres to the SW on the island of Soay (*q.v.*). The correlation is supported by the fact that the nearest outcrops of the Leac Stearnan Member on Skye (a kilometre north of Soay) are along strike from the Camusunary beds and have a similar dip.

Soay

The stratigraphy obtained by Clough, who mapped the island for the Geological Survey (Clough & Harker 1904), is shown in Figure 114. A geological map of the island, based on the original mapping, forms part of the one-inch to the mile Minginish sheet 70, published by the British Geological Survey in 1913. The mapped units probably form part of the Applecross Formation and are here given member status.

The Leac Stearnan Member at the base of the sequence consists of pale red or brown, massive fine to medium-grained sandstone. Contorted bedding is typical. There are frequent metre-thick intercalations of flat bedded sandstone with current lineation. One of these sandstones has been quarried just above high water mark east of Loch Coire Doire na Seilg [NG 45241266]. There are also numerous intercalations of grey micaceous sandstone with linguoid ripple lamination throughout. The current lineation directions are almost random but the ripple lamination shows palaeocurrents flowing towards the south or SW like those in the Laimhrig Shale Member of Rum. Trough cross-bedding, where present, also shows palaeocurrents flowing southwestwards. Symmetrical, straight-crested ripples have an azimuth of about 105° like those in the Laimhrig Shale.

Grey siltstones containing millimetre-thick phosphatic laminae have been found at two localities in the Leac Stearnan Member. The first is on the coast 550 m south of Leac Stearnan [NG 45641282]. Details are given in Figure 115. The high persistency factor for these beds ($p = 100$–500) suggests low river gradients and frequent overbank flooding. The second locality is near the cliff top 200 m east of Loch Coire Doire na Seilg [NG 45311267]. The siltstone at this point has been excavated by the sea to form a large cave. Phosphatic material has also been recovered from micaceous grey sandstone on the east coast of Soay about 400 m north of An Dubh laimhrig [NG 47211557]. Phosphatic siltstone clasts frequently occur in the sandstones beneath fine grained intercalations. Phosphate from all these localities has been found to contain microfossils (W. L. Diver, pers. comm.).

Clough's choice of southern Soay as the type area for the Leac Stearnan Member was probably due to the greater frequency of silty and micaceous intercalations as compared with the much thicker section in eastern Soay. The Leac Stearnan Member in the type area is thus more clearly distinguished from the member above. Silty intercalations form about 10% of the type section, which extends from the headland south of Leac Stearnan [NG 458130] southwestwards along the coast to near Loch Coire Doire na Seilg [NG 450124]. The lowest beds of the Leac Stearnan Member, however, occur at

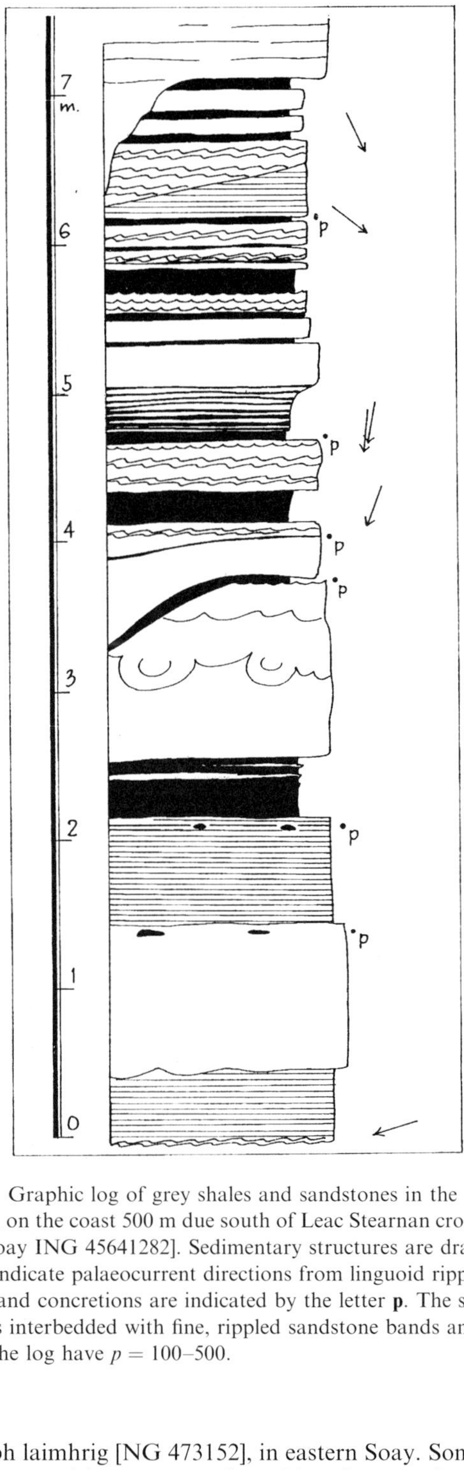

Fig. 115. Graphic log of grey shales and sandstones in the Leac Stearnan Member, on the coast 500 m due south of Leac Stearnan croft house, on the Isle of Soay [NG 45641282]. Sedimentary structures are drawn as seen. Arrows indicate palaeocurrent directions from linguoid ripples. Phosphatic laminae and concretions are indicated by the letter **p**. The shale lithology (black) is interbedded with fine, rippled sandstone bands and flaser. The beds in the log have $p = 100$–500.

An Dubh laimhrig [NG 473152], in eastern Soay. Some 3 m of grey siltstones near high water mark at this point contain siliceous greywacke beds, up to 15 cm thick, each graded upwards and ripple laminated like those in the upper part of the Diabaig Formation at Diabaig. These beds may mark the stratigraphic base of the Leac Stearnan Member, which here is 270 m thick.

The Beinn Bhreac Member, compared by Clough to the Applecross Formation, is much coarser than the Leac Stearnan. The typical lithology is coarse and very coarse pebbly sandstone in which the pink colour of the feldspars contrasts markedly with that of the greenish grey chloritic matrix. Pebbles are commonly a centimetre across and may reach as much as 3 cm. Pebble lithologies are white and greenish yellow quartz, red feldspar porphyry, red, pale red and white quartzite, and quartz–feldspar rock.

Trough cross-bedding and contorted bedding are the typical sedimentary structures of the Beinn Bhreac Member. Trough axes indicate palaeocurrents flowing towards the east ($\theta = 69°$, $n = 65$). Only two intercalations of flat bedded sandstone have been noted. One which is 3 m thick outcrops on the cliff top about 250 m west of Rubh' Aonghais [NG 44091229], where it was once quarried. The erosive base forms a series of shallow troughs, each about 0.5 m deep. The top of the bed is eroded, stepwise, by the overlying sandstone. There is another flat bedded intercalation at the top of the member (see below).

The base of the Beinn Bhreac Member is gradational over 10–20 m. It can be seen on the cliff near Loch Coire Doire na Seilg [NG 450124]. The continuously exposed coast section from here eastwards constitutes the stratotype. The top of the Member is conveniently defined by a grey siltstone bed 0.7 m thick that crops out on the side of a fault gully in the cliff 700 m south of Leac nam Faoileann [NG 42821385]. The siltstone contains small cupiferous nodules at three levels. It is underlain by about a metre of flat bedded sandstone with current lineation, that fills large troughs 2–5 m wide and 0.4 m deep. At the base of each trough there is a coarse lag deposit with reworked phosphatic siltstone clasts containing microfossils. Trough axes and current lineation suggest that the palaeocurrents flowed towards 025° unlike the usual easterly direction in the member. The flat bedding at this locality gives way upwards to ripple lamination, with the ripple crests lying transverse to the trough axes. This sequence of sedimentary structures is like that seen in crevasse splay deposits.

The Beinn Bhreac Member is apparently only 700 m thick, much less than might have been expected by comparison with its probable correlative on Rum, the Scresort Sandstone. This may, perhaps, be due to the major NE trending fault that bisects the outcrop near Doire Chaol. Correlations with Rum are considered in more detail in the description of the Rum sub-area.

The beds above the Beinn Bhreac Member are all fine-grained pale red sandstones with trough cross-bedding and frequent contortions. They were named by Clough after the rock Leac nam Faoileann [NG 426145] at the NW corner of Soay. Black bands rich in heavy minerals are common in the member and are frequently formed into large drop structures. Flat bedded flagstones quarried at the cliff top 650 m south of Leac nam Faoileann [NG 42811393] contain undeformed black bands. Another flat bedded unit 300 m farther north [NG 42761426] also shows a trough-shaped lower boundary like that in the flagstone quarry in the Beinn Bhreac Member. The troughs are about 6 m wide and 0.5 m deep. The few available palaeocurrent readings from this member indicate directions towards the south or SE, roughly at right angles to those in the Beinn Bhreac Member.

The base of the Leac Faoileann Member defined by the siltstone bed is sharp, but the change in grain size that differentiates the member from that beneath starts about 30 m stratigraphically below this level. The type section extends northwards from the siltstone marker bed. The top of the member is not seen.

Rum (Rhum)

The main features of the succession – grey shales beneath red sandstones – were noted by MacCulloch as long ago as 1819. Harker, who mapped the island for the Geological Survey, assigned the shales to the Diabaig Formation and the sandstones to the Applecross (Harker 1908). This classification is maintained and all units established subsequently to Harker's work given member status. Two geological maps of Rum were published in 1994, one by the British Geological Survey, scale 1:50 000, the other by Scottish Natural Heritage, scale 1:20 000 (Emeleus 1994). A comprehensive description of the stratigraphy and sedimentology by Nicholson is contained in the sheet memoir by Emeleus (1997, p. 9–16). Nicholson's members are shown on Figure 114 and on the above-mentioned maps. They replace those proposed by Black & Welsh (1961).

A palaeomagnetic study of samples from a stratigraphic interval about 100 m thick near the middle of the succession shows directions of magnetization like those in the Torridon Group elsewhere (Robinson & McClelland 1987).

The lowest beds of the Torridon Group unconformably overlie Lewisian basement and belong to the Fiachanis Sandstone Member (facies Db1 of the Diabaig Formation), originally mapped as 'basal grit' by Bailey (1945), Hughes (1960), Dunham & Emeleus (1967) and Dunham (1968). Outcrops of the member are found only in the southern and central parts of the island, brought to the surface by Cenozoic uplift within the ring fault. The member consists of coarse and very coarse grey feldspathic, locally epidotic, sandstone. The grey colour is due to contact alteration. There are abundant clasts of angular quartz together with occasional epidosite and rhyolite (Black & Welsh 1961). Maximum pebble size is about a centimetre and clast-supported breccias are absent. Bedding is typically tabular, with trough cross-bedding seen mainly in the coarser sandstones near the basement. The member is about 50 m thick on the type section in southern Rum [NM 37409396 to NM 37509396], underlain by gneiss and overlain by the Laimhrig Shale. It also occurs in central Rum, between Priomh-lochs and Cnapan Breaca where it appears to be 570 m thick (Dunham 1968, plate 25). Just east of Priomh-lochs [NM 37109875] the member is in contact with a fairly smooth unweathered gneiss surface without any marginal breccia. Interbeds of laminated grey siltstone appear towards the top of the member in this area and the contact with the Laimhrig Shale is exposed near the Priomh-lochs [NM 36959905]. The difference in thickness between south and central Rum implies palaeorelief in excess of 500 m but in view of the probable presence of faults in the Fiachanis Sandstone near Priomh-lochs and the absence of any significant breccia the true figure is probably a lot less.

The Laimhrig Shale Member (of the Diabaig Formation) is well exposed along the SE coast of Rum, between Bàgh na h-Uamha [NM 421972] and Dibidil [NM 401925]. It is at least 275 m thick. The base is below sea-level, but the top is well exposed at Bàgh na h-Uamha. Nicholson (*in* Emeleus 1997) recognized three interbedded facies within the shales, like those in the Diabaig Formation shale facies (Db2) at Diabaig:

(1) Silt–mud rhythmites, frequently graded.
(2) Wave-rippled, parallel laminated or massive fine sandstone beds a few centimetres thick. Ripple crests trend consistently 105°. Sand veins descend from these beds into the rhythmites, forming complete polygons 10–20 cm across.
(3) Fine-grained grey sandstone beds 5–100 cm thick, with sharp erosional bases. The beds may be massive, parallel laminated or ripple-drift laminated, with palaeocurrent flow towards the south or SW. The beds become thicker and more abundant towards the top of the member.

The total thickness of the Diabaig Formation in Rum is probably over 500 m, much greater than elsewhere, but nevertheless the facies present are recognizably the same as those in the type area at Diabaig. There is no similarity with the Sleat Group of Skye which, like the Diabaig Formation, overlies Lewisian basement and underlies the Applecross Formation.

The top of the Diabaig Formation and the base of the overlying Allt Mór na h-Uamha Member of the Applecross Formation are conveniently studied on the coast at Bàgh na h-Uamha and in stream sections nearby. The type section along the stream Allt Mór na h-Uamha [NM 42059722 to NM 41089751] is 400 m thick. The lower half of the member is formed of upward-fining cycles metres or tens of metres thick with $p \approx 400$. The lower part of each is pinkish-grey weathering fine-grained sandstone with planar and trough cross-bedding, frequently contorted. Ripple-drift lamination at the tops of the beds shows palaeocurrents that flowed south or southeastwards. The upper half of each cycle is mainly grey siltstone and fine-grained sandstone with parallel and current-ripple lamination. Similar cycles are seen at the base of the Applecross Formation in Skye (Stewart 1966*a*).

The rest of the member consists of about 200 m of pale-grey fine to medium-grained arkose (feldspar $\approx 40\%$) with trough cross-bedding, often contorted. Pebbles are absent.

The Scresort Sandstone Member overlies the Allt Mór na h-Uamha Member on the south side of Loch Scresort, and also in down-faulted blocks north of Bàgh na h-Uamha and at Welshman's Rock. Nicholson designated the north side of Loch Scresort, from Rubha na Roinne westwards, as the type section [NG 42300014 to NM 40859982]. The member is about 2500 m thick and covers most of the northern part of the island. The dominant lithology is medium to coarse-grained, pale red sandstone, with durable pebbles either scattered or concentrated in thin bands. A graphic log of an 18 m thick section of the sandstone published by Nicholson (*in* Emeleus 1997, fig. 5) shows that trough cross-bedding is common and about 25% of the beds contorted. Pebbles are generally 1–3 cm in diameter with a maximum of 6 cm in the stream Allt Rubha na Moine [NG 3820413]. According to Black & Welsh (1961) pebble lithologies include metamorphic quartzite, chert, felsite, porphyrite and quartz mica schist, all of which are typical of the Applecross Formation. Palaeocurrents ($n = 53$) in the lower part of the Scresort Sandstone measured by Welsh (1963) flowed towards the east and NE. According to Nicholson (1993, table 1) the mean direction near Rubha na Roinne was towards the ESE ($\theta = 118°$, $n = 203$), whereas in outcrops along the coast from Rubha na Moine to Kilmory it was southeastwards ($\theta = 132°$, $n = 206$). Welsh used all kinds of cross-beds to get palaeocurrent directions while Nicholson only used trough axes, hence the discordant mean vectors.

Part of the Scresort Sandstone is only sparsely pebbly and was for this reason designated as a separate stratigraphic unit (the Loch nan Eala Arkose) by Black & Welsh (1961). It is exposed on the northern coast from near Creag na h-Iolaire [NG 409024] to Rubha na Moine [NG 387042], where it is about 800 m thick. The palaeomagnetic study by Robinson & McClelland (1987) was conducted on these sandstones, probably because of their relatively fine grain size. Grey shale intercalations just above this sparsely pebbly unit crop out west of Kilmory Glen [NG 355016].

The Sgorr Mhór Sandstone Member, assigned by Nicholson to the Aultbea Formation but here regarded as part of the Applecross Formation, completes the Rum succession. The type section designated by Nicholson is along the coast from Guirdil Bay to Camas na h-Àtha [NG 31490119 to NM 30109962]. The base is gradational over tens of metres into the underlying Scresort Sandstone north of Glen Shellesder, but the top is concealed beneath the sea, giving an exposed thickness of only 175 m. The dominant lithology is fine to medium-grained, pebble-free red sandstone. A graphic log by Nicholson (*in* Emeleus 1997, fig. 6) shows that flat bedding forms 30% of the section and undeformed tabular bedding another 22%. Contorted bedding affects only 35% of the section. Palaeocurrents ($n = 88$) flowed towards the south and SE (Nicholson *in* Emeleus 1997, fig. 6). Nicholson reports frequent black bands formed by concentrations of opaque minerals, that may explain why the member was misidentified as Laimhrig Shale by Black & Welsh (1961) where it crops out at the southern tip of the island. It has been downfaulted in this area by a concealed eastward-dipping listric normal fault, like those at Bàgh na h-Uamha and Welshman's Rock.

Nicholson correlated the Sgorr Mhór Sandstone with the Aultbea Formation because both are fine grained and have frequent contortions and heavy mineral bands. However, the differences are even more striking. The Aultbea in its type area is almost entirely contorted and flat bedding is virtually absent. Furthermore, the Sgorr Mhór Sandstone exposed is less than 200 m thick so that it is uncertain if it forms the base of a major stratigraphic unit or simply a relatively fine-grained unit within the Applecross Formation. The latter alternative is preferred here.

The key elements in the correlation of the Torridon Group between Rum and Soay are:

- coincidence of southerly palaeocurrent directions in both the Allt Mór na h-Uamha Member of Rum and the sedimentologically similar Leac Stearnan Member in Soay;
- lithological similarity of the Scresort Sandstone in Rum and the Beinn Bhreac Member in Soay;
- coincidence of fine grain-size, frequent black bands and southerly directed palaeocurrents in the Sgorr Mhór Sandstone of Rum and the Leac Faoileann Member in Soay. The correlation shown in Figure 114 is based on these considerations.

Canna, Eigg and Hawkes Bank

The Cenozoic rocks of Canna and Eigg, respectively to the west and east of Rum (see Figure 1), overlie Precambrian basement, including red sandstones of the Torridon Group (Harker 1908). At Compass Point on Canna [NG 280056] rounded and subangular blocks of red sandstone up to 10 cm in size occur in Cenozoic conglomerate. Gneiss, schist and epidotic grit are also recorded. At the foot of Sgurr of Eigg [NM 460845] angular blocks of red sandstone up to a metre across form part of a Cenozoic agglomerate. The isotopic composition of the Cenozoic lavas of Eigg also indicates contamination by Torridon Group sandstone (Dickin & Jones 1983).

Seismic and gravity modelling show that the sedimentary basin west of Canna contains about 4 km of Torridonian beneath Triassic and Jurassic sediment (O'Neill & England 1994). Seismic profiling reveals about 6 km of Torridonian in the Sea of the Hebrides basin, SW of Rum (Stein 1988, fig. 11).

Boreholes SH 226 and SH 767 sunk in the sea bed by the British Geological Survey on Hawkes Bank, 20–40 km SW of Rum, proved red sandstone (Binns *et al.* 1974). From the published description of the cores the rock is petrographically like the red Torridon Group sandstone of Rum. Geophysical evidence shows that the outcrop extends southwestwards at least as far as the latitude of Iona, and in pre-Cenozoic times must have been even more extensive. This perhaps explains the presence of 'red Torridonian sandstone' and fossiliferous Durness Limestone in the Triassic conglomerates of Mull (Rast *et al.* 1968).

Iona

Sedimentary rocks unconformably overlying Scourian basement gneisses in the Isle of Iona (Fig. 1) have been studied by Jehu (1922), Bailey & Anderson (1925), Stewart (1962) and Fraser (1977). They were named the Iona Group by Stewart (1969). The British Geological Survey published a 1:50 000 map (Ross of Mull, sheet 43S), incorporating new mapping, in 1999. The following account of the rocks has been prepared in collaboration with Dr F. M. Fraser-Menzies.

The sediments suffered polyphase deformation prior to intrusion of the Ross of Mull granite, dated by Rb–Sr at 414 ± 3 Ma (Halliday *et al.* 1979). Some of this deformation is associated with the Sound of Iona fault that separates the island from relatively high-grade metamorphic rocks above the Moine thrust in the Ross of Mull (Potts *et al.* 1995). The Moine thrust is now deep under the Ross of Mull but to the west has been moved up several kilometres by the Sound of Iona fault so that it structurally overlies Iona. Gravity data show the fault trending NE (not NNE as believed by Potts *et al.*) for nearly 100 km (Rollin 1994). The Sound of Iona fault is thought by Potts *et al.* to be analogous to the Loch Gruinart fault that juxtaposes the Bowmore Group in central Islay against basement gneisses to the west (see Bowmore). It is also analogous to the Camusunary–Skerryvore fault, 25 km to the NW, that downthrows the Lewisian basement 5 km to the SE.

The Iona Group is Torridonian almost by definition (see p. 1) for it is very probably Precambrian (see below) and unconformably overlies Lewisian basement. However, definite lithostratigraphic correlation with the Torridonian is at present impossible.

The northern coast of Iona, opposite Eilean Annraidh, exposes a continuous section of the basal beds of the Iona Group and their unconformable contact with high-grade gneisses [NM 292261]. Deformation at this locality is relatively slight even though to the south the rocks near the contact are intensely sheared. A graphic log for the lowest 165 m of the sequence has been published by Stewart (1962, fig. 10) who divided the strata into units **T1–T12**. The basal conglomerate (unit **T1**) contains sub-angular blocks up to 30 cm across, and occasionally much larger, apparently derived from the basement beneath, in a greyish-green sandy matrix. Clast types include white and purple quartz, red pegmatite and schistose, dark greenish-grey epidotic gneiss, amphibolite, but no quartzite pebbles. Shear planes in some of the breccia clasts fail to cut the matrix (A. L. Harris, pers. comm.), which has been interpreted to mean that there was an active fault scarp nearby (Holdsworth *et al.* 1987; Potts *et al.* 1995).

The basal beds are followed by pink feldspathic conglomerate (unit **T2**) containing clast-supported feldspar and quartz pebbles with diameters up to 0.5 cm, rarely as much as 2.5 cm, set in a deformed green epidotic matrix. Epidote and quartzite pebbles occur rarely. A mylonitized shear zone separates unit **T2** from the coarse grey feldspathic sandstone (unit **T3**) that follows. Units **T2** up to **T11** [NM 294295] consist of coarse grey sandstones and tightly folded dark grey mudstones containing thin sandstone beds (Fraser 1977, figs 2.8 & 2.9). Sedimentary structures such as cross-bedding have been severely deformed.

The coarser sandstones are all potassic arkoses, but abundant albite is recorded in some specimens by Jehu (1922), perhaps derived from unfoliated albite–quartz pegmatites that exist nearby in the basement. Epidotization has affected the Lewisian and the overlying sediments, especially near the unconformity. Jehu records both clastic grains of epidote as well as abundant authigenic epidote. Neither garnet, magnetite nor hematite are mentioned as detrital minerals by Jehu, though the first two occur in the gneisses.

Beds above unit **T11** are all fine-grained grey sandstones, with thin mudstones between, and an exposed thickness of around 300 m. Ripple lamination is typical. Straight-crested ripple forms are occasionally seen on bedding surfaces (Fraser 1977, fig. 2.2) but desiccation cracks have never been noted.

Units **T1** to **T11** represent a small alluvial cone fining upwards into lake deposits, burying an erosion surface that, judging from the slight thickness of the basal breccia, had no more than a few hundred metres of relief. The beds above **T11**, from the small scale of their cross-bedding, may have been deposited on the bottom of a shallow lake by a prograding alluvial fan.

The Iona Group is pre-Devonian because it is cut by the Ross of Mull granite. Northern Scotland was an area of erosion during the Silurian and upper Ordovician (Soper *et al.* 1999); while Cambrian and lower Ordovician sediments are marine shales and limestones, so the continental Iona Group is very probably Precambrian. There are obvious lithological comparisons to be made with the lowest beds of the Stoer, Sleat and Torridon Groups, but this alone, of course, is no basis for correlation. In particular the correlation with the Colonsay Group tentatively advanced by Bentley (1988) should be resisted.

The Colonsay Group, of uncertain age, is built entirely of deltaic deposits and turbidites, sedimentologically like a molasse (Stewart 1969; Stewart & Hackman 1973). The group rests on a shear plane. There has been disagreement over the significance of this shear plane and its possible coincidence with an unconformity (see Muir *et al.* 1994). But the most significant fact is that high-grade gneisses are in contact with the Colonsay Group, in Islay and Colonsay, at stratigraphic levels that are 5 km apart. It is difficult to imagine this as an unconformity surface. In short, the Colonsay Group is very probably allochthonous and is at present virtually impossible to correlate with anything.

Bowmore

The Bowmore sandstone on the Isle of Islay was originally mapped for the Geological Survey by Wilkinson (1907) and correlated with the Torridon Group. It was remapped by Amos (1960) whose work is incorporated in the 1:50 000 North Islay sheet 27, published by the British Geological Survey in 1997. Unfortunately this map covers only the northern part of the outcrop. The southern part falls within sheet 19 which is unrevised and out of print.

There may be an undersea extension of the Bowmore sandstone to the north of Islay, for erratics of pebbly, jasper-bearing sandstone like the Applecross Formation are frequently found on Colonsay, derived from the sea bed to the east (Cunningham-Craig *et al.* 1911, p. 60–61).

The Bowmore sandstone was given group status by Stewart (1969). It can be divided into two units: the Laggan Formation below and the Blackrock Formation above, each about 2400 m thick. Neither the stratigraphic top or the bottom of the Bowmore Group are seen in Islay, for it is tectonically isolated between the Loch Gruinart fault to the west and the Loch Skerrols thrust to the east. However, it is known from gravity and magnetic surveys to be underlain by basement gneisses about 5 km below the present surface (Durrance 1976; Westbrook & Borradaile 1978). Most of the following account of the rocks is based on the work of Amos (1960).

The Laggan Formation is composed mainly of medium-grained sandstone that weathers pale grey or brown. Coarse-grained rocks form the lowest 200 m near Laggan Farm. Feldspars (mainly potassic) form less than 10% of the rock, and matrix about 25%. Detrital zircon and iron ores are frequently concentrated in thin bands. The type section follows the coast from Laggan Point [NR 276553] north to Rubh' an t-Sàile [NR 293591], exposing about 2400 m of strata.

The Blackrock Formation is mainly coarse to very coarse sandstone of greenish-grey colour. Feldspars (mainly potassic) form 15% of the rock and 10–20% of the matrix. Pebbles a centimetre or two in diameter are common, exclusively made of quartz, quartzite, feldspar or jasper. None of the felsite, schist or nordmarkite pebbles reported by Wilkinson (1907) and Green (1924) were found by Amos. Nor has anyone reported porphyry pebbles. The quartzites, though recrystallized, retain the ferruginous pellicles that once surrounded clastic grains. The type section of the formation extends from near Blackrock [NR 306630] west to Uisguintuie [NR 298630]. According to Amos (1960) the Blackrock Formation conformably follows the Laggan Formation west of Laggan Farm.

The Bowmore Group has been strongly deformed, probably during the Caledonian orogeny. Only the feldspar grains retain their detrital shapes – the quartz has been almost entirely recrystallized. The sandstones have been bent into large asymmetrical folds that plunge gently to the east. The folds have a penetrative axial plane cleavage that dips about 30° to the SE. Quartz grains and pebbles elongated in the cleavage plane plunge about 30° to the SE. As a result of the deformation both cross-bedding and contorted bedding are difficult to see and the palaeocurrent directions ill-defined.

Correlation of the Bowmore and Torridon Groups is plausible for they share great thickness and uniformity of grain size and must have formed in similar kinds of basin. The deformation of the Bowmore Group is like that of the Sleat and Torridon Groups in the Kishorn nappe of Skye. The Kishorn nappe is bounded above by the Moine thrust, whereas in Islay the Bowmore Group is covered by the Loch Skerrols thrust, generally thought to be a southern extension of the Moine thrust. The structural position of the rocks in the Sleat of Skye and in central Islay is therefore similar. But the Bowmore Group is very much less feldspathic than either the Sleat or Torridon Groups and lacks their characteristic igneous pebbles. Lithostratigraphic correlation, therefore, remains no more than a possibility.

References

Aldridge, R. J. 1986. Conodont palaeobiogeography and thermal maturation in the Caledonides. *Journal of the Geological Society, London*, **143**, 177–184.

Allen, P. 1949. Wealden petrology: the Top Ashdown Pebble Bed and the Top Ashdown Sandstone. *Quarterly Journal of the Geological Society of London*, **104**, 257–321.

Allen, P. 1991. Provenance research: Torridonian and Wealden. *In*: Morton, A. C., Todd, S. P. & Haughton, P. D. W. (eds) *Developments in sedimentary provenance studies*. Geological Society, London, Special Publications, **57**, 13–21.

Allen, P., Allen, J. R. L., Goldring, R. & Maycock, I. D. 1960. Festoon bedding and 'mud-with-lenticles' lithology. *Geological Magazine*, **97**, 261–263.

Allen, P., Sutton. J. & Watson, J. V. 1974. Torridonian tourmaline-quartz pebbles and the Precambrian crust northwest of Britain. *Journal of the Geological Society, London*, **130**, 85–91.

Allen, P. A. & Allen, J. R. 1990. *Basin analysis: principles and applications*. Blackwell, Oxford.

Amos, B. J. 1960. *The geology of the Bowmore district, Islay*. PhD thesis, Imperial College, London.

Anderson, F. W. & Dunham, K. C. 1966. *The geology of northern Skye*. Memoirs of the Geological Survey of the UK.

Anderton, R. 1980. Distinctive pebbles as indicators of Dalradian provenance. *Scottish Journal of Geology*, **16**, 143–152.

Anderton, R., Gibbons, W. & Nicholson, P. G. 1992. Precambrian. *In*: Cope, C. J. W., Ingham, J. K., Rawson, P. F. (eds) *Atlas of palaeogeography and lithofacies*. Memoirs of the Geological Society of London, **13**, 5–7.

Armstrong, A. 1976. A three-dimensional simulation of slope forms. *Zeitschrift für Geomorphologie, Supplementband*, **25**, 20–28.

Bailey, E. B. 1945. Tertiary igneous tectonics of Rhum (Inner Hebrides). *Quarterly Journal of the Geological Society of London*, **100**, 165–192.

Bailey, E. B. 1954. Relations of Torridonian to Durness Limestone in the Broadford–Strollamus district of Skye. *Geological Magazine*, **91**, 73–78.

Bailey, E. B. 1955. Moine tectonics and metamorphism in Skye. *Transactions of the Edinburgh Geological Society*, **16**, 93–166.

Bailey, E. B. & Anderson, E. M. 1925. *The geology of Staffa, Iona and western Mull*. Memoirs of the Geological Survey of Great Britain.

Baker, V. R. & Kochel, R. C. 1988. Flood sedimentation in bedrock fluvial systems. *In*: Baker, V. R., Kochel, R. C. & Patton, P. C. (eds) *Flood geomorphology*. Wiley, New York, 123–137.

Baltzer, F. 1991. Late Pleistocene and Recent detrital sedimentation in the deep parts of Lake Tanganyika (East African rift). *Special Publications of the International Association of Sedimentologists*, **13**, 147–173.

Barber, A. J., Beach, A., Park, R. G., Tarney, J. & Stewart, A. D. 1978. The Lewisian and Torridonian rocks of north-west Scotland. *Geologists' Association Guide* **21**.

Beach, A. 1976. The interrelations of fluid transport, deformation, geochemistry and heat flow in early Proterozoic shear zones in the Lewisian complex. *Philosophical Transactions of the Royal Society, London*, **A280**, 569–604.

Beacom, L. E., Anderson, T. B. & Holdsworth, R. E. 1999. Using basement-hosted clastic dykes as syn-rifting palaeostress indicators: an example from the basal Stoer Group, northwest Scotland. *Geological Magazine*, **136**, 301–310.

Beadle, L. C. 1981. *The inland waters of tropical Africa: an introduction to tropical limnology*. Longman, London, 2nd ed.

Beckmann, G. E. J. 1976. A palaeomagnetic study of part of the Lewisian complex, north-west Scotland. *Journal of the Geological Society, London*, **132**, 45–59.

Benner, E. & Hoernes, S. 1994. Stable isotope systematics in metasediments of the Scottish Highlands. *Mineralogical Magazine*, **58A**, 73–74.

Bentley, M. 1988. The Colonsay Group. *In*: Winchester, J. A. (ed.) *Later Proterozoic stratigraphy of the northern Atlantic regions*. Blackie, Glasgow, 119–130.

Binns, P. E., McQuillin, R., Kenolty, N. & Cooper, B. 1974. The geology of the Sea of the Hebrides. *Institute of Geological Sciences Report* **73/14**.

Black, G. P. & Welsh, W. 1961. The Torridonian succession of the Ise of Rhum. *Geological Magazine*, **98**, 265–276.

Blair, T. C. & McPherson, J. G. 1994. Alluvial fans and their natural distinction from rivers based on morphology, hydraulic processes, sedimentary processes, and facies assemblages. *Journal of Sedimentary Research*, **A64**, 450–489.

Blatt, H. & Jones, R. L. 1975. Proportions of exposed igneous, metamorphic, and sedimentary rocks. *Bulletin of the Geological Society of America*, **86**, 1085–1088.

Blissenbach, E. 1954. Geology of alluvial fans in semiarid regions. *Bulletin of the Geological Society of America*, **65**, 175–190.

Bluck, B. J. 1986. Upward coarsening sedimentation units and facies lineages, Old Red Sandstone, Scotland. *Transactions of the Royal Society of Edinburgh. Earth Sciences*, **77**, 251–264.

Bluck, B. J. 2001. Caledonian and related events in Scotland. *Transactions of the Royal Society of Edinburgh. Earth Sciences*, **91** (for 2000), 375–404.

Bluck, B. J., Dempster, T. J. & Rogers, G. 1997. Allochthonous metamorphic blocks on the Hebridean passive margin, Scotland. *Journal of the Geological Society, London*, **154**, 921–924.

Blundell, D. J., Hurich, C. A. & Smithson, S. B. 1985. A model for the MOIST seismic reflection profile, N. Scotland. *Journal of the Geological Society, London*, **142**, 245–258.

Boud, R. C. 1974. Aaron Arrowsmith's topographical map of Scotland and John MacCulloch's geological survey. *The Canadian Cartographer*, **11**, 24–34.

Bowes, D. R. 1972. Geochemistry of Precambrian crystalline basement rocks, north-west highlands of Scotland. *Twenty-fourth International Geological Congress*, Montreal, Section **1**, 97–103.

Bowes, D. R., Wright, A. E. & Park, R. G. 1964. Layered intrusive rocks in the Lewisian of the north-west highlands of Scotland. *Quarterly Journal of the Geological Society of London*, **120**, 153–192.

Bowring, S. A. & Erwin, D. H. 1998. A new look at evolutionary rates in deep time: uniting paleontology and high-precision geochronology. *GSA Today*, **8**(9), 1–8.

Brewer, J. A. & Smythe, D. K. 1984. MOIST and the continuity of crustal reflector geometry along the Caledonian–Appalachian orogen. *Journal of the Geological Society, London*, **141**, 105–120.

British Geological Survey 1896. *Applecross, Scotland sheet 81, solid & drift, 1:63 360*. British Geological Survey, Keyworth.

British Geological Survey 1913*a*. *Minginish, Scotland sheet 70, solid and drift, 1:63 360*. British Geological Survey, Keyworth.

British Geological Survey 1913*b*. *Inverbroom, Scotland sheet 92, solid & drift*, British Geological Survey, Keyworth.

British Geological Survey 1923. *Assynt District, Scotland, solid & drift, 1:63 360*. British Geological Survey, Keyworth.

British Geological Survey 1976. *Broadford, Scotland sheet 71W, solid and drift, 1:50 000*. British Geological Survey, Keyworth.

British Geological Survey 1994. *Rum, Scotland sheet 60, solid & drift, 1:50 000*. British Geological Survey, Keyworth.

British Geological Survey 1997. *North Islay, Scotland sheet 27, solid and drift, 1:50 000 provisional series*. British Geological Survey, Keyworth.

British Geological Survey 1998. *Cape Wrath, Scotland sheet 113, solid and drift geology, 1:50 000 provisional series*. British Geological Survey, Keyworth.

British Geological Survey 1998. *Summer Isles, Scotland sheet 101W, solid and drift geology, 1:50 000 provisional series*. British Geological Survey, Keyworth.

British Geological Survey 1999. *Gairloch, Scotland sheet 91 & 100, solid and drift geology, 1:50 000 provisional series*. British Geological Survey, Keyworth.

British Geological Survey 1999. *Ross of Mull, Scotland sheet 43S, solid and drift geology, 1:50 000 provisional series*. British Geological Survey, Keyworth.

British Geological Survey 2002. *Stoer, Scotland sheet 107W, solid and drift geology, 1:50 000 provisional series*. British Geological Survey, Keyworth.

Buchan, K. L., Mertanen, S., Park, R. G., Pesonen, L. J., Elming, S-Å., Abrahamsen, N. & Bylund, G. 2000. Comparing the drift of Laurentia and Baltica in the Proterozoic: the importance of key palaeomagnetic poles. *Tectonophysics*, **319**, 167–198.

Buchan, K. L., Ernst, R. E., Hamilton, M. A., Mertanen, S., Pesonen, L. J. & Elming, S-Å. 2001. Rodinia: the evidence from integrated palaeomagnetism and U–Pb geochronology. *Precambrian Research*, **110**, 9–32.

Bull, W. B. 1962. Relations of alluvial fan size and slope to drainage basin size and lithology in western Fresno County, California. *US Geological Survey Professional Paper* **450-B**, 51–53.

Bull, W. B. 1964. Alluvial fans and near-surface subsidence in western Fresno County, California. *United States Geological Survey Professional Paper* **437-A**.

Bull, W. B. 1972. Recognition of alluvial-fan deposits in the stratigraphic record. *In*: Rigby, K. J. & Hamblin, W. K. (eds) *Recognition of ancient sedimentary environments*. Society of Economic Paleontologists and Mineralogists, Special Publications, **16**, 68–83.

Bullard, E. C., Everett, J. E. & Smith, A. G. 1965. The fit of the continents around the Atlantic. *Philosophical Transactions of the Royal Society of London*, **A258**, 41–51.

Busk, H. G. 1929. *Earth flexures; their geometry and their representation and analysis in geological section with special reference to the problem of oil finding*. Cambridge University Press.

Butler, C. A., Holdsworth, R. E. & Strachan, R. A. 1995. Evidence for Caledonian sinistral strike-slip motion and associated fault zone weakening, Outer Hebrides Fault Zone, NW Scotland. *Journal of the Geological Society, London*, **152**, 743–746.

Butler, R. W. H. 1997. Late Proterozoic rift faults and basement-cover relationships within the Ben More thrust sheet, NW Scotland. *Journal of the Geological Society, London*, **154**, 761–764.

Butler, R. W. H. 2000. A short walk with Archibald Geikie. *Geoscientist*, **10**, 4–6.

Byers, P. N. 1972. *Correlation and provenance of the Precambrian Moine and Torridonian rocks of Morar, Raasay, Rhum, and Skye, northwest Scotland*. PhD thesis, University of Reading.

Campbell, S. E. 1979. Soil stabilization by a prokaryotic desert crust: implications for Precambrian land biota. *Origins of Life*, **9**, 335–348.

Canning, J. C., Henney, P. J., Morrison, M. A., Van Calsteren, P. W. C., Gaskarth, J. W. & Swarbrick, A. 1998. The Great Glen fault: a major vertical lithospheric boundary. *Journal of the Geological Society, London*, **155**, 425–428.

Cannon, W. F. 1994. Closing of the Midcontinent Rift; a far-field effect of Grenvillian compression. *Geology*, **22**, 155–158.

Cartwright, I., Fitches, W. R., O'Hara, M. J., Barnicoat, A. C. & O'Hara, S. 1985. Archaean supracrustal rocks from the Lewisian near Stoer, Sutherland. *Scottish Journal of Geology*, **21**, 187–196.

Cheadle, M. J., McGeary, S., Warner, M. R. & Matthews, D. H. 1987. Extensional structures on the western UK continental shelf: a review of evidence from deep seismic profiling. *In*: Coward, M. P., Dewey, J. F. & Hancock, P. L. (eds) *Continental extension tectonics*. Geological Society, London, Special Publications, **28**, 445–465.

Cliff, R. A. & Rex, D. C. 1989. Evidence for a 'Grenville' event in the Lewisian of the northern Outer Hebrides. *Journal of the Geological Society, London*, **146**, 921–924.

Cloud, P. 1968. Atmospheric and hydrospheric evolution on the primitive earth. *Science*, **160**, 729–736.

Cloud, P. & Germs, A. 1971. New pre-Paleozoic nannofossils from the Stoer Formation (Torridonian), northwest Scotland. *Bulletin of the Geological Society of America*, **82**, 3469–3474.

Clough, C. T. & Harker, A. 1904. *The geology of west-central Skye, with Soay*. Memoirs of the Geological Survey of Great Britain.

Cochran, J. R. 1983. Effects of finite rifting times on the development of sedimentary basins. *Earth and Planetary Science Letters*, **66**, 289–302.

Cocks, L. R. M. 2001. Ordovician and Silurian global geography. *Journal of the Geological Society, London*, **158**, 197–210.

Costa, J. E. 1988. Rheologic, geomorphic, and sedimentologic differentiation of water floods, hyperconcentrated flows, and debris flows. *In*: Baker, V. R., Kochel, R. C. & Patton, P. C. (eds) *Flood geomorphology*. Wiley, New York, 113–122.

Cotter, E. 1978. The evaluation of fluvial style, with special reference to the central Appalachian Palaeozoic. *In*: Miall, A. D. (ed.) *Fluvial sedimentology*. Canadian Society of Petroleum Geologists Memoir, **5**, 361–383.

Coward, M. P. & Whalley, J. S. 1979. Texture and fabric studies across the Kishorn nappe, near Kyle of Lochalsh, western Scotland. *Journal of Structural Geology*, **1**, 259–273.

Cowie, J. & McNamara, K. J. 1978. Olenellus (Trilobita) from the Lower Cambrian strata of north-west Scotland. *Palaeontology*, **21**, 615–634.

Crowell, J. C. 1999. Pre-Mesozoic Ice Ages: their bearing on undertanding the climatic system. *Memoir of the Geological Society of America*, **192**.

Cunningham, R. J. H. 1841. Geognostical account of the County of Sutherland. *Transactions of the Highland and Agricultural Society of Scotland*, Series ii, **13**, 73–114.

Cunningham-Craig, E. H., Wright, W. B. & Bailey, E. B. 1911. *The geology of Colonsay and Oronsay with part of the Ross of Mull*. Memoirs of the Geological Survey of Great Britain.

Dallmeyer, R. D., Strachan, R. A., Rogers, G., Watt, G. R. & Friend, C. R. L. 2001. Dating deformation and cooling in the Caledonian thrust nappes of north Sutherland, Scotland: insights from $^{40}Ar/^{39}Ar$ and Rb–Sr chronology. *Journal of the Geological Society, London*, **158**, 501–512.

Dalziel, I. W. D. 1997. Neoproterozoic–Paleozoic geography and tectonics: review, hypothesis, environmental speculation. *Bulletin of the Geological Society of America*, **109**, 16–42.

Dalziel, I. W. D. & Soper, N. J. 2001. Neoproterozoic extension on the Scottish promontory of Laurentia: paleogeographic and tectonic implications. *Journal of Geology*, **109**, 299–317.

Dalziel, I. W. D. , Mosher, S. & Gahagan, L. M. 2000. Laurentia–Kalahari collision and the assembly of Rodinia. *Journal of Geology*, **108**, 499–513.

Davidson, A. 1995. A review of the Grenville orogen in its North American type area. *AGSO Journal of Australian Geology and Geophysics*, **16**, 3–24.

Davison, S. & Hambrey, M. J. 1996. Indications of glaciation at the base of the Proterozoic Stoer Group (Torridonian), NW Scotland. *Journal of the Geological Society, London*, **153**, 139–149.

Davison, S. & Hambrey, M. J. 1997. Discussion on indications of glaciation at the base of the Proterozoic Stoer Group (Torridonian), NW Scotland. *Journal of the Geological Society, London*, **154**, 1087–1088.

Dawes, P. R. 1997. The Proterozoic Thule Supergroup, Greenland and Canada: history, lithostratigraphy and development. *Geology of Greenland Survey Bulletin* 174.

Demicco, R. V. & Kordesch, E. G. 1986. Facies sequences of a semi-arid closed basin: the Lower Jurassic East Berlin Formation of the Hartford Basin, New England, USA. *Sedimentology*, **33**, 107–118.

Dickin, A. P. & Jones, N. W. 1983. Isotopic evidence for the age and origin of pitchstones and felsites, Isle of Eigg, NW Scotland. *Journal of the Geological Society, London*, **140**, 691–700.

Diver, W. L. 1980. Some factors controlling cryptarch distribution in the late Precambrian Torridon Group. *Abstracts of the 5th International Palynological Conference, Cambridge*, 113.

Diver, W. L. & Peat, C. J. 1979. On the interpretation and classification of Precambrian organic-walled microfossils. *Geology*, **7**, 401–404.

Donnellan, N. C. B. 1981. *Geochemical studies of the Torridonian sediments of NW Scotland*. PhD thesis, University of Birmingham.

Downie, C. 1962. So-called spores from the Torridonian. *Proceedings of the Geological Society of London*, **1600**, 127–128.

Drever, J. I. 1997. *The geochemistry of natural waters: surface and groundwater environments*. Prentice Hall, Upper Saddle River, NJ, 3rd ed.

Dudal, R. & Eswaran, H. 1988. Distribution, properties, and classification of vertisols, *In*: Wilding, L. P. & Puentes, R. (eds) *Vertisols, their distribution, properties, classification, and management*. Texas A & M University Soil Management Support Services Technical Monograph **18**, 1–22.

Dunham, A. C. 1968. The felsites, granophyre, explosion breccias and tuffisites of the north-eastern margin of the Tertiary igneous complex of Rhum, Inverness-shire. *Quarterly Journal of the Geological Society of London*, **123**, 327–350.

Dunham, A. C. & Emeleus, C. H. 1967. The Tertiary geology of Rhum, Inner Hebrides. *Proceedings of the Geologists' Association*, **78**, 391–418.

Durrance, E. M. 1976. A gravity survey of Islay, Scotland. *Geological Magazine*, **113**, 251–261.

Dzulynski, S. & Walton, E. K. 1965. *Sedimentary features of flysch and greywackes*. Elsevier, Amsterdam.

Elliott, D. & Johnson, M. R. W. 1980. Structural evolution in the northern part of the Moine thrust belt, NW Scotland. *Transactions of the Royal Society of Edinburgh. Earth Sciences*, **71**, 69–96.

Elmore, R. D. 1983. Precambrian non-marine stromatolites in alluvial fan deposits, the Copper Harbor Conglomerate, upper Michigan. *Sedimentology*, **30**, 829–842.

Emeleus, C. H. 1994. *1:20 000 solid geology map of Rum*. 2nd ed. Scottish Natural Heritage, Edinburgh.

Emeleus, C. H. 1997. *Geology of Rum and the adjacent islands: memoir for 1:50 000 geological sheet 60 (Scotland)*. Memoirs of the British Geological Survey.

Evans, C. R. & Lambert, R. St J. 1974. The Lewisian of Lochinver, Sutherland; the type area for the Inverian metamorphism. *Journal of the Geological Society, London*, **130**, 125–150.

Evans, D., Chesher, J. A., Deegan, C. E. & Fannin, N. G. T. 1982. The offshore geology of Scotland in relation to the IGS shallow drilling program 1970–1978. *Institute of Geological Sciences Report* **81/12**, 1–36.

Eyles, V. A. 1937. John MacCulloch, F.R.S., and his geological map: an account of the first geological survey of Scotland. *Annals of Science*, **2**, 114–129.

Eyles, V. A. 1939. MacCulloch's geological map of Scotland: an additional note. *Annals of Science*, **4**, 107.

Fairhead, J. D. 1986. Geophysical controls on sedimentation within the African Rift Systems. *In*: Frostick, L. E., Renaut, R. W., Reid, I. & Tiercelin, J. J. (eds) *Sedimentation in the African rifts*. Blackwell, Oxford, 19–27.

Fedo, C. M., Nesbitt, H. W. & Young, G. M. 1995. Unraveling the effects of potassium metasomatism in sedimentary rocks and paleosols, with implications for paleoweathering conditions and provenance. *Geology*, **23**, 921–924.

Fedo, C. M., Young, G. M., Nesbitt, H. W. & Hanchar, J. M. 1997. Potassic and sodic metasomatism in the southern province of the Canadian Shield: evidence from the Paleoproterozoic Serpent Formation, Huronian Supergroup, Canada. *Precambrian Research*, **84**, 17–36.

Fermor, L. L. 1951. On a discovery of copper-ore in the Torridonian rocks of Sutherland. *Geological Magazine*, **88**, 215–218.

Fettes, D. J., Mendum, J. R., Smith, D. I. & Watson, J. V. 1992. *Geology of the Outer Hebrides: memoir for 1:100 000 (solid edition) geological sheets, Lewis and Harris, Uist and Barra (Scotland)*. Memoirs of the British Geological Survey.

Fisher, R. V. & Schmincke, H-U. 1984. *Pyroclastic rocks*. Springer-Verlag, Berlin.

Flint, R. F., Sanders, J. E. & Rodgers, J. 1960. Diamictite, a substitute term for symmictite. *Bulletin of the Geological Society of America*, **71**, 1809–1810.

Foley, S. F., Venturelli, G., Green, D. H. & Toscani, L. 1987. The ultrapotassic rocks: characteristics, classification, and constraints for petrogenic models. *Earth-Science Reviews*, **24**, 81–134.

Fraser, F. M. 1977. *The Lewisian and Torridonian geology of Iona*. PhD thesis, University of St Andrews.

Garcia-Pichel, F., López-Córtes, A. & Nübel, U. 2001. Phylogenetic and morphological diversity of cyanobacteria in soil desert crusts from the Colorado Plateau. *Applied and Environmental Microbiology*, **67**, 1902–1910.

Garrels, R. M. 1988. Sediment cycling during Earth history. *In*: Lerman, A. & Meybeck, M. (eds) *Physical and chemical weathering in geochemical cycles*. Kluwer, Dortrecht, 341–355.

Gay, A. L. & Grandstaff, D. E. 1979. Chemistry and mineralogy of Precambrian paleosols at Elliot Lake, Ontario, Canada. *Precambrian Research*, **12**, 349–373.

Geikie, A. 1880. A fragment of primeval Europe. *Nature*, **22**, 400–403.

Geikie, A. 1892. *Report of the Geological Survey and Museum for 1891*.

Geikie, A. 1894. *Report of the Geological Survey and Museum for 1893*.

Geikie, A. 1895. *Report of the Geological Survey and Museum for 1894*.

Geikie, A. 1900. *Summary of progress of the Geological Survey of Great Britain for 1899*, London.

Geikie, A. 1903. *Summary of progress of the Geological Survey of Great Britain for 1902*, London.

Gilbert, J. S. & Lane, S. J. 1994. The origin of accretionary lapilli. *Bulletin of Volcanology*, **56**, 398–411.

Gilligan, A. 1919. The petrography of the Millstone Grit Series of Yorkshire. *Quarterly Journal of the Geological Society of London*, **75**, 251–294.

Glendinning, N. R. W. 1988. Sedimentary structures and sequences within a late Proterozoic tidal shelf deposit: the Upper Morar Psammite Formation of northwestern Scotland, *In*: Winchester, J. A. (ed.) *Later Proterozoic stratigraphy of the northern Atlantic regions*. Blackie, Glasgow, 14–31.

Glennie, K. W. 1970. *Desert sedimentary environments*. Elsevier, Amsterdam.

Gloppen, T. G. & Steel, R. J. 1981. The deposits, internal structure and geometry in six alluvial fan–fan delta bodies (Devonian Norway) – a study in the significance of braiding sequence in conglomerates. *In*: Ethridge, F. G. & Flores, R. M. (eds) *Recent and ancient nonmarine depositional systems: models for exploration*. Special Publication. Society of Economic Paleontologists and Mineralogists, **31**, 49–69.

Godard, A. 1957. La surface prétorridonienne en Écosse. *Revue de Gèographie alpine*, **45**, 135–153.

Godard, A. 1965. *Recherches de géomorphologie en Écosse du nord-ouest*. Strasbourg.

Goodchild, J. G. 1897. Desert conditions in Britain. *Transactions of the Edinburgh Geological Society*, **7**, 203–222.

Goodchild, J. G. 1898. Desert conditions in Britain. *Transactions of the Geological Society of Glasgow*, **11**, 71–104.

Gordon, M. B. & Hempton, M. R. 1986. Collision-induced rifting: the Grenville orogeny and the Keweenawan rift of North America. *Tectonophysics*, **127**, 1–25.

Gower, C. F. 1988. The Double Mer Formation. *In*: Winchester, J. A. (ed.) *Later Proterozoic stratigraphy of the northern Atlantic regions*. Blackie, Glasgow, 113–118.

Gower, C. F., Ryan, A. B. & Rivers, T. 1990. Mid-Proterozoic Laurentia–Baltica: an overview of its geological evolution and a summary of the contributions made by this volume. *In*: Gower, C. F., Rivers, T & Ryan, B. (eds) *Mid-Proterozoic Laurentia–Baltica*. Geological Association of Canada, Special Papers, **38**, 1–20.

Gracie, A. J. 1964. *The Torridonian rocks in the area of Cnoc Mòr an Rudha Bhig in the Coigach Peninsula, Ross-shire*. PhD thesis, University of Reading.

Gracie, A. J. & Stewart, A. D. 1967. Torridonian sediments at Enard Bay, Ross-shire. *Scottish Journal of Geology*, **3**, 181–194.

GREEN, J. F. N. 1924. The structure of the Bowmore–Port Askaig district of Islay. *Quarterly Journal of the Geological Society of London*, **80**, 72–100.

GREGOR, C. B. 1985. The mass-age distribution of Phanerozoic sediments. *In*: SNELLING, N. J. (ed.) *The chronology of the geological record*. Geological Society, London, Memoirs, **10**, 284–289.

GREGORY, J. W. 1915. Moine pebbles in Torridonian conglomerates. *Geological Magazine*, **52**, 447–450.

GREGORY, J. W. 1917. The geology of phosphates and their bearing on the conservation of mineral resources. *Transactions of the Geological Society of Glasgow*, **16**, 115–163.

GUSTAVSON, T. C. 1991. Buried vertisols in lacustrine facies of the Pliocene Fort Hancock Formation, Hueco Bolson, west Texas and Chihuahua, Mexico. *Bulletin of the Geological Society of America*, **103**, 448–460.

HÅKANSON, L. & JANSSON, M. 1983. *Principles of lake sedimentology*. Springer Verlag, Berlin.

HALLIDAY, A. N., AFTALION, M., VAN BREEMEN, O. & JOCELYN, J. 1979. Petrogenetic significance of Rb–Sr and U–Pb isotopic systems in the 400 Ma old British Isles granitoids and their hosts. *In*: HARRIS, A. L., HOLLAND, C. H. & LEAKE, B. E. (eds) *The Caledonides of the British Isles – reviewed*. Scottish Academic Press, Edinburgh, 653–661.

HAMBREY, M. J., FAIRCHILD, I. J., GLOVER, B. W., STEWART, A. D., TREAGUS, J. E. & WINCHESTER, J. A. 1991. The late Precambrian geology of the Scottish Highlands and Islands. *Geologist's Association Guide*, **44**.

HAMILTON, P. J., EVENSEN, N. M., O'NIONS, R. K. & TARNEY, J. 1979. Sm–Nd systematics of Lewisian gneisses: implications for the origin of granulites. *Nature*, **277**, 25–28.

HARDIE, L. A., SMOOT, J. P. & EUGSTER, H. P. 1978. Saline lakes and their deposits: a sedimentological approach. *Special Publications of the International Association of Sedimentologists*, **2**, 7–41.

HARKER, A. 1908. *The geology of the small isles of Inverness-shire*. Memoirs of the Geological Survey of Great Britain.

HARLAND, W. B. 1997. *The geology of Svalbard*. Memoirs of the Geological Society of London, **17**.

HARVEY, A. M. 1989. The occurence and role of arid zone alluvial fans. *In*: THOMAS, D. S. G. (ed.) *Arid zone geomorphology*. Belhaven Press, London, 136–158.

HAY, S. J., HALL, J., SIMMONS, G. & RUSSELL, M. J. 1988. Sealed microcracks in the Lewisian of NW Scotland: a record of 2 billion years of fluid circulation. *Journal of the Geological Society, London*, **145**, 819–830.

HEDDLE, M. F. 1901. *The mineralogy of Scotland*. David Douglas, Edinburgh.

HEIM, A. 1919. *Geologie der Schweiz*. Tauchnitz, Leipzig.

HICKS. H. 1878. On the metamorphic and overlying rocks in the neighbourhood of Loch Maree, Ross-shire. *Quarterly Journal of the Geological Society of London*, **34**, 811–818.

HIGGINS, A. K. 1988. The Krummedal supracrustal sequence in East Greenland *In*: WINCHESTER, J. A. (ed.) *Later Proterozoic stratigraphy of the northern Atlantic regions*. Blackie, Glasgow, 86–96.

HIGGINS, A. K., SMITH, M. P., SOPER, N. J., LESLIE, A. G., RASMUSSEN, J. A. & SØNDERHOLM, M. 2001. The Neoproterozoic Hekla Sund Basin, eastern North Greenland: a pre-Iapetan extensional sequence thrust across its rift shoulders during the Caledonian orogeny. *Journal of the Geological Society, London*, **158**, 487–499.

HINZE, W. J., ALLEN, D. J., BRAILE, L. W. & MARIANO, J. 1997. Middle Proterozoic to Cambrian rifting, central North America. *In*: OJAKANGAS, R. W., DICKAS, A. B. & GREEN, J. C. (eds) *The midcontinent rift system: a major Proterozoic continental rift*. Special Paper of the Geological Society of America, **312**, 7–35.

HITCHEN, K., STOKER, M. S., EVANS, D. & BEDDOE-STEPHENS, B. 1995. *In*: BOLDY, S. A. R. (ed.) *Permian and Triassic rifting in northwest Europe*. Geological Society, London, Special Publications, **91**, 87–102.

HOFFMAN, P. F. 1989. Precambrian geology and tectonic history of North America. *In*: BALLY, A. W. & PALMER, A. R. (eds) *The geology of North America*, **A**. Geological Society of America, Boulder, 447–512.

HOLDSWORTH, R. E., HARRIS, A. L. & ROBERTS, A. M. 1987. The stratigraphy, structure and regional significance of the Moine rocks of Mull, Argyllshire, Scotland. *Geological Journal*, **22**, 83–107.

HOLLAND, H. D. 1984. *The chemical evolution of the atmosphere and oceans*. Princeton University Press, Princeton.

HOLLAND, H. D. 1999. When did the Earth's atmosphere become oxic? A reply. *The Geochemical News*, **100**, 20–22.

HOLLAND, J. G. & LAMBERT, R. ST J. 1973. Comparative major element geochemistry of the Lewisian of the mainland of Scotland. *In*: PARK, R. G. & TARNEY, J. (eds) *The early Precambrian of Scotland and related rocks of Greenland*. Department of Geology, Keele, 51–62.

HOLMES, A. 1960. A revised geological timescale. *Transactions of the Edinburgh Geological Society*, **17**, 184–216.

HOLMES, A., SHILLIBEER, H. A. & WILSON, J. T. 1955. Potassium-argon ages of some Lewisian and Fennoscandian pegmatites. *Nature*, **176**, 390–392.

HORODYSKI, R. J. 1980. Middle Proterozoic shale-facies microbiota from the lower Belt Supergroup, Little Belt Mountains, Montana. *Journal of Paleontology*, **54**, 649–663.

HUBERT, J. F., REED, A. A. & CAREY, P. J. 1976. Paleogeography of the East Berlin Formation, Newark Group, Connecticut valley. *American Journal of Science*, **276**, 1183–1207.

HUBERT, J. F., FESHBACH-MERINEY, P. E. & SMITH, M. A. 1992. The Triassic–Jurassic Hartford rift basin, Connecticut and Massachusetts: evolution, sandstone diagenesis, and hydrocarbon history. *Bulletin of the American Association of Petroleum Geologists*, **76**, 1710–1734.

HUGHES, C. J. 1960. The southern mountains igneous complex, Isle of Rhum. *Quarterly Journal of the Geological Society of London*, **114**, 111–138.

HUNTER, R. E. 1977. Basic types of stratification in small eolian dunes. *Sedimentology*, **24**, 361–387.

HUSELBEE, M. Y. & THOMAS, A. T. 1998. *Olenellus* and conodonts from the Durness Group, NW Scotland, and the correlation of the Durness succession. *Scottish Journal of Geology*, **34**, 83–85.

HUTCHISON, A. R. & OLIVER, G. J. H. 1998. Garnet provenance studies, juxtaposition of Laurentian marginal terranes and timing of the Grampian orogeny in Scotland. *Journal of the Geological Society, London*, **155**, 541–550.

IRVING, E. 1954. *The palaeomagnetism of the Torridonian Sandstone Series*. PhD thesis, University of Cambridge.

IRVING, E. 1957. The origin of the palaeomagnetism of the Torridonian sandstones of north-west Scotland. *Philosophical Transactions of the Royal Society of London*, **A250**, 100–110.

IRVING, E. 1964. *Paleomagnetism and its application to geological and geophysical problems*. Wiley, New York.

IRVING, E. & RUNCORN, S. K. 1957. Analysis of the palaeomagnetism of the Torridonian sandstone series of north-west Scotland. *Philosophical Transactions of the Royal Society of London*, **A250**, 83–99.

JEFFREYS, H. 1970. *The Earth: its origin history and physical constitution*. Cambridge University Press, 5th ed.

JEHU, T. J. 1922. The Archaean and Torridonian formations and the later intrusive igneous rocks of Iona. *Transactions of the Royal Society of Edinburgh*, **53**, 165–187

JOHNSON, M. R. W., KELLEY, S. P., OLIVER, G. J. H. & WINTER, D. A. 1985. Thermal effects and timing of thrusting in the Moine Thrust zone. *Journal of the Geological Society, London*, **142**, 863–873.

JOLLY, W. 1880. Curious impressions in Cambrian sandstone near Loch Maree. *Nature*, **23**, 94–94.

JONES, B. F. & GALEN, E. 1988. Sepiolite and palygorskite. *In*: BAILEY S. W. (ed.), Hydrous phyllosicates. *Reviews in Mineralogy*, **19**, 631–674.

JONES, E. M., RICE, C. M. & TWEEDIE, J. R. 1987. Lower Proterozoic stratiform sulphide deposits in Loch Maree Group, Gairloch, northwest Scotland. *Transactions of the Institution of Mining and Metallurgy, Section B Applied Earth Science*, **96**, B128–140.

JUDD, J. W. 1878. The secondary rocks of Scotland – third paper. *Quarterly Journal of the Geological Society of London*, **34**, 660–743.

KALLIOKOSKI, J. 1986. Calcium carbonate cement (caliche) in Keweenawan sedimentary rocks ($\sim$1.1 Ga), upper peninsula of Michigan. *Precambrian Research*, **32**, 243–259.

KALSBEEK, F., THRANE, K., NUTMAN, A. P. & JEPSEN, H. F. 2000. Late Mesoproterozoic to early Neoproterozoic history of the East Greenland Caledonides: evidence for Grenville orogenesis? *Journal of the Geological Society, London*, **157**, 1215–1225.

KENNEDY, W. Q. 1951. Sedimentary differentiation as a factor in the Moine-Torridonian correlation. *Geological Magazine*, **88**, 257–266.

KINNY, P. D., FRIEND, C. R. L., STRACHAN, R. A., WATT, G. R. & BURNS, I. M. 1999. U–Pb geochronology of regional migmatites in East Sutherland, Scotland: evidence for crustal melting during the Caledonian orogeny. *Journal of the Geological Society, London*, **156**, 1143–1152.

LAILEY, M., STEIN, A. M. & RESTON, T. J. 1989. The Outer Hebrides fault: a major Proterozoic structure in NW Britain. *Journal of the Geological Society, London*, **146**, 253–259.

LAMBIASE, J. J. 1991. A model for tectonic control of lacustrine stratigraphic sequences in continental rift basins. *In*: KATZ, B. (ed.) *Lacustrine exploration: case studies and modern analogues*. American Association of Petroleum Geologists, Memoirs, **50**, 265–276.

LAWSON, D. E. 1965. Lithofacies and correlation within the lower Torridonian. *Nature*, **207**, 706–708.

LAWSON, D. E. 1970. *The Torridonian rocks of the Gairloch area and parts of Wester Ross and Sutherland, northwest Scotland*. PhD thesis, University of Reading.

LAWSON, D. E. 1972. Torridonian volcanic sediments. *Scottish Journal of Geology*, **8**, 345–362.

LAWSON, D. E. 1976. Sandstone boulder conglomerates and a Torridonian cliffed shoreline between Gairloch and Stoer, northwest Scotland. *Scottish Journal of Geology*, **12**, 67–88.

LEE, G. W. 1920. *The Mesozoic rocks of Applecross, Raasay and north-east Skye*. Memoirs of the Geological Survey of Great Britain.

LEEDER, M. 1999. *Sedimentology and sedimentary basins: from turbulence to tectonics*. Blackwell, Oxford.

LEOPOLD, L. B., WOLMAN, M. G. & MILLER, J. P. 1964. *Fluvial processes in geomorphology*. Freeman, San Francisco.

LEVORSEN, A. I. 1960. *Paleogeologic maps*. Freeman, San Francisco.

LEWIS, C. L. E., CARTER, A. & HURFORD, A. J. 1992. Low-temperature effects of the Skye Tertiary intrusions on Mesozoic sediments in the Sea of the Hebrides basin. *In*: PARNELL, J. (ed.) *Basins on the Atlantic seaboard: petroleum geology, sedimentology and basin evolution*. Geological Society, London, Special Publications, **62**, 175–188.

MACCULLOCH, J. 1819. *A description of the western islands of Scotland, including the Isle of Man, comprising an account of their geological structure, with remarks on their agriculture, scenery, and antiques*. Constable, London, 3 vols.

MACCULLOCH, J. 1836. *Memoirs to His Majesty's Treasury respecting a geological survey of Scotland*. Samuel Arrowsmith, London.

MACGREGOR, M. & PHEMISTER, J. 1972. *Geological excursion guide to the Assynt district of Sutherland*. Edinburgh Geological Society.

MCKEE, E. D. 1966. Structures of dunes at White Sands national monument, New Mexico (and comparison with structures of dunes from other selected areas). *Sedimentology*, **7**, 1–69.

MACK, G. H. & JAMES, W. C. 1994. Paleoclimate and the global distribution of paleosols. *Journal of Geology*, **102**, 360–366.

MACKIE, W. 1899. The felspars present in sedimentary rocks as indicators of the conditions of contemporaneous climate. *Transactions of the Edinburgh Geological Society*, **7**, 443–468.

MACKIE, W. 1901. Seventy chemical analyses of rocks (chiefly from the Moray area) with deductions. *Transactions of the Edinburgh Geological Society*, **8**, 33–60.

MACKIE, W. 1923. The source of purple zircons in the sedimentary rocks of Scotland. *Transactions of the Edinburgh Geological Society*, **11**, 200–213.

MACKIE, W. 1926. The apatites in sedimentary rocks as indicators of the amount of atmospheric carbonic acid in the periods of deposit. *Transactions of the Geological Society of Glasgow*, **18**, 407–421.

MACKIE, W. 1928. The heavy minerals in the Torridon sandstones and metamorphic rocks of Scotland, and their bearing on the relative ages of those rocks. *Transactions of the Edinburgh Geological Society*, **12**, 181–182.

MCMANUS, J & BAJABAA, S. 1998. The importance of air escape processes in the formation of dish-and-pillar and tepee structures within modern and Precambrian fluvial deposits. *Sedimentary Geology*, **120**, 337–343.

MCLENNAN, S. M., SIMONETTI, A. & GOLDSTEIN, S. L. 2000. Nd and Pb isotopic evidence for provenance and post-depositional alteration of the Paleoproterozoic Huronian Supergroup, Canada. *Precambrian Research*, **102**, 263–278.

MANGE, M., TURNER, P., INCE, D., PUGH, J. & WRIGHT, D. 1999. A new perspective on the zonation and correlation of barren strata: an integrated heavy mineral and palaeomagnetic study of the Sherwood Sandstone Group, east Irish Sea basin and surrounding areas. *Journal of Petroleum Geology*, **22**, 325–348.

MARTIGNOLE, J. MACHADO, N. & INDARES, A. 1994. The Wakeham terrane: a Mesoproterozoic terrestrial rift in the eastern part of the Grenville Province. *Precambrian Research*, **68**, 291–306.

MAYCOCK, I. D. 1962. *The Torridonian sandstone round Loch Torridon, Wester Ross*. PhD thesis, University of Reading,

MENDUM, J., PARK, G. *ET AL*. 2003. *Lewisian, Torridonian and Moine rocks of Scotland.* Joint Nature Conservancy Committee, Peterborough (in press).

MEYBECK, M. 1976. Total mineral dissolved transport by world major rivers. *Hydrological Sciences Bulletin*, **21**, 265–284.

MEYBECK, M. 1977. Dissolved and suspended matter carried by rivers: composition, time and space variations, and world balance. *In*: GOLTERMAN, H. L. (ed.) *Interaction between sediments and fresh wate*r. Junk, The Hague, 25–32.

MEYBECK, M. 1988. How to establish and use world budgets of riverine materials. *In*: LERMAN, A. & MEYBECK, M. (eds) *Physical and chemical weathering in geochemical cycles*. Kluwer, Dortrecht, 247–272.

MIALL, A. D. 1996. *The geology of fluvial deposits*. Springer, Berlin.

MILLER, H. 1841. *The Old Red Sandstone; or, new walks in an old field*. J. Johnstone, Edinburgh.

MOORBATH, S. & PARK, R. G. 1972. The Lewisian chronology of the southern region of the Scottish mainland. *Scottish Journal of Geology*, **8** (for 1971), 51–74.

MOORBATH, S., STEWART, A. D., LAWSON, D. E. & WILLIAMS, G. E. 1967. Geochronological studies on the Torridonian sediments of north-west Scotland. *Scottish Journal of Geology*, **3**, 389–412.

MOORBATH, S., WELKE, H. & GALE, N. H. 1969. The significance of lead isotope studies in ancient high-grade metamorphic basement complexes, as exemplified by the Lewisian rocks of northwest Scotland. *Earth & Planetary Science Letters*, **6**, 245–256.

MOORE, J. G & PECK, D. L. 1962. Accretionary lapilli in volcanic rocks of the western continental United States. *Journal of Geology*, **70**, 182–193.

MORTON, A. C. & TAYLOR, P. N. 1991. Geochemical and isotopic constraints on the nature and age of basement rocks from Rockall Bank, NE Atlantic. *Journal of the Geological Society, London*, **148**, 631–634.

MOSELEY, F. 1971. A reconnaissance of the Wadi Beihan, South Yemen, with notes on basement control of gully alignment in superficial deposits. *Proceedings of the Geologist's Association, London*, **82**, 61–69.

MUIR, M. D. & SUTTON, J. 1970. Some fossiliferous Pre-Cambrian chert pebbles within the Torridonian of Britain. *Nature*, **226**, 443–445.

MUIR, R. J., FITCHES, W. R., MALTMAN, A. J. & BENTLEY, M. R. 1994. Precambrian rocks of the southern Inner Hebrides–Malin Sea region: Colonsay, west Islay, Inishtrahull and Iona. *In*: GIBBONS, W. & HARRIS, A. L. (eds) *A revised correlation of Precambrian rocks in the British Isles*. Geological Society, London, Special Reports, **22**, 54–58.

MURCHISON, R. I. 1858. The quartz rocks, crystalline limestone, and micaceous schists of the north-western highlands of Scotland proved to be of lower Silurian age, through the recent fossil discoveries of Mr C. W. Peach, with a note on the fossils by J. W. Salter. *Report of the British Association for 1857*, 82.

MURTHY, G., GOWER, C., TUBRETT, M. & PÄTZOLD, R. 1992. Paleomagnetism of Eocambrian Long Range dykes and Double Mer Formation from Labrador, Canada. *Canadian Journal of Earth Sciences*, **29**, 1224–1234.

NAUMOVA, S. N. & PAVLOVSKY, E. V. 1961. The discovery of plant remains (spores) in the Torridonian shales of Scotland. *Doklady Akademii Nauk SSSR*, **141**, 181–182.

NESBITT, H. W. 1992. Diagenesis and metamorphism of weathered profiles, with emphasis on Precambrian paleosols. *In*: MARTINI, I. P. & CHESWORTH, W. (eds) *Weathering, soils and paleosols*, Elsevier, Amsterdam, 127–152.

NESBITT, H. W. & YOUNG, G. M. 1982. Early Proterozoic climates and plate motions inferred from major element chemistry of lutites. *Nature*, **299**, 715–718.

NESBITT, H. W. & YOUNG, G. M. 1984. Prediction of some weathering trends of plutonic and volcanic rocks based on thermodynamic and kinetic considerations. *Geochimica et Cosmochimica Acta*, **48**, 1523–1534.

NESBITT, H. W., YOUNG, G. M., MCLENNAN, S. M. & KEAYS, R. R. 1996. Effects of chemical weathering and sorting on the petrogenesis of siliclastic sediments, with implications for provenance studies. *Journal of Geology*, **104**, 525–542.

NICHOLSON, P. G. 1992. *Sedimentation of the fluvial 'Torridonian' Applecross Formation, NW Scotland.* PhD thesis, University of Glasgow.

NICHOLSON, P. G. 1993. A basin reappraisal of the Proterozoic Torridon Group, northwest Scotland. *In*: FROSTICK, L. E. & STEEL, R. J. (eds) *Tectonic controls and signatures in sedimentary successions*. International Association of Sedimentologists, Special Publications, **20**, 183–202.

NICOL, J. 1857*a*. On the red sandstone and conglomerate, and the superposed quartz-rocks, limestones, and gneisses of the north-west coast of Scotland. *Quarterly Journal of the Geological Society of London*, **13**, 17–39.

NICOL, J. 1857*b*. On the newer red sandstone, and on some other geological phenomena, near Greinord, in Ross-shire. *Quarterly Journal of the Geological Society of London*, **14**, 167–171.

NICOL, J. 1866. *The geology and scenery of the north of Scotland: being two lectures given at the Philosophical Institution, Edinburgh, with notes and an appendix*. Oliver & Boyd, Edinburgh.

NIGGLI, P. 1954. *Rocks and mineral deposits*. W. H. Freeman, San Francisco.

OJAKANGAS, R. W., MOREY, G. B. & GREEN, J. C. 2001. The Mesoproterozoic Midcontinent rift system, Lake Superior region, USA. *Sedimentary Geology*, **141–142**, 421–442.

OLSEN, P. E. & GORE, P. J. W. 1989. Tectonic, depositional, and paleoecological history of early Mesozoic rift basins, Eastern North America. *28th International Geological Congress, Washington, DC, Guidebook* **T-351**.

O'NEILL, P. S. & ENGLAND, R. W. 1994. The structure of the Sea of the Hebrides basin: an integrated gravity and seismic model. *Scottish Journal of Geology*, **30**, 1–9.

O'NIONS, R. K., HAMILTON, P. J. & HOOKER, P. J. 1983. A Nd isotope investigation of sediments related to crustal development in the British Isles. *Earth & Planetary Science Letters*, **63**, 229–240.

OWEN, G. 1995. Soft-sediment deformation in upper Proterozoic Torridonian sandstones (Applecross Formation) at Torridon, northwest Scotland. *Journal of Sedimentary Research*, **A65**, 495–504.

OWEN, G. 1996*a*. Anatomy of a water-escape cusp in Upper Proterozoic Torridon Group sandstones, Scotland. *Sedimentary Geology*, **103**, 117–128.

OWEN, G. 1996*b*. Experimental soft-se!diment deformation: structures formed by the liquefaction of unconsolidated sands and some ancient examples. *Sedimentology*, **43**, 279–293.

PARK, R. G., CLIFF, R. A., FETTES, D. J. & STEWART, A. D. 1994. Precambrian rocks in northwest Scotland west of the Moine Thrust: the Lewisian Complex and the Torridonian. *In*: GIBBONS, W. & HARRIS, A. L. (eds) *A revised correlation of Precambrian rocks in the British Isles*. Geological Society, London, Special Reports, **22**, 6–22.

PARK, R. G., STEWART, A. D. & WRIGHT, D. T., 2002. The Hebridean terrane. *In*: TREWIN, N. H. (ed.) *Geology of Scotland*. Geological Society, London, 4th ed. (in press).

PARRY, J. H. 1957. The problem of reversed magnetizations and its study by magnetic methods. *Advances in Physics*, **6**, 299–305.

PAVLOVSKY, YE. V. 1958. Precambrian and Lower Palaeozoic history of the Scottish Highlands and the role of abyssal fractures. *Izvestiya Akademii Nauk SSSR. Seriya geologicheskaya*, **7** (in translation), 1–16.

PEACH, B. N. & HORNE, J. 1892. The *Olenellus* zone in the north-west highlands of Scotland. *Quarterly Journal of the Geological Society of London*, **48**, 227–242.

PEACH, B. N. & HORNE, J. 1914. *Guide to the geological model of the Assynt mountains*. Geological Survey & Museum, Edinburgh.

PEACH, B. N. & HORNE, J. 1930. *Chapters on the geology of Scotland*. Oxford University Press, London.

PEACH, B. N., HORNE, J. GUNN, W., CLOUGH, C. T. HINXMAN, L. W. & TEALL, J. J. H. 1907. *The geological structure of the north-west Highlands of Scotland*. Memoirs of the Geological Survey of Great Britain.

PEACH, B. N., HORNE, J. GUNN, W., CLOUGH, C. T. HINXMAN, L. W., TEALL, J. J. H. & CADELL, H. M. 1888. Report on the recent work of the Geological Survey in the northwest highlands of Scotland based on the field notes and maps of Messers B. N. Peach, J. Horne, W. Gunn, C. T Clough, L. Hinxman and H. M. Cadell. *Quarterly Journal of the Geological Society of London*, **44**, 378–441.

PEACH, B. N., HORNE, J., WOODWARD, H. B., CLOUGH, C. T., HARKER, A. & WEDD, C. B. 1910. *The geology of Glenelg, Lochalsh and south-east part of Skye*. Memoirs of the Geological Survey of Great Britain.

PEAT, C. J. 1984. Comments on some of Britain's oldest microfossils. *Journal of Micropalaeontology*, **3**, 65–71.

PEAT, C. J. & DIVER, W. 1982. First signs of life on Earth. *New Scientist*, **95**, 776–781.

PEAT, C. J. & LLOYD, B. J. 1974. Direct observation of rock macerates. *Nature*, **251**, 294–295.

PENCK, A. 1897. Geomorphologische Probleme aus Nordwest-Schottland. *Zeitschrift der Gesselschaft für Erdkunde, Berlin*, **32**, 146–191.

PETTIJOHN, F. J. 1975. *Sedimentary rocks*. Harper & Row, New York, 3rd ed.

PETTIJOHN, F. J., POTTER, P. E. & SIEVER, R. 1987. *Sand and sandstone*. Springer Verlag, New York, 2nd ed.

PHEMISTER, J. 1936. *British regional geology: Scotland – the northern Highlands*. Department of Scientific & Industrial Research, Geological Survey & Museum, 1st ed.

PHEMISTER, J. 1948. *British regional geology: Scotland – the northern Highlands*. Department of Scientific & Industrial Research, Geological Survey & Museum, 2nd ed.

PHEMISTER, J. 1960. British regional geology: Scotland – the northern highlands. Department of Scientific & Industrial Research, Geological Survey & Museum, 3rd ed.

PIPER, J. D. A. 1992. Palaeomagnetic properties of a Precambrian metamorphic terrane: the Lewisian complex of the Outer Hebrides, NW Scotland. *Tectonophysics*, **201**, 17–48.

PIPER, J. D. A. & POPPLETON, T. J. 1991. Palaeomagnetic conglomerate tests on basal Stoer Group sediments, NW Scotland. *Scottish Journal of Geology*, **27**, 97–106.

PLANT, J. A. 1984. Regional geochemical maps of the United Kingdom. *NERC News Journal*, **3**, 1 & 5–7.

POPPLETON, T. J. & PIPER, J. D. A. 1989. The palaeomagnetism of a Precambrian metamorphic boundary in the Loch Torridon area, NW Scotland. *Journal of the Geological Society, London*, **146**, 799–808.

Potts, G. J. 1990. A palaeomagnetic study of recumbently folded and thermally metamorphosed Torridon Group sediments, Eishort anticline, Skye, Scotland. *Journal of the Geological Society, London*, **147**, 999–1007.

Potts, G. J., Hunter, R. H., Harris, A. L. & Fraser, F. M. 1995. Late-orogenic extensional tectonics at the NW margin of the Caledonides in Scotland. *Journal of the Geological Society, London*, **152**, 907–910.

Prave, A. R. 1999. The Neoproterozoic Dalradian Supergroup of Scotland: an alterative hypothesis. *Geological Magazine*, **136**, 609–617.

Prave, A. R., Oliver, G. & Parrish, R. 2000. An objective reappraisal of the Neoproterozoic tectonic evolution of Scotland. *Geological Society of London, Abstracts Annual Meeting of the Tectonic Studies Group, Manchester*.

Prave, A. R., Oliver, G., Stephens, E., Parrish, R. & Pringle, M. 2001. The Scottish Highlands: a revisionist Neoproterozoic tectonostratigraphic template. *Geological Society of London, Abstracts Annual Meeting of the Tectonic Studies Group, Leeds*.

Rachocki, A. H. 1981. *Alluvial fans: an attempt at an empirical approach*. Wiley, Chichester.

Rainbird, R. H., Hamilton, M. A. & Young, G. M. 2001. Detrital zircon geochronology and provenance of the Torridonian, NW Scotland. *Journal of the Geological Society, London*, **158**, 15–27.

Ramsay, A. C. 1871*a*. Physical relations of the New Red Marl, Rhaetic Beds, and Lower Lias. *Quarterly Journal of the Geological Society of London*, **27**, 189–199.

Ramsay, A. C. 1871*b*. Red rocks of England of older date than the Trias. *Quarterly Journal of the Geological Society of London*, **27**, 241–256.

Ramsay, J. G. 1969. The measurement of strain and displacement in orogenic belts. *In*: Kent, P. E., Satterthwaite, G. E. & Spencer, A. M. (eds) Time and place in orogeny, Geological Society, London, Special Publications, **3**, 43–79.

Ramsay, J. G. & Huber, M. I. 1983. *The techniques of modern structural geology, vol. 1: strain analysis*. Academic Press, London.

Rasmussen, B. & Buick, R. 1999. Redox state of the Archean atmosphere: evidence from detrital heavy minerals in *c*. 3250–2750 Ma sandstones from the Pilbara craton, Australia. *Geology*, **27**, 115–118.

Rast, N., Diggens, J. N. & Rast, D. E. 1968. Triassic rocks of the Isle of Mull: their sedimentation, facies, structure and relationship to the Great Glen fault and the Mull caldera. *Proceedings of the Geological Society of London*, **1645**, 299–304.

Rayner, D. H. 1981. *The stratigraphy of the British Isles*. Cambridge University Press, 2nd ed.

Rejebian, V. A., Harris, A. G. & Huebner, J. S. 1987. Conodont color and textural alteration: an index to regional metamorphism, contact metamorphism, and hydrothermal alteration. *Bulletin of the Geological Society of America*, **99**, 471–479.

Retallack, G. J. 1990. *Soils of the past: an introduction to paleopedology*. Harper Collins, London.

Retallack, G. J. & Mindszenty, A. 1994. Well preserved late Precambrian paleosols from northwest Scotland. *Journal of Sedimentary Research*, **A64**, 264–281.

Retallack, G. J. & Mindszenty, A. 1995. Well preserved late Precambrian paleosols from northwest Scotland – reply. *Journal of Sedimentary Research*, **A65**, 445.

Rickmers, W. R. 1913. *The Duab of Turkestan: a physiographic sketch and account of some travels*. Cambridge University Press.

Roberts, A. M. & Holdsworth, R. E. 1999. Linking onshore and offshore structures: Mesozoic extension in the Scottish Highlands. *Journal of the Geological Society, London*, **156**, 1061–1064.

Robinson, M. A. & McClelland, E. A. 1987. Palaeomagnetism of the Torridonian of Rhum, Scotland: evidence for limited uplift of the Central Intrusive Complex. *Earth and Planetary Science Letters*, **85**, 473–487.

Rodd, J. A. 1983. *The sedimentology and geochemistry of the type Diabaig Formation in the Upper Proterozoic Torridon Group of Scotland*. PhD thesis, University of Reading.

Rodd, J. A. & Stewart, A. D. 1992. Geochemistry, weathering and diagenesis of the Diabaig Formation (Torridon Group) in NW Scotland. *Scottish Journal of Geology*, **28**, 27–35.

Roest, W. R. & Srivistava, S. P. 1989. Sea-floor spreading in the Labrador Sea: a new reconstruction. *Geology*, **17**, 1000–1003.

Rogers, G., Krogh, T. E., Bluck, B. J. & Kwok, Y. Y. 1990. Provenance ages of the Torridonian sandstone of NW Scotland using single grain U–Pb zircon analysis. *Geological Society of Australia Abstracts*, **27**, 84.

Rogers, G., Hyslop, E. K., Strachan, R. A., Paterson, B. A. & Holdsworth, R. E. 1998. The structural setting and U–Pb geochronology of Knoydartian pegmatites in W. Inverness-shire: evidence for Neoproterozoic tectonothermal events in the Moine of NW Scotland. *Journal of the Geological Society, London*, **155**, 685–696.

Rollin, K. E. 1994. Geophysical correlation of Precambrian rocks of northern Britain. *In*: Gibbons, W. & Harris, A. L. (eds) *A revised correlation of Precambrian rocks in the British Isles*. Geological Society, London, Special Reports, **22**, 65–74.

Rollinson, H. R. 1996. Tonalite–trondhjemite–granodiorite magmatism and the genesis of Lewisian crust during the Archaean. *In*: Brewer, T. S. (ed.) *Precambrian crustal evolution in the north Atlantic region*. Geological Society, London, Special Publications, **112**, 25–42.

Ronov, A. B., Mikhailovskaya, M. S. & Solodkova, I. I. 1966. Evolution of the chemical and mineralogical composition of arenaceous rocks. *In*: Vinogradov, A. P. (ed.) *Chemistry of the Earth's crust*, vol. 1 Israel Program for scientific translations, Jerusalem, 212–266.

Rye, R. & Holland, H. D. 1998. Paleosols and the evolution of atmospheric oxygen: a critical review. *American Journal of Science*, **298**, 621–673.

Salter, J. W. 1859. Durness Limestone fossils described. *Quarterly Journal of the Geological Society of London*, **15**, 374–381.

Sanders, I. S. & Johnston, J. D. 1989. The Torridonian Stac Fada Member: an extrusion of fluidised peperite? *Transactions of the Royal Society of Edinburgh. Earth Sciences*, **80**, 1–4.

Sanders, I. S. & Johnston, J. D. 1990. Reply to 'The Torridonian Stac Fada Member: a discussion'. *Transactions of the Royal Society of Edinburgh. Earth Sciences*, **81**, 249–250.

Schlische, R. W. & Olsen, P. E. 1990. Quantitative filling model for continental extension basins with applications to the early Mesozoic rifts of eastern North America. *Journal of Geology*, **98**, 135–155.

Schumm, S. A. 1968. Speculations concerning paleohydrologic controls of terrestrial sedimentation. *Bulletin of the Geological Society of America*, **79**, 1573–1588.

Schwarz, E. J. 1971. Magnetochemical aspects of the heating of red and green beds. *Geological Survey of Canada Paper* **70-63**, 31–34.

Sedgwick, A. & Murchison, R. I. 1828. On the old conglomerates, and other secondary deposits of the north coasts of Scotland. *Proceedings of the Geological Society of London*, **1**, 77–80.

Sedgwick, A. & Murchison, R. I. 1835. On the structure and the relations of the deposits contained between the Primary rocks and the Oolitic Series in the north of Scotland. *Transactions of the Geological Society of London*, 2nd Series, **3**, 125–160.

Selley, R. C. 1963. *Sedimentation of the Torridonian rocks of Raasay, Scalpay and adjacent islets, Inverness-shire*. PhD thesis, Imperial College, University of London.

Selley, R. C. 1964. The penecontemporaneous deformation of heavy mineral bands in the Torridonian sandstone of northwest Scotland. *In*: Van Straaten, L. M. J. U. (ed.) *Deltaic and shallow maine deposits*. Elsevier, Amsterdam, 362–367.

Selley, R. C. 1965*a*. Diagnostic characters of fluviatile sediments of the Torridonian formation (Precambrian) of northwest Scotland. *Journal of Sedimentary Petrology*, **35**, 366–380.

Selley, R. C. 1965*b*. The Torridonian succession on the islands of Fladday, Raasay, and Scalpay, Inverness-shire. *Geological Magazine*, **102**, 361–369.

Selley, R. C. 1966. Petrography of the Torridonian rocks of Raasay and Scalpay, Inverness-shire. *Proceedings of the Geologists' Association*, **77**, 293–314.

Selley, R. C. 1969. Torridonian alluvium and quicksands. *Scottish Journal of Geology*, **5**, 328–346.

Selley, R. C. 1970. Studies of sequence in sediments using a simple mathematical device. *Quarterly Journal of the Geological Society of London*, **125**, 557–581.

Selley, R. C. 1996. *Ancient sedimentary environments and their subsurface diagnosis*. Chapman & Hall, London, 4th ed.

Selley, R. C. , Shearman, D. J., Sutton, J. & Watson, J. 1963. Some underwater disturbances in the Torridonian of Skye and Raasay. *Geological Magazine*, **100**, 224–243.

Sheraton, J. W., Skinner, A. C. & Tarney, J. 1973. The geochemistry of the Scourian gneisses of the Assynt district. *In*: Park, R. G. & Tarney, J. (eds) *The early Precambrian of Scotland and related rocks of Greenland*. Department of Geology, Keele, 13–30.

Smith, J. V. & Brown, W. L. 1988. *Feldspar minerals, volume 1 crystal structures, physical, chemical, and microtextural properties*. Springer-Verlag, Berlin.

Smith, N. D. 1970. The braided stream depositional environment: comparison of the Platte River with some Silurian clastic rocks, north-central Appalachians. *Bulletin of the Geological Society of America*, **81**, 2993–3014.

Smith, N. D. 1971. Transverse bars and braiding in the lower Platte River, Nebraska. *Bulletin of the Geological Society of America*, **82**, 3407–3420.

Smith, R. L., Stearn, J. E. F. & Piper, J. D. A. 1983. Palaeomagnetic studies of the Torridonian sediments, NW Scotland. *Scottish Journal of Geology*, **19**, 29–45. (And supplementary publication SUP 24015).

Smoot, J. P. 1991. Sedimentary facies and depositional environments of early Mesozoic Newark Supergroup basins, eastern North America. *Palaeogeography, Palaeoclimatology, Palaeoecology*, **84**, 369–423.

Sønderholm, M. & Tirsgaard, H. 1993. Lithostratigraphic framework of the Upper Proterozoic Eleornore Bay Supergroup of east and north-east Greenland. *Grønlands Geologiske Undersøgelse Bulletin*, **167**.

Soper, N. J. 1994. Was Scotland a Vendian RRR junction? *Journal of the Geological Society, London*, **151**, 570–582.

Soper, N. J. & Anderton, R. 1984. Did the Dalradian slides originate as extensional faults? *Nature*, **307**, 357–360.

Soper, N. J. & England, R. W. 1995. Vendian and Riphean rifting in NW Scotland. *Journal of the Geological Society, London*, **152**, 11–14.

Soper, N. J., Harris, A. L. & Strachan, R. A. 1998. Tectonostratigraphy of the Moine Supergroup: a synthesis. *Journal of the Geological Society, London*, **155**, 13–24.

Soper, N. J., Ryan, P. D. & Dewey, J. F. 1999. Age of the Grampian orogeny in Scotland and Ireland. *Journal of the Geological Society, London*, **156**, 1231–1236.

Spalletta, C. & Vai, G. B. 1984. Upper Devonian intraclast parabreccias interpreted as seismites. *Marine Geology*, **55**, 133–144.

Steel, R. J. & Wilson, A. C. 1975. Sedimentation and tectonism (?Permo-Triassic) on the margin of the North Minch Basin, Lewis. *Journal of the Geological Society, London*, **131**, 183–202

Steel, R. J., Mæhle, S., Nilsen, H., Røe, S. L. & Spinnangr, Å. 1977. Coasening-upward cycles in the alluvium of the Hornelen basin (Devonian) Norway: sedimentary response to tectonic events. *Bulletin of the Geological Society of America*, **88**, 1124–1134.

Stein, A. M. 1988. Basement controls upon basin development in the Caledonian foreland, NW Scotland. *Basin Research*, **1**, 107–119.

Stein, A. M. 1992. Basin development and petroleum potential in The Minches and Sea of the Hebrides basins. *In*: Parnell, J. (ed.) *Basins on the Atlantic seaboard: petroleum geology, sedimentology and basin evolution*. Geological Society, London, Special Publications, **62**, 17–20.

Steiner, M. B. 1983. Detrital remanent magnetization in hematite. *Journal of Geophysical Research*, **88**, 6523–6539.

Stewart, A. D. 1962. On the Torridonian sediments of Colonsay and their relationship to the main outcrop in north-west Scotland. *Liverpool and Manchester Geological Journal*, **3**, 121–156.

Stewart, A. D. 1963. On certain slump structures in the Torridonian sandstones of Applecross. *Geological Magazine*, **100**, 205–218.

Stewart, A. D. 1966*a*. On the correlation of the Torridonian between Rhum and Skye. *Geological Magazine*, **103**, 432–439.

Stewart, A. D. 1966*b*. An unconformity in the Torridonian. *Geological Magazine*, **103**, 462–465.

Stewart, A. D. 1969. *Torridonian rocks of Scotland reviewed*. American Association of Petroleum Geologists, Memoirs, **12**, 595–608.

Stewart, A. D. 1972. Pre-Cambrian landscapes in northwest Scotland. *Geological Journal*, **8**, 111–124.

Stewart, A. D. 1975. 'Torridonian' rocks of western Scotland *In*: Harris, A. L., Shackleton, R. M., Watson, J., Downie, C., Harland, W. B. & Moorbath, S. (eds) *A correlation of the Precambrian rocks in the British Isles*. Geological Society, London, Special Reports, **6**, 43–51.

Stewart, A. D. 1977. Quantitative limits to palaeogravity. *Journal of the Geological Society, London*, **133**, 281–291.

Stewart, A. D. 1980. Limits to palaeogravity since the late Precambrian. *Nature*, **271**, 153–155.

Stewart, A. D. 1982. Late Proterozoic rifting in NW Scotland: the genesis of the 'Torridonian'. *Journal of the Geological Society, London*, **139**, 413–420.

Stewart, A. D. 1988*a*. The Stoer Group, Scotland. *In*: Winchester, J. A. (ed.) *Later Proterozoic stratigraphy of the northern Atlantic regions*. Blackie, Glasgow, 97–103.

Stewart, A. D. 1988*b*. The Sleat and Torridon Groups, *In*: Winchester, J. A. (ed.) *Later Proterozoic stratigraphy of the northern Atlantic regions*. Blackie, Glasgow, 104–112.

Stewart, A. D. 1990. The Torridonian Stac Fada Member: a discussion. *Transactions of the Royal Society of Edinburgh. Earth Sciences*, **81**, 247.

Stewart, A. D. 1991*a*. Geochemistry, provenance and climate of the Upper Proterozoic Stoer Group in Scotland. *Scottish Journal of Geology*, **26** (for 1990), 89–97.

Stewart, A. D. 1991*b*. Geochemistry, provenance and palaeoclimate of the Sleat and Torridon Groups in Skye. *Scottish Journal of Geology*, **27**, 81–95.

Stewart, A. D. 1991*c*. Torridonian. *In*: Craig, G. Y. (ed.) *Geology of Scotland*. Geological Society, London, 3rd edn, 65–85.

Stewart, A. D. 1993*a*. Late Proterozoic and late Palaeozoic movement on the Coigach fault in NW Scotland. *Scottish Journal of Geology*, **29**, 21–28.

Stewart, A. D. 1993*b*. The ratio of mechanical to chemical denudation in alluvial systems, derived from geochemical mass balance. *Transactions of the Royal Society of Edinburgh, Earth Sciences*, **84**, 73–78.

Stewart, A. D. 1995*a*. Well preserved late Precambrian paleosols from northwest Scotland – discussion. *Journal of Sedimentary Research*, **A65**, 444.

Stewart, A. D. 1995*b*. Possible sources of clastic sediment in the late Proterozoic Torridon Group, from geochemical mass balance. *Transactions of the Royal Society of Edinburgh. Earth Sciences*, **85**, 303–309.

Stewart, A. D. 1997. Discussion of the evidence for glaciation in the Proterozoic Stoer Group of Scotland. *Journal of the Geological Society, London*, **154**, 375–376.

Stewart, A. D. & Donnellan, N. C. B. 1992. Geochemistry and provenance of red sandstones in the Upper Proterozoic Torridon Group of Scotland. *Scottish Journal of Geology*, **28**, 143–153.

Stewart, A. D. & Hackman, B. D. 1973. Precambrian sediments of western Islay. *Scottish Journal of Geology*, **9**, 185–201.

Stewart, A. D. & Irving, E. 1974. Palaeomagnetism of Precambrian sedimentary rocks from NW Scotland and the apparent polar wandering path of Laurentia. *Geophysical Journal of the Royal Astronomical Society*, **37**, 51–72.

Stewart, A. D. & Parker, A. 1979. Palaeosalinity and environmental interpretation of red beds from the late Precambrian ('Torridonian') of Scotland. *Sedimentary Geology*, **22**, 229–241.

Stone, P., Plant, J. A., Mendum, J. R. & Green, P. M. 1999. A regional geochemical assessment of some terrane relationships in the British Caledonides. *Scottish Journal of Geology*, **35**, 145–156.

Sutton, J. 1962. Torridonian microfossils. *Geological Magazine*, **99**, 379.

Sutton, J. 1963. Some events in the evolution of the Caledonides. *In*: Johnson, M. R. W. & Stewart, F. H. (eds) *The British Caledonides*, Oliver & Boyd, Edinburgh, 249–269.

Sutton, J. & Watson, J. V. 1951. The pre-Torridonian metamorphic history of the Loch Torridon and Scourie areas in the north-west highlands, and its bearing on the chronological classification of the Lewisian. *Quarterly Journal of the Geological Society of London*, **106**, 241–308.

Sutton, J. & Watson, J. V. 1960. Sedimentary structures in the Epidotic Grits of Skye. *Geological Magazine*, **97**, 106–122.

Sutton, J. & Watson, J. V. 1963. Structures in shallow-water Precambrian sediments from northwest Scotland. *Sedimentology*, **2**, 207–214.

Sutton, J. & Watson, J. V. 1964. Some aspects of Torridonian stratigraphy in Skye. *Proceedings of the Geologists' Association*, **75**, 251–289.

Svenningsen, O. M. 2001. Onset of seafloor spreading in the Iapetus Ocean at 608 Ma: precise age of the Sarek dyke swarm, northern Swedish Caledonides. *Precambrian Research*, **110**, 241–254.

Tarney, J. 1973. The Scourie dyke suite and the nature of the Inverian event in Assynt. *In*: Park, R. G. & Tarney, J. (eds) *The early Precambrian of Scotland and related rocks of Greenland*. Department of Geology, Keele, 105–118.

Tarney, J. & Weaver, B. L. 1987. Geochemistry of the Scourian complex: petrogenesis and tectonic models. *In*: Park, R. G. & Tarney, J. *Evolution of the Lewisian and comparable Precambrian high grade terrains*. Geological Society, London, Special Publications, **27**, 45–56.

Taylor, S. R. & McLennan, S. M. 1985. *The continental crust: its composition and evolution. An examination of the geochemical record preserved in sedimentary rocks*. Blackwell, Oxford.

Thorpe, R. S., Potts, P. J. & Sarre, M. B. 1977. Rare earth evidence concerning the origin of granites of the Isle of Skye, northwest Scotland. *Earth & Planetary Science Letters*, **36**, 111–120.

Timofeev, B. V. 1966. *Mikropaleofitologicheskoe issledovanie drevnikh svit*. Akademiya Nauk SSSR, Isdatelskvo Nauka, Moscow.

Torsvik, T. H. & Sturt, B. A. 1987. On the origin and stability of remanence and the magnetic fabric of the Torridonian red beds, NW Scotland. *Scottish Journal of Geology*, **23**, 23–38.

Torsvik, T. H. & Rehnström, E. F. 2001. Cambrian palaeomagnetic data from Baltica: implications for true polar wander and Cambrian palaeogeography. *Journal of the Geological Society, London*, **158**, 321–329.

Turnbull, M. J. M., Whitehouse, M. J. & Moorbath, S. 1996. New isotopic age determinations for the Torridonian, NW Scotland. *Journal of the Geological Society, London*, **153**, 955–964.

Turner, P. 1979. The palaeomagnetic evolution of continental red beds. *Geological Magazine*, **116**, 289–301.

Upfold, R. L. 1984. Tufted microbial (cyanobacterial) mats from the Proterozoic Stoer Group, Scotland. *Geological Magazine*, **121**, 351–355.

Van de Kamp, P. C. & Helmold, K. P. 1991. Compositions of sediments derived from granitic basement rocks in glacial to humid tropical climates (abstract). *Geological Society of America, Annual Meeting*, San Diego.

Van de Kamp, P. C. & Leake, B. E. 1997. Mineralogy, geochemistry, provenance and sodium metasomatism of Torridonian rift basin clastic rocks, NW Scotland. *Scottish Journal of Geology*, **33**, 105–124.

Van Houten, F. B. 1968. Iron oxides in red beds. *Bulletin of the Geological Society of America*, **79**, 399–416.

Vance, D., Strachan, R. A. & Jones, K. A. 1998. Extensional versus compressional settings for metamorphism: garnet chronometry and pressure–temperature–time histories in the Moine Supergroup, northwest Scotland. *Geology*, **26**, 927–930.

Walther, J. 1894. *Einleitung in die Geologie als historische Wissenschaft: III Lithogenesis der Gegenwart*. Verlag Gustav Fischer, Jena.

Watson, J. 1977. The Outer Hebrides: a geological perspective. *Proceedings of the Geologists' Association*, **88**, 1–14.

Watt, G. R., Kinny, P. D. & Friderichsen, J. D. 2000. U–Pb geochronology of Neoproterozoic and Caledonian tectonothermal events in the East Greenland Caledonides. *Journal of the Geological Society, London*, **157**, 1031–1048.

Weaver, B. L. & Tarney, J. 1980. Rare earth geochemistry of Lewisian granulite-facies gneisses, northwest Scotland: implications for the petrogenesis of the Archaean lower continental crust. *Earth & Planetary Science Letters*, **51**, 279–296.

Weaver, B. L. & Tarney, J. 1981. The Scourie dyke suite: petrogenesis and geochemical nature of the Proterozoic sub-continental mantle. *Contributions to Mineralogy & Petrology*, **78**, 175–188.

Weil, A. B., Van der Voo, R., Niocaill, C. M. & Meert, J. G. 1998. The Proterozoic supercontinent Rodinia: paleomagnetically derived reconstructions for 1100 to 800 Ma. *Earth & Planetary Science Letters*, **154**, 13–24.

Welsh, W. 1963. Sedimentary features of the Rhum Torridonian. *Transactions of the Edinburgh Geological Society*, **19**, 486–492.

Westbrook, G. K. & Borradaile, G. J. 1978. The geological significance of magnetic anomalies in the region of Islay. *Scottish Journal of Geology*, **14**, 213–224.

Whittaker, A., Cope, J. C. W. *et al.* 1991. A guide to stratigraphical procedure. *Journal of the Geological Society, London*, **148**, 813–824.

Wilkinson, S. B. 1907. *The geology of Islay, including Oronsay and portions of Colonsay and Jura*. Memoirs of the Geological Survey of Great Britain.

Williams, G. E. 1966*a*. *The Precambrian Torridonian sediments of the Cape Wrath district, north-west Scotland*. PhD thesis, University of Reading.

Williams, G. E. 1966*b*. Palaeogeography of the Torridonian Applecross Group. *Nature*, **209**, 1303–1306.

Williams, G. E. 1966*c*. Planar cross-stratification formed by the lateral migration of shallow streams. *Journal of Sedimentary Petrology*, **36**, 742–746.

Williams, G. E. 1968. Torridonian weathering, and its bearing on Torridonian palaeoclimate and source. *Scottish Journal of Geology*, **4**, 164–184.

Williams, G. E. 1969*a*. Characteristics and origin of a Precambrian pediment. *Journal of Geology*, **77**, 183–207.

Williams, G. E. 1969*b*. *Petrography and origin of pebbles from Torridonian strata (late Precambrian), northwest Scotland*. American Association of Petroleum Geologists, Memoirs, **12**, 609–629.

Williams, G. E. 1970. Origin of disturbed bedding in Torridon Group sandstones. *Scottish Journal of Geology*, **6**, 409–414.

Williams, G. E. 1971. Flood deposits of sand-bed ephemeral streams of central Australia. *Sedimentology*, **17**, 1–40.

Williams, G. E. 2001. Neoproterozoic (Torridonian) alluvial fan succession, northwest Scotland, and its tectonic setting and provenance. *Geological Magazine*, **138**, 471–494.

Williams, G. E. & Schmidt, P. W. 1997. Palaeomagnetic dating of sub-Torridon Group weathering profiles, NW Scotland: verification of Neoproterozoic palaeosols. *Journal of the Geological Society, London*, **154**, 987–997.

Winchester, J. A. 1988. Later Proterozoic environments and tectonic evolution in the northern Atlantic lands. *In*: Winchester, J. A. (ed.) *Later Proterozoic stratigraphy of the northern Atlantic regions*. Blackie, Glasgow, 253–270.

Windley, B. F. 1995. *The evolving continents*, 3rd. edn, Wiley, Chichester.

Young, G. M. 1999*a*. Some aspects of the geochemistry, provenance and palaeoclimatology of the Torridonian of NW Scotland. *Journal of the Geological Society, London*, **156**, 1097–1111.

Young, G. M. 1999*b*. A geochemical investigation of palaeosols developed on Lewisian rocks beneath the Torridonian Applecross Formation, NW Scotland. *Scottish Journal of Geology*, **35**, 107–118.

YOUNG, G. M. 2002. Stratigraphy and geochemistry of volcanic debris flows in the Stac Fada Member of the Stoer Group, Torridonian, NW Scotland. *Transactions of the Royal Society of Edinburgh. Earth Sciences* (in press).

ZHANG ZHONGYING 1982. Upper Proterozoic microfossils from the Summer Isles, NW Scotland. *Palaeontology*, **25**, 443–460.

ZHANG ZHONGYING, DIVER, W. L. & GRANT, P. R. 1981. Microfossils from the Aultbea Formation, Torridon Group, on Tanera Beg, Summer Isles. *Scottish Journal of Geology*, **17**, 149–154.

Index

Page numbers in *italics* refer to Figures and page numbers in **bold** refer to Tables

PLATE CAPTIONS

Plate 1. The Torridonian outcrop in NW Scotland, showing component groups and formations.

Plate 2. The Torridonian outcrop in NW Scotland showing the sub-areas described in the Directory, together with the locations of figured maps and sections.